Forest Soil Respiration under Climate Changing

Forest Soil Respiration under Climate Changing

Special Issue Editors

Robert Jandl
Mirco Rodeghiero

MDPI • Basel • Beijing • Wuhan • Barcelona • Belgrade

MDPI

Special Issue Editors
Robert Jandl
Austrian Forest Research Center (BFW)
Austria

Mirco Rodeghiero
Research and Innovation Centre, Fondazione Edmund Mach San Michele all'Adige,
Italy

Editorial Office
MDPI
St. Alban-Anlage 66
Basel, Switzerland

For citation purposes, cite each article independently as indicated on the article page online and as indicated below:

ISBN 978-3-03897-178-8 (Pbk)
ISBN 978-3-03897-179-5 (PDF)

Contents

About the Special Issue Editors

Robert Jandl, Univ Lecturer, Dr is a forest ecologist at the Austrian Research Center for Forests and is coordinating the research activities related to climate change. His main research interest is currently the role of forest soils in climate change mitigation and the carbon dynamics in forest ecosystems of the temperate zone. Robert Jandl is a member of the commission for Climate and Airquality of the Austrian Academy of Sciences and a Board Member of the Austrian Center for Climate Change (CCCA).

Mirco Rodeghiero, Dr, Forest Ecology PhD is a researcher at Fondazione Edmund Mach (San Michele all'Adige, Italy). His main research activity is focused on the effects of climate change on soil carbon and nitrogen dynamics by combining physiological and pedological measurements. In particular he investigated the main determinants of soil carbon dioxide efflux in forest ecosystems. He coordinated the soil sampling campaign for the Italian National Forest Inventory and was involved in the major European integrative projects for the study of soil carbon dynamics.

Preface to "Forest Soil Respiration under Climate Changing"

Soil respiration is a process of prime relevance for understanding the carbon cycle in forest ecosystems and for properly comprehending the role of forests in climate change mitigation. The process is divided into two components: (i) autotrophic soil respiration, i.e. the efflux of CO_2 from the respiration of tree roots, and (ii) heterotrophic soil respiration, i.e. the efflux of CO_2 due to respiration of soil microorganisms. A third component, the respiration of mycorrhizae, is still debated and it is not yet clear whether it should be accounted for in autotrophic or heterotrophic soil respiration, respectively, or whether it should be treated as a third component.

The rate of soil respiration is controlled by environmental factors. Expectedly, the strongest driver is soil temperature, followed by soil moisture. The relevance of either factor depends on site properties. Two papers are reinforcing this view. An asset of the paper compilation is the collection of case studies where other factors besides temperature and soil moisture are evidently greatly affecting the rate of soil respiration. The characteristics of the forest stand such as tree density, stand age, and tree species and additional soil properties such as aggregate stability are influencing soil respiration.

The book gives guidance on the current state of knowledge and helps identifying knowledge gaps for future research endeavours.

Robert Jandl, Mirco Rodeghiero
Special Issue Editors

forests

MDPI

Article

The Role of Respiration in Estimation of Net Carbon Cycle: Coupling Soil Carbon Dynamics and Canopy Turnover in a Novel Version of 3D-CMCC Forest Ecosystem Model

Sergio Marconi [1,2,*], Tommaso Chiti [3,4], Angelo Nolè [5], Riccardo Valentini [2,4] and Alessio Collalti [2,6]

1   School of Natural Resources and Environment, 103 Black Hall, University of Florida,
     Gainesville, FL 32611, USA
2   Foundation Euro-Mediterranean Center on Climate Change—Impacts on Agriculture,
     Forest and Ecosystem Services (IAFES), 01100 Viterbo, VT, Italy; rik@unitus.it (R.V.);
     alessio.collalti@cmcc.it (A.C.)
3   Department for Innovation in Biological, Agro-Food and Forest Systems (DIBAF), University of Tuscia,
     01100 Viterbo, VT, Italy; tommaso.chiti@unitus.it
4   Far Eastern Federal University (FEFU), Ajax St., Vladivostok, 690920 Russky Island, Russia
5   School of Agricultural, Forestry, Food and Environmental Sciences (SAFE), University of Basilicata,
     85100 Potenza, Italy; angelo.nole@gmail.com
6   CNR-ISAFOM National Research Council of Italy, Institute for Agriculture and Forestry Systems in the
     Mediterranean, 87036 Rende, CS, Italy
*   Correspondence: sergio.marconi@weecology.org; Tel.: +1-352-745-9685

Academic Editors: Robert Jandl and Mirco Rodeghiero
Received: 16 March 2017; Accepted: 15 June 2017; Published: 21 June 2017

Abstract: Understanding the dynamics of organic carbon mineralization is fundamental in forecasting biosphere to atmosphere net carbon ecosystem exchange (NEE). With this perspective, we developed 3D-CMCC-PSM, a new version of the hybrid process based model 3D-CMCC FEM where also heterotrophic respiration (R_h) is explicitly simulated. The aim was to quantify NEE as a forward problem, by subtracting ecosystem respiration (R_{eco}) to gross primary productivity (GPP). To do so, we developed a simplification of the soil carbon dynamics routine proposed in the DNDC (DeNitrification-DeComposition) computer simulation model. The method calculates decomposition as a function of soil moisture, temperature, state of the organic compartments, and relative abundance of microbial pools. Given the pulse dynamics of soil respiration, we introduced modifications in some of the principal constitutive relations involved in phenology and littering sub-routines. We quantified the model structure-related uncertainty in NEE, by running our training simulations over 1000 random parameter-sets extracted from parameter distributions expected from literature. 3D-CMCC-PSM predictability was tested on independent time series for 6 Fluxnet sites. The model resulted in daily and monthly estimations highly consistent with the observed time series. It showed lower predictability in Mediterranean ecosystems, suggesting that it may need further improvements in addressing evapotranspiration and water dynamics.

Keywords: forest ecosystem; Fluxnet; soil respiration; net ecosystem exchange; phenology

---

1. Introduction

Global concerns over increasing levels of greenhouse gas concentrations, particularly carbon dioxide (CO_2), have pushed research efforts to better investigate biogeochemical carbon (C) flux dynamics and patterns between atmosphere and biosphere, and to upscale C flux estimates from site-specific to regional, continental and global scales. Increased atmospheric concentration of CO_2, combined with increasing temperatures and size variations of ecosystem C pools, are responsible for year-to-year terrestrial ecosystem carbon flux perturbations, through the variation of both photosynthetic and respiration rates [1].

In the last decades, the eddy covariance (EC) technique has provided long-term continuous measurements of net carbon ecosystem exchange (NEE), water vapor and energy, within the global network of EC flux towers (FLUXNET) distributed over major terrestrial ecosystems. The availability of EC measure of NEE contributed to quantify and to determine seasonal and inter-annual variability of ecosystem C budgets at EC tower site-specific scale [2–4]. Observed NEE does not directly quantify the two major components of ecosystem C flux balance represented by ecosystem respiration (R_{eco}) and gross primary productivity (GPP). Thus, flux partitioning algorithms have been developed to partition eddy covariance NEE into photosynthetic uptake and respiratory release [5,6]. At the same time, EC flux measurements provide key information for the parameterization, calibration and validation of process-based forest ecosystem models (FEMs) contributing to large-scale estimates of main ecosystem C pools.

The implementation of both forest process-based models (PBMs) [7–12] and functional–structural tree models (FSTMs) [13–16], based on the widely used light use efficiency (LUE) approach [17], has contributed to understanding and upscaling the main physiological processes supporting ecosystem C uptake. Although most of forest ecosystem models provide reliable estimates of forest growth, limitations for NEE estimates are related to the uncertainty in R_{eco} estimation [18,19]. Hence, the implementation of biogeochemical models integrating soil respiration models and FSTMs is a great opportunity to reliably estimate NEE [20–22].

Although soil respiration and soil organic matter (SOM) decomposition depend mainly on abiotic factors, such as temperature and soil moisture [23–25], a key role is played by soil organic carbon (SOC) stock size, microbial pools [26,27], and dynamics of SOC supply to soil with littering [28]. Leaf and fine root senescence is of primary importance in determining dynamics in heterotrophic respiration [29,30], and exhibits seasonal patterns and intra-seasonal pulses [31,32]. These pulses are mostly driven by phenological transitions through stages of dormancy, active growth, and senescence. For example, supply of dead leaves in soil is strongly dependent on a tree's leaffal strategy. This is also true when predicting tree respiration, which is directly related to their growth and nitrogen content, which depends on spring phenology. Unfortunately, processes involving budburst and senescence are still partly obscure. PBMs usually represent these processes simplistically; these simplifications may lead some terrestrial ecosystem models to result in biased predictions [33–35].

For these reasons, we developed the 3D-CMCC-PSM (3D-CMCC-Phenology and soil Model), a new version of the hybrid process-based model 3D-CMCC FEM proposed by Collalti et al. [36,37], where (1) spring phenology was directly taking into account tradeoffs between growth and Non Structural carbon (NSC) demand; (2) phenological transitions and supply of Fresh organic matter (FOM) to soil were more explicitly represented; and (3) heterotrophic respiration (R_h) was explicitly simulated to dynamically quantify stock changes of 7 different SOC pools mediated by the amount of active microbial C pools, following the rationale proposed in the DNDC [38]. The aim of this study was to: (1) test the performance of the modified model version, comparing model NEE estimates against independent time series for 6 Fluxnet sites, representing different forests in different climatic areas, distributed over a wide latitudinal gradient amongst European EC sites; and (2) quantify uncertainty associated to 3D-CMCC-PSM constitutive relations structure and parameterization.

2. Materials and Methods

2.1. Study Area and Data

Eddy covariance data were collected from FLUXNET [39]. We chose the 6 sites to represent a climatic and longitudinal transect through Europe (Figure 1), so that the model could be tested on different critical boundary conditions (Table 1). We used EC data from different time series from daily to annual, processed using the method described in [40]. Gross primary production (GPP) and ecosystem respiration (R$_{eco}$) were partitioned using [6]. Information about forest structure and total SOC at the beginning of the simulation was collected from literature (e.g., [41,42]) and PIs information. Sites were chosen to represent 3 diverse forest ecosystems, dominated by different species composition from deciduous broadleaved, DBF (i.e., *Fagus sylvatica* L.), evergreen broadleaved, EBF (i.e., *Quercus ilex* L.), and evergreen needle leaved, ENF (i.e., *Pinus sylvestris* L. and *Picea abies* L.), representing the most common European forest species from boreal to mediterranean ecoregions across Europe.

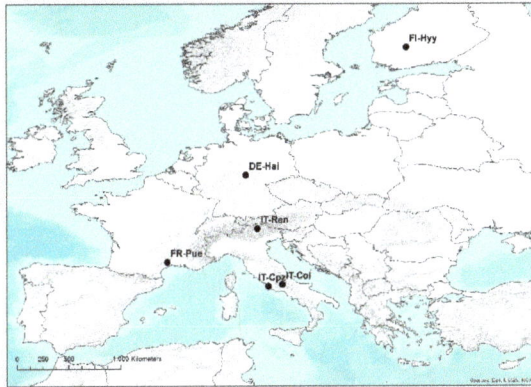

Figure 1. Location of the 6 Fluxnet sites used to evaluate 3D-CMCC-PSM.

2.2. Model Description

The 3D-CMCC FEM is a stand-scale process-based model (PBM) designed to simulate C and water cycle in natural and managed forest ecosystems (for a full description see [36,37]). Several eco-physiological processes were modeled at species-specific level, and at a variable temporal scale (from daily to annual) depending on the process to simulate (Figure 2). Model outputs were generally represented at hectare scale, while processes were simulated at different spatial scales from cellular (e.g., stomatal conductance), to canopy (e.g., transpiration), to individual tree, up to stand level.

Carbon assimilation was modeled for sun and shaded leaves using the LUE approach [43]: potential C assimilation was constrained by environmental and stand structural (e.g., tree age) scalars [44]. Autotrophic respiration (R$_A$) was explicitly modeled as the sum of growth and maintenance respiration (R$_G$ and R$_M$, respectively). The first was computed as a fixed ratio of new growth tissues (30%) and the latter was based on nitrogen content in stems, branches, leaves, fine and coarse roots, non-structural carbon (NSC), and fruit tree pools. carbon allocation among these pools was controlled by species-specific parameters, phenology, light and water availability. Water cycle was modeled calculating the daily balance between precipitation, canopy transpiration, evaporation, soil evaporation, and runoff. Meteorological variables used to force the model were: global solar radiation (MJ m^{-2} day^{-1}), maximum and minimum air temperature (°C), relative humidity (%), and daily cumulated precipitation (mm day^{-1}). To be initialized, the model required knowing stand structural characteristics such as: species composition, stand density, diameter at breast height (DBH), tree height and age. Soil initialization required the estimation of total organic carbon (TOC) in the different SOC pools, as described in the following section.

Table 1. Site descriptions and stand initialization data. Mean temperature (T) and Precipitation (prec.) are annual averages collected at site.

Site Name	Coords (Lat°/Lon°)	Climate	Species Composition	Mean Annual T (°C)	Mean Annual Prec. (mm year^{-1})	Mean DBH (cm)	Tree Height (m)	Stand Age (years)	Stand Density (trees ha^{-1})
Collelongo (ITCol)	41.8/13.5	Temperate	*Fagus sylvatica* L. (DBF)	6.3	1180	20.27	19.84	100	900
Hainich (DEHai)	51.0/10.4	Temperate	*Fagus sylvatica* L. (DBF)	8.3	720	30.8	23.1	120	334
Hyytiälä (FIHyy)	61.8/24.2	Boreal	*Pinus sylvestris* L. (ENF)	3.8	709	10.3	6.5	39	1796
Renon (ITRen)	46.5/11.4	Temperate	*Picea abies* L. (ENF)	4.7	809	16.98	11.32	50	767
Castelporziano (ITCpz)	41.7/12.3	Mediterranean	*Quercus ilex* L. (EBF)	15.6	780	16	12.5	45	458
Puechabon (FRPue)	43.7/3.6	Mediterranean	*Quercus ilex* L. (EBF)	13.5	883	7	6	59	6149

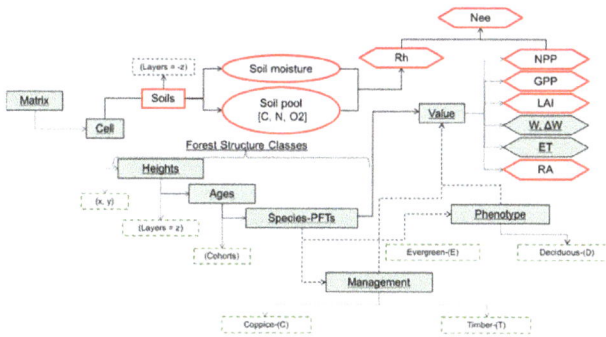

Figure 2. 3D-CMCC version of PSM main flowchart modified from [36]. Red-circled boxes represent the pools and variables introduced or modified by 3D-CMCC-PSM.

2.3. Model Improvements

2.3.1. Soil Carbon Dynamics

The most recent 3D-CMC FEM model version (v.5.1, [37]) lacks in representing SOC dynamics, preventing any estimation of NEE. With that perspective, we developed a simplified version of the method described in [38] to quantify, dynamically, changes of 7 different SOC pools mediated by the amount of active microbial C pool (Figure 3).

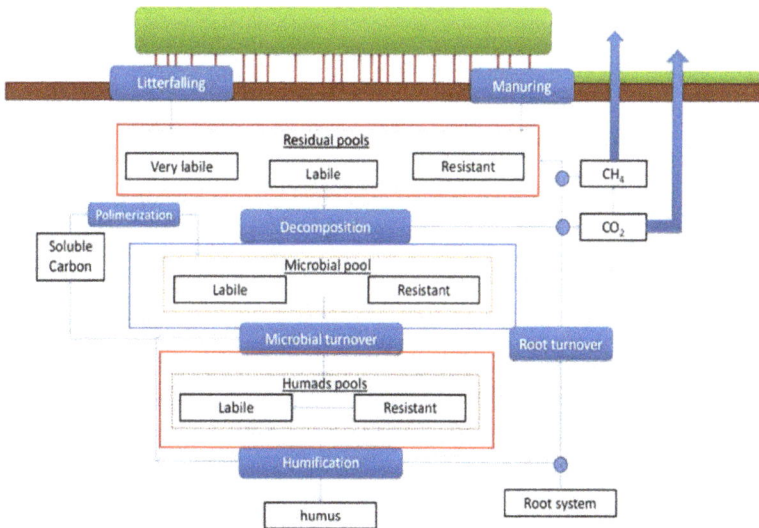

Figure 3. Soil Carbon dynamics in 3D-CMCC-PSM. The three macro pools are highlighted by red boxes (dead C pool) and blue box (live C pool, i.e., microbial). Blue-filled boxes represent the processes simulated by the soil model.

The litter C decomposed by microbial activity is partly mineralized as CO_2, partly stored into microbial metabolic biomass (labile), partly in structural microbial biomass (resistant), and partly transformed in other organic compounds [38]. SOC stability is also related to its chemical recalcitrance, its accessibility, and interaction with clays [45]. We divided SOC in 7 pools to take these differences into account. Humic pool (we use the term Humads, to be consistent with [38]) was divided into a more labile (labile Humads, which stands for Humic acid), an intermediate (resistant Humads, which stands for Fulvic acid) and a more resistant sub-pool (Humus, which stands for Humine).

We assumed that microbial C use efficiency was different for different pools, but constant within each pool [44]. Microbial activity is strictly related to micro-environmental conditions too. Given these premises we modeled C dynamics among soil pools using the following two partial differential equations:

$$\begin{cases} \frac{\partial C_p}{\partial t} = C_L(t) - \beta C_p(t)B(t) \\ \frac{\partial B}{\partial t} = \beta C_p(t)B(t) - \alpha\, B(t) \end{cases}, \tag{1}$$

with $\frac{\partial C_p}{\partial t}$ being the CO_2 efflux produced from a specific C pool decomposition, $C_L(t)$ is the amount of new carbon entering the specific soil pool, $C_p(t)$ the amount of carbon in that C pool, $B(t)$ the microbial biomass competing for $C(t)$. α and β respectively represented the microbial turnover and SOM consumption factors. α was treated as a constant value, as in [38].

The consumption factor β was estimated as a function of both SOM stability and micro-environmental conditions:

$$\begin{cases} \beta\left(C_{pool}\right) = \mu_T\,\mu_M\mu_A k_{pool} \\ \mu_T = -0.014\,T^2 + 0.099\,T + 0.02 \\ \mu_M = -2.85\,\theta^3 + 1.49\,\theta^2 + 1.77\,\theta - 0.03 \\ \mu_D = 0.6\,z^{-0.136}(-0.02\,clay_\% + 0.03) \\ \mu_A = \mu_D\frac{\log\left(0.14\,clay_\%^{-1}\right)}{2.3026} + 1 \end{cases} \tag{2}$$

where μ_T represented the temperature factor, μ_M the moisture factor, μ_A the accessibility, k_{pool} the recalcitrance, and μ_D a clay dependent depth factor [46]. The k_{pool} factor was treated as a specific parameter depending on each compartment's biomass. T represented temperature in Celsius degrees, θ stands for water moisture, z the mean depth of the soil, $clay_\%$ is the percentage of clays in soil. Humads were decomposed using the same rationale, but had slower rates. Inert organic matter (IOM) was calculated following [47].

2.3.2. Deciduous Phenology

Similarly to [37] the phenology scheme was constrained in 3 and 5 sub-phases respectively for evergreen and deciduous species (Table 2). These phases were driven by photoperiod, thermal sum, and maximum leaf biomass (resulting from maximum attainable Leaf Area Index, LAI, $m^2\,m^{-2}$).

3D-CMCC FEM represents leaves development as a by-product of leaf biomass pool dynamics [48]. Despite being a reasonable simplification, such method leads to non-negligible excesses in growth respiration and NSC consumption during budburst. Such unrealistic demand could eventually consume all the available carbon, causing the death of the tree. For this reason, we proposed a modification of budburst phenology, to explicitly simulate the dynamic tradeoffs between demand in NSC, increase in foliar biomass, and maturation of progressively self-sufficient leaves.

Table 2. Description of the different phenological phases for deciduous and evergreen species used in 3D-CMCC PSM.

Deciduous		Evergreen	
Phase	Trigger	Phase	Trigger
Bud Burst	GDD threshold	Bud Burst	GDD treshold
Leaf Development	PeakLai/2	PeakLai	Pipe Model
PeakLai	Pipe Model	Leaffall	Daylength Treshold
Leaffall	Daylength Treshold		
Unvegetative	[49]		

We based the tradeoff function on the hypothesis that new leaf demand of NSC is higher during budburst, and gets progressively lower when maturing [50]. Idealistically, the total NSC mass (B_R) demanded by maturing shoots and leaves could be represented by the linear differential equation (ODE):

$$\frac{\delta B_R}{\delta t} - R_{NSC}(t) \cdot B_R(t) = 0 \tag{3}$$

where $R_{NSC}(t)$ is the instantaneous proportion of NSC demand. We assumed that C request per leaf exponentially decreases with maturity, while the total demand increases by the increasing number of leaf primordia. For simplicity, we assumed that the two components resulted in the linear reduction of $R_{NSC}(t)$. Being a fraction, the maximum value of the integral of $R_{S\&F\&R}(t)$ is equal to 1. Being the first vegetative day t_0, and the last day of budburst (BB_T) the last possible one to reach complete leaf development, the domain of the function was [t_0, BB_T]:

$$1 = \int_0^{BB_T} \frac{R_{NSC_{MAX}} \cdot t}{BB_T} \delta t \ \rightarrow \ R_{NSC_{MAX}} = \frac{2}{BB_T} \ \rightarrow \ R_{NSC}(t) = \frac{2t}{BB_T{}^2}, \tag{4}$$

resolving the ODE, and substituting $R_{S\&F\&R}(t)$ it gives:

$$B_R(t) = \frac{B_0 \cdot e^{\frac{2t+2}{BB_T{}^2}} \cdot e^{-t^2/BB_T{}^2}}{BB_T}, \tag{5}$$

where BB_T is the parameter used in [37], $e^{\frac{2t+2}{BB_T{}^2}}$ is the biomass dependent and $e^{\frac{-x^2}{BB_T{}^2}}$ the maturity dependent factors (expressed in days). Graphically, the equation represents a skewed function of the amount of NSC allowed to be used by the trees of the specific class; the faster the leaves reach maturity, the more daily specific allocation is allowed (Figure 4).

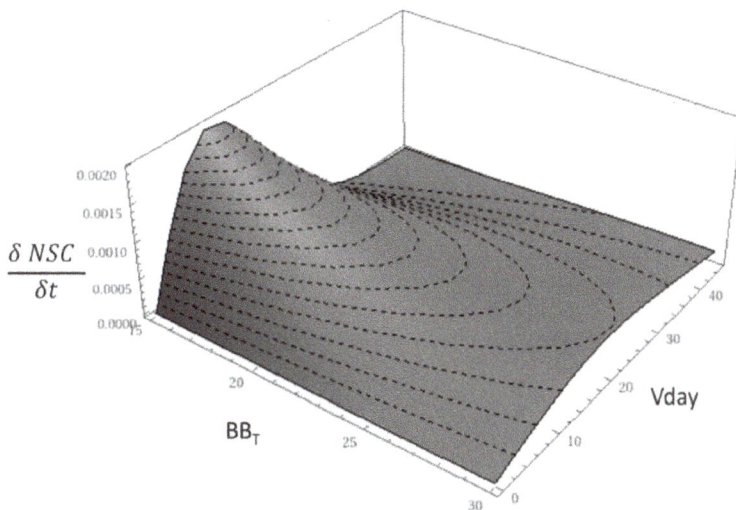

Figure 4. Graphic representation of the C tradeoff function. The axes represent respectively bud burst days (BB_T), vegetative days (Veg_day), and the fraction of total NSC invested in leaf development ($\frac{\partial NSC}{\partial t}$). The shorter BB_T, the higher the maximum NSC fraction.

Another major modification we introduced involves fall leaf yellowing and senescence. Falling leaves contribute to half of annual litter production. Correctly estimating their timing has a strong impact on both GPP and Rh dynamics [32]. The 5.1 version of 3D-CMCC-FEM represents senescence by linearly decreasing leaves biomass of a predefined fraction until their pool is emptied. However, such method systematically overestimated leaf turnover at the beginning of the leaffall phase, and is poorly flexible in calculating its duration. For these reasons, linear loss of leaf biomass was replaced with a sigmoid function. Assuming for hypothesis that all leaves fall by the end of senescence season, the sigmoid function was:

$$\frac{\partial L}{\partial t} = LAI(t) - Max\left(0, \frac{\alpha(h,a,sp)}{1+e^{\frac{t-\beta(t)}{\gamma(t)}}}\right),$$ (6)

where $\alpha(t)$, $\beta(t)$, $\gamma(t)$ are three parameters with biological or physical meaning. In fact:

$$\begin{cases} \alpha(h,a,sp) = LAI_{0+\delta t}(h,a,sp) \\ \beta(s) = t_0(s) + \frac{\Delta t(s)}{2} \\ \gamma(s) \cong \frac{\Delta t(s)}{In(0.1) - In(10(\beta(s)-0.1))} \end{cases},$$ (7)

where $L_{0+\delta t}$ stands for LAI value at peak of green, $t_0(s)$ is the first day of senescence (triggered by a species-specific day-length parameter), $\Delta t(s)$ is the length of the senescence period (days). $\Delta t(s)$ was calculated using [49], as a function of temperature and photoperiod:

$$\begin{cases} R_{sen}(t) = (T_{Max}(s) - T(t)) \cdot \left(\frac{D_L(t)}{D_L(0)}\right)^2 \\ \Delta t(s) = \sum_{R_{sen}(t)=0}^{R_{sen}(t)<Y_c} R_{sen}(t) \end{cases},$$ (8)

where T_{Max} is the maximum temperature at which senescence is effective, $T(t)$ is daily average Temperature, $D_L(0)$ photoperiod at the first day of senescence, $D_L(t)$ photoperiod at the ith day of the year.

2.3.3. Evergreen Phenology

Evergreen canopy turnover was modified from [37]. 3D-CMCCFEM v.5.1 assumes that evergreen leaf turnover is constant throughout the year, and that annual leaf turnover is equal to leaf biomass produced the year before [37,48]. However, leaf turnover seems to be concentrated in specific seasonal windows: (1) consistently to leaf emergence (spring); and (2) approaching of photosynthetic inefficient season (fall) [51]. For this reason, this approach may affect the ability in estimating leaf biomass and GPP intra-annual variability. To better represent leaf turnover dynamics, we developed a new framework where competition for light dynamically affected leaf turnover. We assumed the canopy to be a population of leaves optimized to intercept the highest amount of light. Since leaves cannot move from their position in the canopy, they get partially shaded by new emergent leaves when aging. Conceptually, leaves can be assumed to be in competition for one resource, light, and their turnover can be predicted by using a competition for one resource scheme as in [52]:

$$\begin{cases} \frac{\partial B_i(t)}{\partial t} B_i(t)^{-1} = f_i(R) - f_i(m_i) \\ f_i(R) = \frac{r_i \cdot R}{(R+K_i)} \\ R_i^* = \frac{m_i \cdot Y_i}{(r_i - m_i)} \end{cases},$$ (9)

In this model, R and R_i^* are respectively the concentration of available light in J m^{-2} day^{-1} (i.e., resource that leaves are competing for), and the light needed to survive in a progressively more shaded canopy. r_i is the max photosynthetic rate (gC m^{-2} day^{-1}), m_i is the maintenance respiration in gC m^{-2} day^{-1}, Y_i carbon yield.

We assumed for hypothesis that:

(1) Older leaves live in the shaded portions of the canopy, where light transmitted is reduced following Lambert Beer's exponential decay equation. For this reason, we expect an exponential reduction in absorbed photosynthetically active radiation (APAR);

(2) An age dependent quasi-exponential decay in leaf quantum yield efficiency [53]. These decays impact on the reduction of r_i;

(3) Nitrogen content in older leaves is often lower than in young ones, because of its transfer from portions of the crown with low productivity to portions more exposed to light [53]. Since maintenance respiration is proportional to nitrogen content, we expect an exponential reduction in m_i;

(4) Y_i was assumed to be constant as in [17] because of the joint effect of reduction in respiration rate and quantum yield efficiency.

We assumed that the three components of the Equation (9) may have the following shapes:

$$\begin{cases} m_i = m_0 \cdot e^{-k_\eta \cdot t} + m_m \\ r_i = \left(\alpha_0 \cdot e^{-k_\alpha \cdot t} + \alpha_m \right) \cdot \left(\lambda_0 \cdot e^{-k_\lambda \cdot t} + \lambda_m \right) \\ k_{i=const} \end{cases} , \qquad (10)$$

where t is time, the independent variable. m_0, α_0, λ_0 are MR, quantum yield and APAR at first day of budburst, the second year of life. m_m, α_m, λ_m are the theoretical minimum values of MR, quantum yield and APAR to let a proportion of leaves survive in a non-shading context. k_η, k_α, k_λ are the exponential parameters for the three functions.

Based on these hypotheses Equation (10) becomes:

$$R_i^* = \frac{\left(m_0 \cdot e^{-k_\eta \cdot t} + m_m \right) \cdot k_i}{\left(\alpha_0 \cdot e^{-k_\alpha \cdot t} + \alpha_m \right) \cdot \left(\lambda_0 \cdot e^{-k_\lambda \cdot t} + \lambda_m \right) - \left(m_0 \cdot e^{-k_\eta \cdot t} + m_m \right)}, \qquad (11)$$

Assuming that $k_\eta = k_\alpha = k_\lambda$, and that the denominator of Equation (9) has to be greater than zero for the leaves in the ith generation to survive, R* is a sigmoid positive function. Knowing R*, we can calculate the amount of live leaf biomass of the ith generation $(_{(i)})$ as the inversion of R* (i.e., $S_i^*(t)$). We simplified the theoretical model using the following function:

$$S_i^*(t) = \frac{1}{2}t^2 - \frac{2BF_{LS(i)} + 1}{2}t + BF_{LS(i)}, \qquad (12)$$

where t is the number of days since leaves of the ith generation have emerged. According to this model the theoretical maximum age of each generation $(BF_{LS(i)})$ should correspond to the year in which the R* almost reaches its asymptote. We used Equation (12) to quantify leaf turnover for each generation. About 60% of leaf turnover happened in early spring, when new leaves emerged: the amount of biomass lost every day was proportional to new leaf biomass production the same day. The rest of annual turnover happened in fall, when each leaf biomass generation was reduced of a constant value calculated from Equation (12). No foliar re-sprouting was simulated in fall, even though there are evidences of it for *Quercus ilex* [54]. Leaf biomass reduction was determined by linearly decreasing each Bi to the quantity predicted by the specific parabolic decay for the end of the year.

2.3.4. Production of Fresh Organic Matter

At a relatively fine temporal scale (i.e., daily time step) timing of litter formation and FOM production may be fundamental in correctly estimating Rh [55]. Littering for woody tissues followed the rationale of BIOME-BGM family [48]. Partitioning of leaves and fine root turnover followed [56]. FOM coming from any plant biomass pool to the soil was added to the litter pool. Litter pool was consisting of three sub-pools: metabolic very labile C, structural labile C, and structural resistant C. FOM was divided into the three litter sub pools according to its original C:N ratio, as in [38]:

$$
\begin{cases}
\frac{\partial}{\partial t} C_{i+1} = \frac{\left(CN_{lt}^{-1} - CN_i^{-1}\right)}{\left(CN_i^{-1} - CN_{i+1}^{-1}\right)} \cdot C_{lt} \\[2mm]
\frac{\partial}{\partial t} C_i = \frac{\partial}{\partial t} C_{lt} - \frac{\partial}{\partial t} C_{i+1} \\[2mm]
CN_i < CN_{lt} < CN_{i+1}
\end{cases} \qquad (13)
$$

where C_{lt} is the FOM entering litter from each structural C compartments of the plant. FOM C and N were distributed to the litter sub pools with closest C:N. C_{i+1}, represents the pool with higher recalcitrance. When CN_{lt} was higher than any litter sub pool, all the new C was added to the structural resistant pool; otherwise, if CN_{lt} was lower than the CN of the metabolic pool, all its C and N were added to the very labile sub pool. Litter C dynamically moved from a pool to another. Microbes absorbed and partially immobilized litter C in their biomass, and released it again in the soil, during the humification processes [57].

2.3.5. Optimization

We introduced a new calibration scheme to provide an optimized parameterization, and quantify uncertainty related to the equations used in 3D-CMCC-PSM to simulate NEE, and choice of a subset of 45 species-specific physiological parameters (Table S1). Calibration was performed on each site independently, using a training dataset composed by 3 years of EC daily NEE time-series. We chose 2000–2002 for DEHai, 2001–2003 for FIHyy, ITCpz, and FRPue; 2001–2004 for ITCol; 2006–2008 for ITRen. Because of the strong autocorrelation characterizing time-series, we didn't randomly split our dataset, and decided to use sub-time series of contiguous years [58]. We preferred to use the beginning of each simulation for training, before the global recession following the Financial Crisis of 2007–2008. We decided to do so to (1) facilitate comparison among different sites' simulations; (2) have the calibration start when vegetation structure was known from literature; and (3) reduce the risk for parameters to be fitted over a changing environment instead of eco-physiological properties.

To sample the parameterization space (the realistic values of each physiological parameter of the species simulated in each site), we randomly extracted 1000 parameter-set combinations from prior distributions. Prior distributions were assumed to be the same among individuals of the same species, different across species. We assumed each parameter to follow a truncated normal distribution, to avoid any possibility to have non-realistic negative values. Average and variance were estimated by using values found in literature, as in [37]. We used the same averaged value as in [37] for those parameters whose observations in literature were less than 3, because we didn't have enough information to calculate sample standard deviation. The optimization was performed by choosing the parameterization set maximizing the objective function QF through:

$$
QF = \left(\frac{\sum_i^n Y_i^{obs} \cdot Y_i^{sim} - \frac{\sum Y_i^{obs} \cdot \sum Y_i^{sim}}{n}}{\sqrt{\sum \left(Y_i^{obs}\right)^2 - \frac{\left(\sum Y_i^{obs}\right)^2}{n}} \sqrt{\sum \left(Y_i^{sim}\right)^2 - \frac{\left(\sum Y_i^{sim}\right)^2}{n}}} \right)^2 \times \left[1 - \frac{\sum_i^n \left(Y_i^{obs} - Y_i^{sim}\right)^2}{\sum_i^n \left(Y_i^{obs} - \overline{Y}^{obs}\right)^2} \right], \qquad (14)
$$

where Y^{obs} represents EC daily NEE, Y^{sim} Modeled NEE for the same day, $\overline{Y}^{obs}$ the average EC daily NEE over the train time series. The first part of the RHS of the equation represents the square of the Pearson Correlation coefficient (R), the second the Nash-Sutcliffe Efficiency index (NSE).

2.3.6. Validation Analysis

Results were compared to Eddy Covariance data on long-term daily, monthly and annual averages, over the full series of testing years (~5 years). For the validation, we used 2003–2007 for DEHai, 2004–2008 for FRPue and ITCpz, 2004–2011 for FIHyy, 2005–2012 for ITCol, 2009–2011 for ITRen. Then we evaluated how the model performed in the different seasons aggregating values for months of the same season.

To evaluate the model efficiency, we calculated daily, monthly, and seasonal: (1) R; (2) NSE; (3) root mean square error (RMSE); and (4) mean absolute bias (MAB). Each statistic was considered differently informative [59] as summarized in Table 3. The model's ability in representing observed anomalies was determined by analyzing inter annual variability (IAVs) following [60] and [37].

Table 3. Statistics used for Model's results validation against Eddy Covariance data.

Statistics	Formulation	Use and Ranges
Pearson Coefficient	$r = \dfrac{\sum_i^n Y_i^{obs} \cdot Y_i^{sim} - \frac{\sum Y_i^{obs} \cdot \sum Y_i^{sim}}{n}}{\sqrt{\sum (Y_i^{obs})^2 - \frac{(\sum Y_i^{obs})^2}{n}} \sqrt{\sum (Y_i^{sim})^2 - \frac{(\sum Y_i^{sim})^2}{n}}}$	Estimation of model's measure of correlation with EC data [0;1]
Nash Sutcliffe efficiency	$NSE = 1 - \dfrac{\sum_i^n (Y_i^{obs} - Y_i^{sim})^2}{\sum_i^n (Y_i^{obs} - \overline{Y}^{obs})^2}$	Estimation of model's predictability $[-\infty;1]$
Root Mean Square Error	$RMSE = \sqrt{\dfrac{\sum_i^n (Y_i^{obs} - Y_i^{sim})^2}{n}}$	Estimation of model's accuracy gC m^{-2} day^{-1} [0; ∞]
Mean Absolute Bias	$MAB = \frac{1}{n} \sum_i^n \dfrac{|Y_i^{obs} - Y_i^{sim}|}{\sigma(Y_i^{obs})}$	Estimation of model's bias gC m^{-2} day^{-1} [0; ∞]

3. Results

3.1. Evaluation of Daily, Seasonal, and Annual NEE Estimations

To evaluate 3D-CMCC-PSM NEE predictions, we compared predicted (MD) daily and monthly NEE time series to EC daily data. The analyses were performed only on the test data (i.e., portions of the series which have not been used for calibration) to avoid any effect of overfitting. The model showed high correlations with observed EC data at all sites for both daily and monthly fluxes, apart from ITCpz site (Table 4). Excluding ITCpz, R ranged at all sites from 0.65 to 0.84 for daily, and 0.59 to 0.97 for monthly scale. Beech-dominated Deciduous Forests (DBF) performed better than Conifer species (ENF) and evergreen Mediterranean broadleaved forests (EBF). ENF and EBF in FRPue performed similarly on daily scale, for all the statistics used. However, ENF predictability significantly increased on monthly scale (R ranging between 0.92 and 0.97), while EBF performed worse (R 0.42 in ITCpz, and 0.59 in FRPue). RMSE on average was 1.92 gC m^{-2} day^{-1}. MAE ranged between 0.96 and 1.78 gC m^{-2} day^{-1}, and on average it decreased almost twice on monthly timescale. MAB showed similar behavior for DBF and ENF. It ranged between 0.39 and 0.56 gC m^{-2} day^{-1} (0.50 on average) for daily time series. Mediterranean forests resulted as the ones with highest MAB, and showed no significant reduction when predictions were aggregated on monthly scale. Differently from the other simulations, even NSE just improved slightly for ITCpz, and even reduced for FRPue simulation.

Table 4. Daily and Monthly Validation statistics calculated on the test-set. As stated in Table 3, R and NSE are dimensionless; RMSE and MAB are gC m^{-2} day^{-1}.

	DEHai	ITCol	FIHyy	ITRen	ITCpz	FRPue	Mean
			Daily NEE				
R	0.84	0.76	0.67	0.65	0.24	0.65	0.64
NSE	0.67	0.5	0.34	0.21	-0.26	0.35	0.3
RMSE	1.84	2.7	1.48	2.32	1.8	1.39	1.92
MAB	0.39	0.5	0.53	0.56	1.15	0.76	0.65
			Monthly NEE				
R	0.93	0.92	0.96	0.97	0.42	0.59	0.8
NSE	0.81	0.76	0.9	0.87	0.12	0.2	0.61
RMSE	1.15	1.58	0.45	0.72	1.24	1	1.02
MAB	0.28	0.32	0.21	0.24	1.25	0.86	0.53

Daily results were aggregated in seasonal series to evaluate seasonal predictability. Daily NEE were averaged to give a time series of mean seasonal NEE (gC m^{-2} day^{-1}), one value per season, for the duration of the test dataset. Seasonal aggregations showed that 3D-CMCC-PSM poorly performed in predicting seasonal fluxes. NSE was generally negative in summer, with the exclusion of DEHai and FRPue. 3D-CMCC-PSM generally best reproduced NEE dynamics in fall (R ranging between 0.22 and 0.89). ENF ecosystems showed consistently higher correlation in spring predictions, with R of 0.65 and MAB of 0.62 gC m^{-2} day^{-1} on average. In the case of evergreen stands, 3D-CMCC-PSM consistently showed poor performance in summer. Expectedly, DBF performed the worst in winter (Table 5). NSE on average resulted positive only in fall for both DBF and EBF, and spring, for ENF stands.

Table 5. Seasonal validation statistics calculated on the test-sets and aggregated by ecosystem type. As stated in Tables 3 and 4, R and NSE are dimensionless; RMSE and MAB are gC m^{-2} day^{-1}. MAM stands for March, April, and May. JJA stands for June, July, and August. SON stands for September, October, and November. DJF stands for December, January, and February. DBF stands for Deciduous Broadleaf Forests, EBF for Evergreen Broadleaf Forests, ENF for Evergreen Needle leaf Forests.

	R	NSE	RMSE	MAB
		DBF		
MAM	0.43	−0.36	2.77	0.72
JJA	0.36	−0.02	2.96	1.17
SON	0.82	0.58	1.9	0.58
DJF	0.2	−0.93	0.7	1.58
		EBF		
MAM	0.65	0.28	1.83	0.62
JJA	0.11	−0.92	2.76	1.01
SON	0.51	0.03	1.56	0.74
DJF	0.38	−1.59	0.51	0.82
		ENF		
MAM	0.18	−0.41	1.95	1.2
JJA	0.32	−0.34	1.58	1
SON	0.45	−0.19	1.3	0.79
DJF	0.47	−6.13	1.45	1.4

We used Taylor diagrams [61] to graphically summarize how closely Daily, Monthly and Annual NEE patterns matched EC observations (Figure 5). 3D-CMCC-PSM performance was generally satisfactory. Daily simulations resulted in all sites but ITCpz being within the ±1 normalized standard deviation region. Monthly scale predictions were more consistent with EC data, especially for BDF and ENF sites. It resulted in all 4 simulations falling within ±0.5 normalized standard deviation from the reference point, and $R > 0.9$. Again, 3D-CMCC-PSM performed worst in EBF, with FRPue still inside ±1 normalized standard deviation region, and ITCpz falling outside the ±1.5 normalized SD region. The consistently worse predictability in ITCpz and FRPue confirm a systematic weakness in 3D-CMCC to represent fluxes for these sites as already described in [37]. Model performance on annual scale showed a different pattern, mostly because of some sites' consistent biases in seasonal NEE, and the difference in NEE magnitude. Delay in spring phenology, and the consistent underestimation of summer NEE, resulted in significant underestimation and scarce predictability of ITCol annual NEE ($R < 0.2$). ITCPz and FRPue resulted among the sites with higher annual predictability, partly because of the low seasonal variance in NEE, partly because winter and spring bias tends to compensate each other.

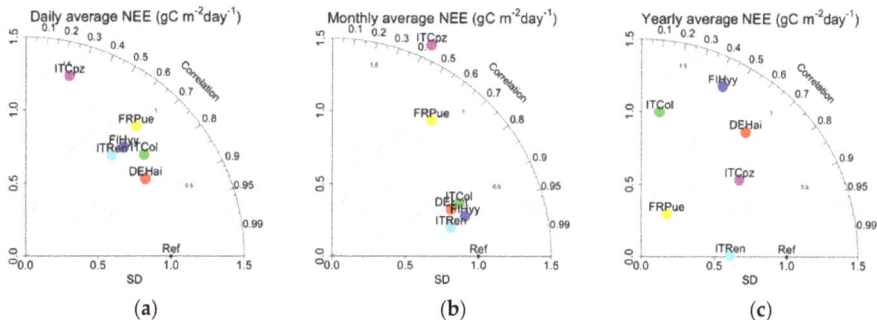

Figure 5. Taylor diagrams representing 3D-CMCC-PSM performance in (**a**) daily, (**b**) monthly and (**c**) annual NEE estimation for the test-set. ITCol and DEHai represent DBF (red and green dots); ITRen and FIHyy ENF (blue and turquoise). ITCpz and FRPue represent EBF (yellow and magenta dots). The closest a simulation lied to the "Ref" point, the better 3D-CMCC-PSM represented NEE patterns. X- and Y- axes represent NEE standard deviation (SD): the closest to 1, the better the performance. Simulations with $R \geq 0.9$, and difference in SD with EC NEE less than 0.5 5 gC m^{-2} day^{-1} showed very good performance. Simulations with $0.75 \leq R < 0.9$, and difference in SD with EC NEE between 0.5 and 1 gC m^{-2} day^{-1} showed good performance. Simulations with $0.35 \leq R < 0.75$, and difference in SD with EC NEE between 1 and 1.5 gC m^{-2} day^{-1} showed sufficiently good performance.

3.2. Anomalies and Parameters Related Uncertainty

Figure 6 shows uncertainty associated to random choice of parameters. Overall, uncertainty was expectedly higher in summer and fall. Such increase was particularly clear for deciduous forests, which not only showed wider NEE standard deviation, but also had optimal modeled NEE falling outside standard deviation area.

DBF sites showed also high uncertainty in estimating the first vegetative day, suggesting that a better representation of winter dormancy effects on budburst dates may significantly improve model's predictability. Uncertainty was generally lower in Mediterranean sites, probably because model's performance was generally lower, and fitting parameters to data would have little effect on performance. 3D-CMCC-PSM uncertainty was generally low for ENF for most of the year, but was generally high when temperatures were higher. Higher uncertainty for warmer days was generally found in DBF sites too, suggesting that 3D-CMCC-PSM was expectedly sensitive to high temperatures for both photosynthesis [62] and respiration [23]; because of such variability, calibrating parameters on data resulted in a significant boost in model's predictions.

NEE inter-annual variability was generally underestimated by the model. Nevertheless, 3D-CMCC-PSM correctly reproduced 81% of the sign of the anomalies, and residual difference in magnitude was usually less than 0.3 gC m^{-2} day^{-1}. Highest difference in magnitude occurred in ITCol (difference in residuals higher than 0.5 gC m^{-2} day^{-1} in 5 years out of 12). Highest difference was shown in ITCpz, where the sign was correctly reproduced only once out of 8 years, and having more than 1gC m^{-2} day^{-1} of residual difference (Figure 7).

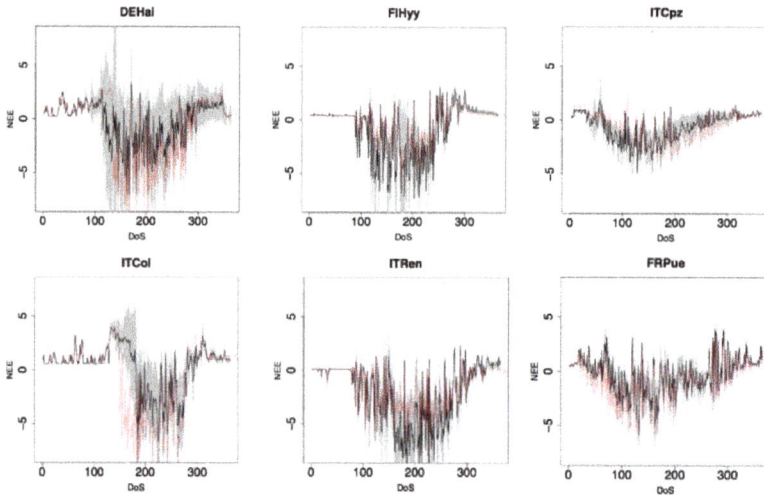

Figure 6. Model structure-related uncertainty in estimating NEE (gC m^{-2} day^{-1}) per DoS (Day of Simulation) by a random choice of parameter values from prior distributions. Data represent 300 1-year simulations from randomly extracted parameterization-sets. Average daily simulations (black lines) and standard deviation (grey areas). Red dotted lines represent daily NEE simulation for the optimized parameterization set (Table S2). First column represents DBF sites, the second ENF, the third EBF.

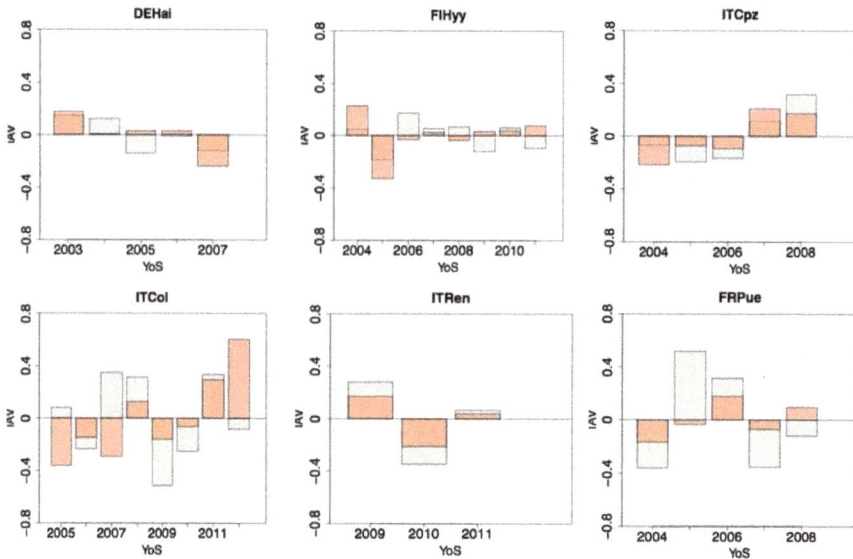

Figure 7. Inter-annual variability (IAV) for the test time-series (gC m^{-2} month^{-1}). Observed IAV in gray boxes, simulated IAV in orange. First column represents DBF sites, the second ENF, the third EBF.

3.3. Comparison with the 5.1 Version of 3D-CMCC-FEM

We compared the results of 3D-CMCC-FEM 5.1 [37] and 3D-CMCC-PSM for daily and monthly GPP for ITRen, FRPue, DEHai and FIHyy. We used the same non-optimized parameterization set for both versions. Except for FR-Pue, 3D-CMCC-PSM showed lower RMSE and higher R for daily GPP. Daily NSE too was generally higher in 3D-CMCC-PSM, except for FIHyy. Monthly aggregated predictions were consistently outperforming those of 3D-CMCC-FEM 5.1 (Table 6).

Table 6. Comparisons between 3D-CMCC-FEM 5.1 (5.1) and 3D-CMCC-PSM (PSM) versions, using the same parameterization for 4 out of 6 sites (ITRen, FRPue, DEHai, FIHyy). As stated in Table 3, R and NSE are dimensionless; RMSE is gC m^{-2} day^{-1}. Bold values represent best performing version.

	Version	ITRen	FRPue	DEHai	FIHyy	Avg
Daily	5.1	0.81	**0.82**	0.92	0.91	**0.87**
R	PSM	**0.88**	0.64	**0.93**	0.91	0.85
Monthly	5.1	0.95	0.64	0.97	0.96	0.89
R	PSM	**0.96**	**0.84**	**0.98**	0.96	**0.94**
Daily	5.1	0.61	−0.54	0.84	**0.87**	0.45
NSE	PSM	**0.72**	**0.09**	**0.96**	0.76	**0.63**
Monthly	5.1	0.91	−0.11	0.94	0.91	0.66
NSE	PSM	0.91	**0.56**	**0.98**	**0.92**	**0.84**
Daily	5.1	2.09	**1.52**	1.85	1.56	1.76
RMSE	PSM	**1.59**	1.96	**1.91**	**1.57**	1.76
Monthly	5.1	0.97	**1.01**	1.07	0.91	0.99
RMSE	PSM	**0.82**	1.09	**0.82**	**0.93**	**0.92**

3.4. Daily and Monthly Reco

We evaluated Reco (ecosystem respiration) by comparing modeled (MD) and observed (EC) daily and monthly time series. Daily R ranged between 0.45 in ITCpz and 0.9 in FIHyy (Table 7). RMSE was of 1.28 gC m^{-2} day^{-1} on average, and MEB ranged between 0.43 and 1 gC m^{-2} day^{-1}. Most of the bias happened in summer, where Reco was generally overestimated, especially in ITCol and ITCpz. NSE was positive in any case but ITCpz. It was generally lower in DBF (0.32), and higher in FIHyy (0.60) and FRPue (0.57). Reco at monthly timescale was strongly improved in predictability, especially for ITRen, whose R increased to 0.67 and NSE to 0.84. Monthly predictions showed improvements for the other simulations too, improving R and NSE of about 0.07. Since most of the bias occurred in summer, monthly predictions showed no dramatic improvements for neither RMSE nor MEB. The only exception was ITRen, whose RMSE reduced of about 0.4 gC m^{-2} day^{-1}, suggesting that daily Reco may be noisy.

Table 7. Daily and Monthly Validation statistics for R$_{eco}$ calculated on the test-set. As stated in Table 3, R and NSE are dimensionless; RMSE and MAB are gC m^{-2} day^{-1}.

	DEHai	ITCol	FIHyy	ITRen	ITCpz	FRPue	Mean
			Daily Reco				
R	0.79	0.71	0.90	0.67	0.45	0.86	0.73
NSE	0.32	0.32	0.60	0.34	−0.43	0.57	0.29
RMSE	1.29	1.83	1.18	1.03	1.65	0.70	1.28
MEB	0.63	1.00	0.43	0.62	0.98	0.48	0.69
			Monthly Reco				
R	0.86	0.79	0.96	0.84	0.54	0.93	0.82
NSE	0.40	0.37	0.66	0.69	−0.40	0.67	0.40
RMSE	1.10	1.67	1.04	0.64	1.49	0.54	1.08
MEB	0.60	0.99	0.41	0.50	1.04	0.43	0.66

4. Discussion

4.1. 3D-CMCC-PSM Predictability in Estimating NEE

In general, the inclusion of a simplistic SOC routine resulted in a reliable estimation of daily and monthly NEE patterns. While daily and monthly patterns are consistent with EC data, seasonal patterns showed non-negligible misrepresentations, which resulted in negative NSE in most of the cases. This inconsistency may be driven by the strong seasonality in both R_{eco} and GPP [63], which positively affects correlation between EC data and MD results.

NEE patterns during summer and fall were much more consistent with measured ones, than winter and spring patterns. During these seasons the biases appeared mostly affected by estimation of R_{eco}. The scarcity of the model in representing EBF C fluxes was especially attributable to GPP predictions. 3D-CMCC-PSM and 3D-CMCC FEM inability in predicting GPP in ITCpz and FRPue sites, denoted the necessity to better represent the relations between Mediterranean forests and environmental factors [64,65]. In FRPue the model well reproduced spring, summer and fall NEE. On the other hand, it showed a bias of around 1 gC m^{-2} day^{-1} in winter, suggesting it was missing some particularly important seasonal processes. For example, evergreen phenology still didn't consider secondary or continuous growth. Thus, species like *Quercus ilex*, which exhibit secondary gem sprouts in fall [54], have fresh leaves and mild temperatures to guarantee photosynthetic activity in fall and winter, partly explaining 3D-CMCC FEM and 3D-CMCC-PSM systematic underestimation.

ITCpz showed the same pattern. However, differently from FRPue, it poorly performed also in early spring and summer, especially for GPP patterns. This poor performance was expected, because of the physical characteristics of the site. In fact, 3D-CMCC FEM soil water dynamics routine was still simplistic, and to date, such other similar models have not included any effect of water table dynamics. On the contrary, ITCpz is characterized with the presence of a shallow groundwater table, which seems to reduce water stress in early summer [37]. Moreover, we used average daily meteo data collected from the Eddy Covariance stations, and initialized simulations using average structural information found in literature. Uncertainty about those data was potentially high, and could have dramatically affected 3D-CMCC-PSM results.

Summer NEE misrepresentation in DBF was probably affected by the assumption that LAI and photosynthetic capacity reach their maximum in early summer, at the same time. On the contrary, maximum photosynthetic capacity may be reached in late summer, and varies across the canopy. Without taking this into account, GPP could be overestimated up to 40% [66]. Notwithstanding, comparing model outputs with published works [67,68], these defects are common also for other PBMs.

Seasonal patterns showed that the model consistently misrepresented NEE in winter, suggesting that R_{eco} still needs to be improved. Especially for DBF sites (e.g., DEHai), winter R_{eco} was mostly driven by RH. RH was exponentially affected by soil temperatures and especially moisture [69], which are calculated by the model, and could be over-fluctuating in winter. Moreover, EC data are prone to random noise [70], whose relative impact on performance metrics may be relatively larger.

Interestingly, annual predictions suggested reasonably high performance of 3D-CMCC-PSM, despite these seasonal inconsistencies. This suggests that biases are usually consistent within a season, but have different signs across seasons (Figure 8), resulting in a compensating effect at coarser time scale.

It was not always possible to individually validate the different components of respiration. Since EC Reco was not measured but inferred, evaluation metrics should be interpreted as a general ability of 3D-CMCC-PSM in predicting it. Despite daily Reco being noisy, 3D-CMCC-PSM could reproduce respiration processes well enough, at least at monthly timescale.

NPP:GPP ranged between 0.37 and 0.62, and was consistent with literature [71–73]. It was generally higher in DBF (0.49 in DEHai, 0.55 in ITCol), lower in EBF (0.38 in FRPue, 0.41 in ITCpz). These results matched with those of [71] who showed that the ratio between NPP and GPP (CUE) is generally 0.53, ranging from 0.23 to 0.83. CUE was relatively low in FIHyy, where Reco predictions were overestimated: therefore, poor predictability was probably ascribable to excesses in RA.

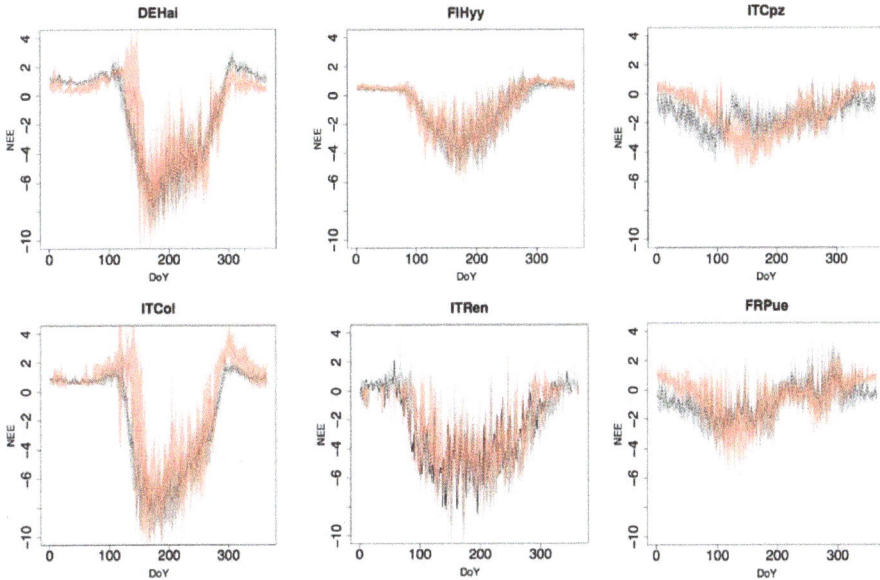

Figure 8. Patterns in daily NEE (gC m^{-2} day^{-1}) per DoY (Day of Year) calculated from test-sets on a site level. Observed EC average patterns (black dotted line) and standard deviation (gray area). Simulated average patterns (red dotted line) and standard deviation (orange area). First column represents DBF sites, the second ENF, the third EBF.

Ecosystem respiration was overestimated during summer, causing NEE systematic overestimation at each simulation, especially in ITCol. Except for FRPue and ITCol, RA grew about 1 gC m^{-2} day^{-1} higher than RH, suggesting that it may be the principal driver of biased summer Reco (Figure 9). This misbehavior may be related to the method used to estimate maintenance respiration, an exponential relationship between respiration, moisture, and temperature [61].

Lack of data to validate the SOC dynamics reduced the spectrum of speculations, which could be statistically analyzed. SOC didn't change its quantity in ten years; this result was consistent with the theoretical stability of the SOC, an indicator which rarely changes within 10 years if no strong disturbance events (e.g., land use changes) have occurred [74]. Litter C was highly fluctuating within a year, but its quantity was stable if compared at the end of each year. This suggested that the model realistically represented litter turnover and decomposition, since residues were degraded into humus labile substances about within a year [75]. Microbial biomass was highly variable, as expected. However, the magnitude of change was too broad throughout the simulations. These results may be related to the use of 5% as the initial active microbial biomass for each site, a value that may be far from the equilibrium for different soils. Moreover, tradeoffs within microbial growth and between the environmental conditions may be scarcely represented. As a matter of fact, 3D-CMCC-PSM simulated the soil as having the same physical-chemical structure throughout the profile. This implied that microbes could find the same amount of C, O$_2$, and living space, with no depth limitation.

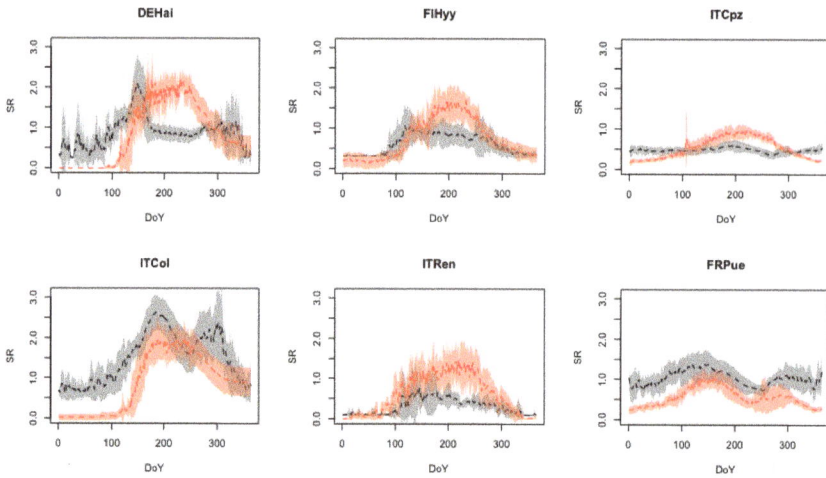

Figure 9. Patterns in daily soil respiration (SR) heterotrophic and autotrophic components (gC m^{-2} day^{-1}). Patterns per DoY (Day of Year) calculated from test-sets on a site level. Microbial respiration average patterns (black-dotted lines) and standard deviation (gray areas). Autotrophic root respiration average patterns (red-dotted lines) and standard deviation (orange areas). First column represents DBF sites, the second ENF, the third EBF.

4.2. 3D-CMCC-PSM Uncertainty in Estimating NEE

We analyzed 300 random parameterization-sets per site to quantify model assumptions and uncertainty. The model showed different behaviors in different sites, but expectedly consistent across species behaving in a similar functional way. These results may suggest that using functional traits combinations to provide physiological parameters, instead of fixed species-specific ones, may produce still reliable and more general predictions, particularly useful in case of larger spatial/temporal simulations [76–78]. Using a species-level parameterization, in fact, may result in a too-fine "resolution" because: (1) it would require excessive computational resources and a finely-detailed parameterization, usually inaccessible on a broad scale [79]; and (2) the model's rationale in predicting forest structure is mainly driven by competition for resources. However, there are not explicit tradeoffs, positive interactions between different tree cohorts, or intra-specific trait variability, which are fundamental to forecast forest ecosystem structure on long-run simulations [80]. Having fixed species-specific parameters throughout a century would potentially result that only a very reduced amount of species would dominate the different cohorts on landscape to regional scale.

According to Figures 6–8, strong uncertainties still reside in timing for the different phenological phases. The biggest source of uncertainty in deciduous stands was driven by the amount of degree days needed to begin the vegetative period. The use of a site-specific thermal sum (GDD) to activate vegetation period is widely used, but proven to be very site sensitive, and not very effective for a regional generalization [81]. On the other hand, the processes triggering bud-burst timing are still partly unknown. Moreover, those models proposing a process-oriented promoter-inhibition rationale are generally over-complex and not prone to spatial generalization [82]. A possible solution in this context is to use remotely-sensed data to train a latitudinal-explicit regression, constraining GDD estimation.

3D-CMCC-PSM showed high uncertainty also in catching the beginning of the senescence phase. The new phenological scheme didn't reduce such uncertainty, since it was still using a photoperiod threshold as the senescence phase trigger [37]. Another strong source of uncertainty in summer GPP may be held by the over-simplicity of soil structure and thus of the soil water routine.

Forests **2017**, *8*, 220

As shown in [37], EC data are prone to high uncertainty. We focused on NEE fluxes to reduce the uncertainty cascade related to NEE partitioning. The next natural step will be reframing the model with a hierarchical Bayesian fashion, to quantify error propagation and parameter uncertainty from the posterior distribution [83].

Daily R_{eco} estimation was affected by the cascade of uncertainties related to the calculation of R_A and heterotrophic respiration, calculated independently. R_A routine may strongly be influenced by uncertainties in R_M estimation, which often resulted in R_A overestimation. The R_M was in fact simulated by a set of empirical relations, which involve the use of a fixed non-acclimating Q10 factor, whose generality is known to be inaccurate [84]. Moreover, the rationale of Ryan's R_M calculation [85] is affected by uncertainty in estimating daily increment of N pools, generally estimated by forest ecosystem models as a fixed proportion of daily C increment.

5. Conclusions

Soil respiration has a key role in determining NEE in a deterministic fashion [86]. In general, this work showed how the inclusion of a simplistic soil carbon routine allowed prediction of trends and variability of NEE across the most diffuse European forest ecosystems. Modifications in the phenology scheme produced slight improvements in predicting GPP. However, they were still limited by the correct estimation of bud burst timing, leaf senescence starting point and duration. The use of an optimized parameter-set improved the model's performance only for those sites where the bio-geophysical processes were correctly reproduced. As a matter of fact, we showed how Mediterranean terrestrial forests, which showed lacks in representing some biological and/or physical processes, performed significantly worse than the other modeled sites, regardless of the use of optimized parameters.

In conclusion, we think that 3D-CMCC-PSM can reliably estimate NEE and R_{eco} dynamics in a forest ecosystem, especially in scaling up daily results to monthly NEE averages. We think that 3D-CMCC-PSM is a solid basis to further explore the effects of soil structure on carbon and Water dynamics, especially in Mediterranean systems, and be used as a tool for predicting forest growth and ecosystem services, and address questions related to future scenario forecasting.

Supplementary Materials: The following are available online at www.mdpi.com/1999-4907/8/6/220/s1, Table S1: Mean and standard deviation of the priors used to sample parameters used for the optimization, Table S2: Values of the parameters used for the 6 simulations, produced by the oprimization scheme.

Acknowledgments: S.M. was partially supported by the Gordon and Betty Moore Foundation's Data-Driven Discovery Initiative through Grant GBMF4563 to Ethan P. White. This work was partially supported by the Italian Ministry of Education, University and Research and the Italian Ministry of Environment and Protection of Land and Sea under the project GEMINA (n. 232/2011). We finally thanks the University of Tuscia for additional funds provided to the research. A.C. was partially supported by the European Union's Horizon 2020 research and innovation program under grant agreement N. 641816 (CRESCENDO project; http://crescendoproject.eu/) and by the "ALForLab" (PON03PE_00024_1) project co-funded by the National Operational Program for Research and Competitiveness (PON R&C) 2007-2013, through the European Regional Development Fund (ERDF) and national resource (Revolving Fund—Cohesion Action Plan (CAP) MIUR. This work used eddy covariance data acquired and shared by the FLUXNET community, including these networks: AmeriFlux, AfriFlux, AsiaFlux, CarboAfrica, CarboEuropeIP, CarboItaly, CarboMont, ChinaFlux, Fluxnet-Canada, GreenGrass, ICOS, KoFlux, LBA, NECC, OzFlux-TERN, TCOS-Siberia, and USCCC. The ERA-Interim reanalysis data are provided by ECMWF and processed by LSCE. The FLUXNET eddy covariance data processing and harmonization was carried out by the European Fluxes Database Cluster, AmeriFlux Management Project, and Fluxdata project of FLUXNET, with the support of CDIAC and ICOS Ecosystem Thematic Center, and the OzFlux, ChinaFlux and AsiaFlux offices. We acknowledge Ethan P. White and Monia Santini for providing critical comments, Carlo Trotta and Giorgio Matteucci for insights about the use of Eddy Covariance data, and Sergio Noce for his help in generating GIS-based maps.

Author Contributions: S.M. conceived the project, designed the experiments, co-developed the model code, performed the simulations, analyzed the data, and wrote the manuscript. A.C. contributed in designing the experiment, co-developing the model code, and writing the manuscript. T.C., A.N. and R.V. contributed in writing the manuscript.

Conflicts of Interest: The authors declare no conflict of interest.

Forests 2017, 8, 220

References

1. Baldocchi, D.; Ryu, Y.; Keenan, T. Terrestrial Carbon Cycle Variability. F1000Res 2016, 5. [CrossRef] [PubMed]
2. Göckede, M.; Foken, T.; Aubinet, M.; Aurela, M.; Banza, J.; Bernhofer, C.; Bonnefond, J.M.; Brunet, Y.; Carrara, A.; Clement, R.; et al. Quality control of CarboEurope flux data & ndash; Part 1: Coupling footprint analyses with flux data quality assessment to evaluate sites in forest ecosystems. Biogeosciences 2008, 5, 433–450. [CrossRef]
3. Xiao, J.; Zhuang, Q.; Law, B.E.; Chen, J.; Baldocchi, D.D.; Cook, D.R.; Oren, R.; Richardson, A.D.; Wharton, S.; Ma, S.; et al. A continuous measure of gross primary production for the conterminous U.S. derived from MODIS and AmeriFlux data. Remote Sens. Environ. 2010, 114, 576–591. [CrossRef]
4. Tang, X.G.; Liu, D.W.; Song, K.S.; Munger, J.W.; Zhang, B.; Wang, Z.M. A new model of net ecosystem carbon exchange for the deciduous-dominated forest by integrating MODIS and flux data. Ecol. Eng. 2011, 37, 1567–1571. [CrossRef]
5. Reichstein, M.; Falge, E.; Baldocchi, D.D.; Papale, D.; Valentini, R.; Aubinet, M.; Berbigier, P.; Bernhofer, C.; Buchmann, N.; Gilmanov, T.; et al. On the separation of net ecosystem exchange into assimilation and ecosystem respiration: Review and improved algorithm. Glob. Chang. Biol. 2005, 11, 1–16. [CrossRef]
6. Lasslop, G.; Migliavacca, M.; Bohrer, G.; Reichstein, M.; Bahn, M.; Ibrom, A.; Jacobs, C.; Kolari, P.; Papale, D.; Vesala, T.; et al. On the choice of the driving temperature for eddy-covariance carbon dioxide flux partitioning. Biogeosciences 2012, 9, 5243–5259. [CrossRef]
7. Potter, C.S.; Randerson, J.T.; Field, C.B.; Matson, P.A.; Vitousek, P.M.; Mooney, H.A.; Klooster, S.A. Terrestrial ecosystem production: A process model based on global satellite and surface data. Glob. Biogeochem. Cycles 1993, 74, 811–841. [CrossRef]
8. Prince, S.D.; Goward, S.N. Global primary production: A remote sensing approach. J. Biogeogr. 1995, 22, 815–835. [CrossRef]
9. Mäkelä, A.; Landsberg, J.; Ek, A.R.; Burk, T.E.; Ter-Mikaelian, M.; Agren, G.I.; Oliver, C.D.; Puttonen, P. Process-based models for forest ecosystem management: Current state of the art and challenges for practical implementation. Tree Physiol. 2000, 20, 289–298. [CrossRef] [PubMed]
10. Yuan, W.P.; Liu, S.G.; Zhou, G.S.; Zhou, G.Y.; Tieszen, L.L.; Baldocchi, D.; Bernhofer, C.; Gholz, H.; Goldstein, A.H.; Goulden, M.L.; et al. Deriving a light use efficiency model from eddy covariance flux data for predicting daily gross primary production across biomes. Agric. For. Meteorol. 2007, 1433, 189–207. [CrossRef]
11. Yuan, W.; Cai, W.; Xia, J.; Chen, J.; Liu, S.; Dong, W.; Merbold, L.; Law, B.; Arain, A.; Beringer, J.; et al. Global comparison of light use efficiency models for simulating terrestrial vegetation gross primary production based on the LaThuile database. Agric. For. Meteorol. 2014, 192–193, 108–120. [CrossRef]
12. Cai, W.W.; Yuan, W.P.; Liang, S.L.; Zhang, X.T.; Dong, W.J.; Xia, J.Z.; Fu, Y.; Chen, Y.; Liu, D.; Zhang, Q. Improved estimations of gross primary production using satellite-derived photosynthetically active radiation. J. Geophys. Res. Biosci. 2014, 119. [CrossRef]
13. Lacointe, A. Carbon allocation among tree organs. A review of basic processesand representation in functional-structural tree models. Ann. For. Sci. 2000, 57, 521–533. [CrossRef]
14. Sievänen, R.; Nikinmaa, E.; Nygren, P.; Ozier-Lafontaine, H.; Perttunen, J.; Hakula, H. Components of functional-structural tree models. Ann. For. Sci. 2000, 57, 399–412. [CrossRef]
15. Lu, M.; Nygren, P.; Perttunen, J.; Pallardy, S.G.; Larsen, D.R. Application of the functional-structural tree model LIGNUM to growth simulation of short-rotation eastern cottonwood. Silva Fenn. 2011, 45, 431–474. [CrossRef]
16. Nikinmaa, E.; Sievänen, R.; Hölttä, T. Dynamics of leaf gas exchange, xylem and phloem transport, water potential and carbohydrate concentration in a realistic 3-D model tree crown. Ann. Bot. 2014, 114, 653–666. [CrossRef] [PubMed]
17. Monteith, J.; Moss, C.J. Climate and the efficiency of crop production in Britain. Philos. Trans. R. Soc. Lond. 1977, 281, 277–294. [CrossRef]
18. Trumbore, S. Carbon respired by terrestrial ecosystems-recent progress and challenges. Glob. Chang. Biol. 2006, 12, 141–153. [CrossRef]
19. Mitchell, S.; Beven, K.; Freer, J. Multiple sources of predictive uncertainty in modeled estimates of net ecosystem CO$_2$ exchange. Ecol. Model. 2009, 220, 3259–3270. [CrossRef]

20

20. Yuan, F.; Arain, M.A.; Barr, A.G.; Black, T.A.; Bourque, C.P.A.; Coursolle, C.; Margolis, H.A.; Mccaughey, J.H.; Wofsy, S.C. Modeling analysis of primary controls on net ecosystem productivity of seven boreal and temperate coniferous forests across a continental transect. *Glob. Chang. Biol.* **2008**, *14*, 1765–1784. [CrossRef]

21. Xu, B.; Yang, Y.; Li, P.; Shen, H.; Fang, J. Global patterns of ecosystem carbon flux in forests: A biometric databased synthesis. *Glob. Biogeochem. Cycles* **2014**, *28*, 962–973. [CrossRef]

22. Zobitz, J.M.; Moore, D.J.P.; Sacks, W.J.; Monson, R.K.; Bowling, D.R.; Schimel, D.S. Integration of process-based soil respiration models with whole-ecosystem CO_2 measurements. *Ecosystems* **2008**, *11*, 250–269. [CrossRef]

23. Lloyd, J.; Taylor, J.A. On the temperature dependence of soil respiration. *Funct. Ecol.* **1994**, *8*, 315–323. [CrossRef]

24. Kirschbaum, M.U.F. Soil respiration under prolonged soil warming: Are rate reductions caused by acclimation or substrate loss? *Glob. Chang. Biol.* **2004**, *10*, 1870–1877. [CrossRef]

25. Xu, J.; Chen, J.; Brosofske, K.; Li, Q.; Weintraub, M.; Henderson, R.; Wilske, B.; John, R.; Jensen, R.; Li, H.; et al. Influence of timber harvesting alternatives on forest soil respiration and its biophysical regulatory factors over a 5-year period in the Missouri Ozarks. *Ecosystems* **2011**, *14*, 1310–1327. [CrossRef]

26. Schimel, J.P.; Weintraub, M.N. Soil organic matter does not break itself down the implications of exoenzyme activity on microbial carbon and nitrogen limitation in soil: A theoretical model. *Soil Biol. Biochem.* **2003**, *35*, 549–563. [CrossRef]

27. Tian, H.; Lu, C.; Yang, J.; Banger, K.; Huntzinger, D.N.; Schwalm, C.R.; Michalak, A.M.; Cook, R.; Ciais, P.; Hayes, D.; et al. Global patterns and controls of soil organic carbon dynamics as simulated by multiple terrestrial biosphere models: Current status and future directions. *Glob. Biogeochem. Cycles* **2015**, *29*, 775–792. [CrossRef] [PubMed]

28. Chapin, S., III; McFarland, J.; David McGuire, A.; Euskirchen, E.S.; Ruess, R.W.; Kielland, K. The changing global carbon cycle: Linking plant-soil carbon dynamics to global consequences. *J. Ecol.* **2009**, *97*, 840–850.

29. Hibbard, K.A.; Law, B.E.; Reichstein, M.; Sulzman, J. An analysis of soil respiration across northern hemisphere temperate ecosystems. *Biogeochemistry* **2005**, *73*, 29–70. [CrossRef]

30. Knohl, A.; Schulze, E.D.; Kolle, O.; Buchmann, N. Large carbon uptake by an unmanaged 250-year-old deciduous forest in Central Germany. *Agric. For. Meteorol.* **2003**, *118*, 151–167. [CrossRef]

31. Facelli, J.M.; Pickett, S.T.A. Plant litter: Its dynamics and effects on plant community structure. *Bot. Rev.* **1991**, *57*, 1–32. [CrossRef]

32. Richardson, A.D.; Anderson, R.S.; Arain, M.A.; Barr, A.G.; Bohrer, G.; Chen, G.; Chen, J.M.; Ciais, P.; Davis, K.J.; Desai, A.R.; et al. Terrestrial biosphere models need better representation of vegetation phenology: Results from the North American Carbon Program Site Synthesis. *Glob. Chang. Biol.* **2012**, *18*, 566–584. [CrossRef]

33. Kucharik, C.J.; Barford, C.C.; El Maayar, M.; Wofsy, S.C.; Monson, R.K.; Baldocchi, D.D. A multiyear evaluation of a Dynamic Global Vegetation Model at three AmeriFlux forest sites: Vegetation structure, phenology, soil temperature, and CO_2 and H_2O vapor exchange. *Ecol. Model.* **2006**, *196*, 1–31. [CrossRef]

34. Ryu, S.R.; Chen, J.; Noormets, A.; Bresee, M.K.; Ollinger, S.V. Comparisons between PnET-Day and eddy covariance based gross ecosystem production in two Northern Wisconsin forests. *Agric. For. Meteorol.* **2008**, *148*, 247–256. [CrossRef]

35. Richardson, A.D.; Keenan, T.F.; Migliavacca, M.; Ryu, Y.; Sonnentag, O.; Toomey, M. Climate change, phenology, and phenological control of vegetation feedbacks to the climate system. *Agric. For. Meteorol.* **2013**, *169*, 156–173. [CrossRef]

36. Collalti, A.; Perugini, L.; Santini, M.; Chiti, T.; Nolè, A.; Matteucci, G.; Valentini, R. A process-based model to simulate growth in forests with complex structure: Evaluation and use of 3D-CMCC Forest Ecosystem Model in a deciduous forest in Central Italy. *Ecol. Model.* **2014**, *272*, 362–378. [CrossRef]

37. Collalti, A.; Marconi, S.; Ibrom, A.; Trotta, C.; Anav, A.; D'andrea, E.; Grünwald, T. Validation of 3D-CMCC Forest Ecosystem Model (v. 5.1) against eddy covariance data for 10 European forest sites. *Geosci. Model Dev.* **2016**, *9*, 479–504. [CrossRef]

38. Zhang, Y.; Li, C.; Zhou, X.; Moore, B. A simulation model linking crop growth and soil biogeochemistry for sustainable agriculture. *Ecol. Model.* **2002**, *151*, 75–108. [CrossRef]

39. FLUXNET-Fluxdata. Available online: http://fluxnet.fluxdata.org (accessed on 3 June 2014).

40. Papale, D.; Reichstein, M.; Aubinet, M.; Canfora, E.; Bernhofer, C.; Longdoz, B.; Kutsch, W.; Rambal, S.; Valentini, R.; Vesala, T.; et al. Towards a standardized processing of Net Ecosystem Exchange measured with eddy covariance technique: Algorithms and uncertainty estimation. *Biogeosciences* **2006**, *3*, 571–583. [CrossRef]

41. Mencuccini, M.; Bonosi, L. Leaf/sapwood area ratios in Scots pine show acclimation across Europe. *Can. J. For. Res.* **2001**, *31*, 442–456. [CrossRef]

42. Pilegaard, K.; Mikkelsen, T.N.; Beier, C.; Jensen, N.O.; Ambus, P.; Ro-Poulsen, H. Field Measurements of Atmosphere—Biosphere Interactions in a Danish Beech Forest. *Boreal Environ. Res.* **2003**, *8*, 315–333.

43. Landsberg, J.J.; Waring, R.H. A generalised model of forest productivity using simplified concepts of radiation-use efficiency, carbon balance and partitioning. *For. Ecol. Manag.* **1997**, *95*, 209–228. [CrossRef]

44. Molina, J.A.E.; Clapp, C.E.; Shaffer, M.J.; Chichester, F.W.; Larson, W.E. NCSOIL, A Model of Nitrogen and Carbon Transformations in Soil: Description, Calibration, and Behavior1. *Soil Sci. Soc. Am. J.* **1983**, *47*, 85. [CrossRef]

45. Sollins, P.; Homann, P.; Caldwell, B.A. Stabilization and destabilization of soil organic matter: Mechanisms and controls. *Geoderma* **1996**, *74*, 65–105. [CrossRef]

46. Tamon, F.; Kobayashi, K.; Li, C.; Yagi, K.; Hasegawa, T. Revising a Process-based Biogeochemistry Model (DNDC) to Simulate Methane Emission from Rice Paddy Fields under Various Residue Management and Fertilizer. *Glob. Chang.* **2008**, *14*, 382–402. [CrossRef]

47. Coleman, K.; Jenkinson, D.S. "ROTHC-26.3." A Model for the Turnover of Carbon in Soils, No. November 1999. Available online: http://www.uni-kassel.de/~w_dec/Modellierung/wdec-rothc_manual.pdf (accessed on 1 October 2014).

48. Running, S.W.; Raymond Hunt, E., Jr. *Generalization of a Forest Ecosystem Process Model for Other Biomes, BIOME-BCG, and an Application for Global-Scale Models*; Academic Press, Inc.: London, UK, 1993; pp. 141–158.

49. Delpierre, N.; Dufrêne, E.; Soudani, K.; Ulrich, E.; Cecchini, S.; Boé, J.; François, C. Modelling Interannual and Spatial Variability of Leaf Senescence for Three Deciduous Tree Species in France. *Agric. For. Meteorol.* **2009**, *149*, 938–948. [CrossRef]

50. Keel, S.G.; Schädel, C. Expanding leaves of mature deciduous forest trees rapidly become autotrophic. *Tree Physiol.* **2010**, *30*, 1253–1259. [CrossRef] [PubMed]

51. Chabot, B.F.; Hicks, D.J. The Ecology of Leaf Life Spans. *Ann. Rev. Ecol. Syst.* **1982**, *13*, 229–259. Available online: http://www.jstor.org/stable/2097068 (accessed on 6 October 2015). [CrossRef]

52. Tilman, D. Constraints and tradeoffs: Toward a predictive theory of competition and succession. *Oikos* **1990**, *58*, 3–15. [CrossRef]

53. Waring, R.H.; Landsberg, J.J.; Williams, M. Net Primary Production of Forests: A Constant Fraction of Gross primary production? *Tree Physiol.* **1998**, *18*, 129–134. [CrossRef] [PubMed]

54. Praciak, A.; Pasiecznik, N.; Sheil, D.; van Heist, M.; Sassen, M.; Correia, C.S.; Dixon, C.; Fyson, G.; Rushford, K.; Teeling, C. *The CABI Encyclopedia of Forest Trees*; CABI: Oxfordshire, UK, 2013.

55. Kuzyakov, Y. Priming effects: Interactions between living and dead organic matter. *Soil Biol. Biochem.* **2010**, *42*, 1363–1371. [CrossRef]

56. Thornton, P. *Biome BGC Version 4.2: Theoretical Framework of Biome-BGC*; Technical Documentation: Missoula, MT, USA, 2010.

57. Liang, B.; Lehmann, J.; Sohi, S.P.; Thies, J.E.; O'Neill, B.; Trujillo, L.; Luizão, F.J. Black carbon affects the cycling of non-black carbon in soil. *Org. Geochem.* **2010**, *41*, 206–213. [CrossRef]

58. James, G.; Witten, D.; Hastie, T.; Tibshirani, R. *An Introduction to Statistical Learning*; Springer: New York, NY, USA, 2013; Volume 6.

59. Walther, B.A.; Moore, J.L. The concepts of bias, precision and accuracy, and their use in testing the performance of species richness estimators, with a literature review of estimator performance. *Ecography* **2005**, *28*, 815–829. [CrossRef]

60. Keenan, T.F.; Baker, I.; Barr, A.; Ciais, P.; Davis, K.; Dietze, M.; Dragoni, D.; Gough, C.M.; Grant, R.; Hollinger, D.; et al. Terrestrial Biosphere Model Performance for Inter-Annual Variability of Land-Atmosphere CO_2 Exchange. *Glob. Chang. Biol.* **2012**, *18*, 1971–1987. [CrossRef]

61. Taylor, K. Summarizing multiple aspects of model performance in a single diagram. *J. Geophys. Res.* **2001**, *106*, 7183–7192. [CrossRef]

62. Booth, B.B.B.; Jones, C.D.; Collins, M.; Totterdell, I.J.; Cox, P.M.; Sitch, S.; Huntingford, C.; Betts, R.A.; Harris, G.R.; Lloyd, J. High sensitivity of future global warming to land carbon cycle processes. *Environ. Res. Lett.* **2012**, *7*, 024002. [CrossRef]

63. Zhao, Y.; Ciais, P.; Peylin, P.; Viovy, N.; Longdoz, B.; Bonnefond, J.M.; Rambal, S.; Klumpp, K.; Olioso, A.; Cellier, P.; et al. How errors on meteorological variables impact simulated ecosystem fluxes: A case study for six French sites. *Biogeosciences* **2012**, *9*, 2537–2564. [CrossRef]

64. Doblas-Miranda, E.; Martínez-Vilalta, J.; Lloret, F.; Álvarez, A.; Ávila, A.; Bonet, F.J.; Ferrandis, P. Reassessing global change research priorities in mediterranean terrestrial ecosystems: How far have we come and where do we go from here? *Glob. Ecol. Biogeogr.* **2015**, *24*, 25–43. [CrossRef]

65. Huber-Sanwald, E.; Huenneke, L.F.; Jackson, R.B.; Kinzig, A.; Leemans, R.; Lodge, D.M.; Walker, B.H. Biodiversity global biodiversity scenarios for the year 2100. *Science* **2000**, *287*, 1770–1774.

66. Muraoka, H.; Koizumi, H. Photosynthetic and structural characteristics of canopy and shrub trees in a cool-temperate deciduous broadleaved forest: Implication to the ecosystem carbon gain. *Agric. For. Meteorol.* **2005**, *134*, 39–59. [CrossRef]

67. Kramer, K.; Leinonen, I.; Bartelink, H.H.; Berbigier, P.; Borghetti, M.; Cienciala, E.; Dolman, A.J.; Froer, O.; Gracia, C.; Granier, A.; et al. Evaluation of Six Process-Based Forest Growth Models Using Eddy-Covariance Measurements of CO_2 and H_2O Fluxes at Six Forest Sites in Europe. *Glob. Chang. Biol.* **2002**, 213–230. [CrossRef]

68. Keenan, T.; Garcia, R.; Sabate, S.; Gracia, C.A. Process Based Forest Modelling: A Thorough Validation and Future Prospects for Mediterranean Forests in a Changing World. *Cuad. Soc. Esp. Cienc. For.* **2007**, *92*, 81–92.

69. Zhou, W.; Hui, D.; Shen, W. Effects of soil moisture on the temperature sensitivity of soil heterotrophic respiration: A laboratory incubation study. *PLoS ONE* **2014**, *9*, e92531. [CrossRef] [PubMed]

70. Hollinger, D.Y.; Richardson, A.D. Uncertainty in eddy covariance measurements and its application to physiological models. *Tree Physiol.* **2005**, *25*, 873–885. [CrossRef] [PubMed]

71. DeLucia, E.H.; Drake, J.E.; Thomas, R.B.; Gonzalez-Meler, M.A. Forest carbon use efficiency: Is respiration a constant fraction of gross primary production? *Glob. Chang. Biol.* **2007**, *13*, 1157–1167. [CrossRef]

72. Ito, A.; Oikawa, T. Global mapping of terrestrial primary productivity and light-use efficiency with a process-based model. In *Global Environmental Change in the Ocean and on Land*; TERRAPUB: Tokyo, Japan, 2004; pp. 343–358.

73. Sabaté, S.; Gracia, C.A.; Sánchez, A. Likely effects of climate change on growth of *Quercus ilex*, *Pinus halepensis*, *Pinus pinaster*, *Pinus sylvestris* and *Fagus sylvatica* forests in the Mediterranean region. *For. Ecol. Manag.* **2002**, *162*, 23–37. [CrossRef]

74. Chiti, T.; Papale, D.; Smith, P.; Dalmonech, D.; Matteucci, G.; Yeluripati, J.; Valentini, R. Predicting changes in soil organic carbon in mediterranean and alpine forests during the Kyoto Protocol commitment periods using the CENTURY model. *Soil Use Manag.* **2010**, *26*, 475–484. [CrossRef]

75. Dickinson, C.H. *Biology of Plant Litter Decomposition*; Elsevier: Amsterdam, The Netherlands, 2012; Volume 2.

76. Yang, Y.; Zhu, Q.; Peng, C.; Wang, H.; Xue, W.; Lin, G.; Li, S. A novel approach for modelling vegetation distributions and analysing vegetation sensitivity through trait-climate relationships in China. *Sci. Rep.* **2016**, *6*. [CrossRef] [PubMed]

77. Yang, Y.; Zhu, Q.; Peng, C.; Wang, H.; Chen, H. From plant functional types to plant functional traits: A new paradigm in modelling global vegetation dynamics. *Prog. Phys. Geogr.* **2015**, *39*, 514–535. [CrossRef]

78. Messier, J.; McGill, B.J.; Lechowicz, M.J. How do traits vary across ecological scales? A case for trait-based ecology. *Ecol. Lett.* **2010**, *13*, 838–848. [CrossRef] [PubMed]

79. Claussen, M.; Mysak, L.A.; Weaver, A.J.; Crucifix, M.; Fichefet, T.; Loutre, M.-F.; Weber, S.L.; Alcamo, J.; Alexeev, V.A.; Berger, A.; et al. Earth System Models of Intermediate Complexity: Closing the Gap in the Spectrum of Climate System Models. *Clim. Dyn.* **2002**, *18*, 579–586.

80. Lavorel, S.; Díaz, S.; Cornelissen, J.H.C.; Garnier, E.; Harrison, S.P.; McIntyre, S.; Urcelay, C. Plant functional types: Are we getting any closer to the Holy Grail? In *Terrestrial Ecosystems in a Changing World*; Springer: Berlin/Heidelberg, Germany, 2007; pp. 149–164.

81. Schaber, J.; Badeck, F.W. Physiology-Based Phenology Models for Forest Tree Species in Germany. *Int. J. Biometeorol.* **2003**, *47*, 193–201. [CrossRef] [PubMed]

82. Linkosalo, T.; Lappalainen, H.K.; Hari, P. A comparison of phenological models of leaf bud burst and flowering of boreal trees using independent observations. *Tree Physiol.* **2008**, *28*, 1873–1882. [CrossRef] [PubMed]

83. Uusitalo, L.; Lehikoinen, A.; Helle, I.; Myrberg, K. An overview of methods to evaluate uncertainty of deterministic models in decision support. *Environ. Model. Softw.* **2015**, *63*, 24–31. [CrossRef]

84. Wythers, K.R.; Reich, P.B.; Bradford, J.B. Incorporating temperature-sensitive Q10 and foliar respiration acclimation algorithms modifies modeled ecosystem responses to global change. *J. Geophys. Res. Biogeosci.* **2013**, *118*, 77–90. [CrossRef]

85. Ryan, M.G.; Gower, S.T.; Hubbard, R.M. Woody Tissue Maintenance Respiration of Four Conifers in Contrasting Climates. *Oecologia* **1995**, *101*, 133–140. [CrossRef] [PubMed]

86. Valentini, R.; Matteucci, G.; Dolman, A.J.; Shulze, E.D.; Rebmann, C.; Moors, E.J.; Granier, A.; Lindroth, A. Respiration as the main determinant of Carbon balance in European forests. *Nature* **2000**, *404*, 851–865. [CrossRef] [PubMed]

forests

MDPI

Communication

Effect of Soil Moisture on the Response of Soil Respiration to Open-Field Experimental Warming and Precipitation Manipulation

Guanlin Li [1], Seongjun Kim [1], Seung Hyun Han [1], Hanna Chang [1] and Yowhan Son [1,2,*]

[1] Department of Environmental Science and Ecological Engineering, Korea University, Seoul 02841, Korea; guanlin.li1207@gmail.com (G.L.L.); dao1129@hanmail.net (S.K.); aryian@naver.com (S.H.H.); wkdgkssk59@naver.com (H.C.)

[2] Department of Biological and Environmental Sciences, Qatar University, Doha 2713, Qatar

* Correspondence: yson@korea.ac.kr; Tel.: +82-2-3290-3015

Academic Editors: Robert Jandl and Mirco Rodeghiero
Received: 22 December 2016; Accepted: 23 February 2017; Published: 25 February 2017

Abstract: Soil respiration (R_S, Soil CO_2 efflux) is the second largest carbon (C) flux in global terrestrial ecosystems, and thus, plays an important role in global and regional C cycling; moreover, it acts as a feedback mechanism between C cycling and global climate change. R_S is highly responsive to temperature and moisture, factors that are closely related to climate warming and changes in precipitation regimes. Here, we examined the direct and interactive effects of climate change drivers on R_S of *Pinus densiflora* Sieb. et Zucc. seedlings in a multifactor climate change experiment involving atmospheric temperature warming (+3 °C) and precipitation manipulations (−30% and +30%). Our results indicated that atmospheric temperature warming induced significant changes in R_S ($p < 0.05$), enhancing R_S by an average of 54.6% and 59.7% in the control and elevated precipitation plots, respectively, whereas atmospheric temperature warming reduced R_S by 19.4% in plots subjected to lower rates of precipitation. However, the warming effect on R_S was influenced by soil moisture. On the basis of these findings, we suggest that atmospheric temperature warming significantly influenced R_S, but the warming effect on R_S may be weakened by warming-induced soil drying in water-limited environments.

Keywords: soil respiration; climate change; warming effect; soil moisture

---

1. Introduction

Mean air temperatures and precipitation regimes across regional and global scales have been altered as a result of global climate change, and are expected to continue to change, engendering and exacerbating regional drought conditions, especially in mid-latitude regions [1–3]. These changes are likely to have significant impacts on soil respiration (R_S, soil CO_2 efflux), one of the largest fluxes in the global carbon (C) cycle [4]. As a critical process in the C cycle in terrestrial ecosystems, R_S plays an important role in regulating CO_2 flux from soil to the atmosphere [5]. Any potential change in R_S could, therefore, greatly affect the atmospheric CO_2 concentration, and subsequently affect climate change feedbacks [6]. Thus, a better understanding of the changes in R_S under present and future climate change would help guide projections of terrestrial C fluxes in the warming world.

Soil temperature and soil moisture, both of which are highly responsive to changes in air temperature and precipitation, are two of the primary abiotic drivers that regulate R_S [6,7]. As such, an increasing number of studies is focusing on the responses of R_S to climate warming (i.e., air and/or soil temperature warming) [6,8–11] and alterations in precipitation patterns [2,6,12,13]. These studies have revealed that not all ecosystems respond in a manner similar to these global climate change drivers. For instance, Feng et al. [14], in a review of studies examining the effects of major global change drivers on R_S across China, noted that R_S response to these drivers differed among ecosystem types (e.g., forest, grassland, tundra). Similarly, Zhong et al. [10], in a review of studies focusing on how warming affects R_S on the Tibetan Plateau, found large variations among regions. Moreover, R_S response to global climate change drivers may also be dependent on the experimental treatment level (e.g., elevated temperature level, precipitation change) or experimental period [14].

R_S is a combined flux that consists of two biotic processes: autotrophic respiration, which originates from plant roots and the associated rhizosphere community, and heterotrophic respiration, which originates from soil microbes and fauna [6]. Because these two processes are sensitive to climate conditions, climate warming and transformations in precipitation patterns may also affect R_S indirectly by modifying autotrophic and heterotrophic respiration [12]. Nevertheless, to what degree the combination of global warming and changes in precipitation patterns will alter R_S, and the roles that other abiotic and biotic factors will play, remain unknown. Hence, in order to forecast global C cycling in the warming world, it is necessary to understand the impact of these changes on the regulation of R_S.

The aim of this study was to examine R_S response to changes in temperature and precipitation by exposing *Pinus densiflora* Sieb. et Zucc. seedlings to various temperatures, precipitation amounts, and combinations of the two. Based on previous studies that indicated that warming increased the root collar diameter, above- and below-ground biomass of *P. densiflora* seedlings, and soil microbial activity [9,15–17], we first hypothesized that warming would enhance R_S by increasing both autotrophic and heterotrophic respiration. Given that shifts in precipitation can undoubtedly change soil conditions (i.e., temperature and moisture), which will in turn change the activities of root and soil microbial activity [18], we then hypothesized that elevated precipitation would also enhance R_S and that reduced precipitation would reduce R_S. Moreover, a combination of warming and precipitation manipulation may aggravate the changes in soil conditions and the responses of *P. densiflora* seedlings and soil microbes. Thus, we also hypothesized that the response of R_S to warming would vary under different precipitation manipulations.

2. Materials and Methods

2.1. Experimental Design

The experiment was conducted on the grounds of Korea University, located in Seoul, South Korea (37°35′36″ N, 127°1′31″ E). We chose *P. densiflora* seedlings for this experiment because this species is one of the representative temperate coniferous trees in South Korea [8]. Mean air temperature and annual precipitation were 13.6 °C and 792.1 mm, respectively, in 2015, and 12.5 °C and 1450.5 mm, respectively, from 1981 to 2010 (Korea Meteorological Administration, 2016). In April 2013, a total of 18 experimental plots (1.5 m × 1.5 m with a 50 cm buffer between the plots) containing 45 2-year-old *P. densiflora* seedlings were established in the study site. The soil at this site is classified as loamy sand (80% sand, 14% clay, 6% silt) [19].

The experimental treatment system was established in April 2013 and consisted of six different treatments with three replicates: two levels of atmospheric warming (control (C) and +3 °C (W)) were crossed with three levels of precipitation (control (P0), −30% (P−), and +30% (P+)). The six treatments consisted of (1) atmospheric temperature control and precipitation control (C*P0), or the "ambient" treatment; (2) atmospheric warming and precipitation control (W*P0); (3) atmospheric temperature control and reduced 30% precipitation (C*P−); (4) atmospheric warming and reduced 30% precipitation (W*P−); (5) atmospheric temperature control and elevated 30% precipitation (C*P+); and (6) atmospheric warming and elevated 30% precipitation (W*P+). An infrared heater (FTE-1000; Mor Electric Heating Instrument Inc., Grand Rapids, MI, USA) was used to elevate the air temperatures in the warming plots. The infrared heaters were set at a height of 60 cm above the *P. densiflora* seedling canopy in warmed plots; dummy heaters (without warming lamps) were set at the same height in non-warmed plots. A transparent panel was used to reduce natural precipitation in the decreased precipitation plots, and an automatic pump and drip-irrigation system was used to elevate precipitation in the plots with higher precipitation levels. These changes in air temperature and precipitation were designed to simulate climate change conditions expected in Korea over the next 50 years, based on RCP 8.5 climate-change scenarios.

2.2. Field Measurements

Soil respiration was measured in the morning between 9:00 and 12:00 on 19 June, 19 August, and 20 October, 2015 using a closed-chamber system with a portable diffusion-type, non-dispersive infrared (NDIR) CO_2 sensor (GMP343, Vaisala CARBOCAP, Helsinki, Finland) and a polyacrylics chamber (10 cm in diameter, 12 cm in height). R_S measurements and calculations were based on the methodology described by Noh et al. [9]. Briefly, CO_2 concentrations in the chamber were recorded every 5 s for 300 s using a handheld controller and logger (MI-70, Vaisala CARBOCAP, Helsinki, Finland) coupled with the NDIR CO_2 sensor; the first 30 s of data after the placement of the chamber were excluded from subsequent analyses. R_S was calculated using the equation (Equation (1)):

$$R_S = \frac{dCO_2}{dt} \times \frac{PV}{ART},\tag{1}$$

where P is the atmospheric pressure, V is the volume of the headspace gas within the chamber, A is the soil surface area enclosed by the chamber, R is the gas constant, and T is the air temperature (K).

Air temperature was measured using infrared temperature sensors (SI-111, Campbell Scientific, Logan, UT, USA), soil temperature was measured at a depth of 5 cm using temperature sensors (107-L34, Campbell Scientific, Logan, UT, USA), and soil moisture was measured at a depth of 10 cm using reflectometer probes (CS616, Campbell Scientific, Logan, UT, USA) ($n = 18$). Air temperature, soil temperature, and soil moisture were logged every 30 min, and the data were recorded using a data logger (CR3000, Campbell Scientific, Logan, UT, USA).

2.3. Statistical Analysis

Repeated measures ANOVA was used to test the effects of warming, precipitation manipulation, and their interaction on soil temperature, moisture, and R_S. We used the relative R_S between the warming plots and non-warmed plots to assess the effect of warming on R_S under different precipitation regimes. Fisher's least significant difference (LSD) test was used to analyze differences in air temperature, soil temperature, soil moisture, and R_S among the treatments. In addition, covariance and linear regression analyses were used to assess the relationships between R_S and soil moisture. All statistical analyses were performed with SAS v.9.3 (SAS Systems, Cary, NC, USA), and significance was set at $p \leq 0.05$. All associated data were available in Tables A1 and A2.

3. Results

3.1. Soil Temperature and Moisture

During the study period, only atmospheric warming and sampling month had significant effects on soil temperature ($p < 0.05$ and $p < 0.01$; Table 1). Warming increased air temperatures around the canopy surface by 3.09 °C, 2.34 °C, and 2.89 °C on average in the control, reduced, and elevated precipitation plots, respectively (Table 2), but the differences in soil temperature induced by atmospheric warming were lower than the differences in air temperature among the precipitation treatments. Warming increased soil temperatures by 0.63 °C, 0.28 °C, and 0.44 °C on average in the control, reduced, and elevated precipitation plots, respectively (Table 2). Although only warming and month had significant effects on soil moisture (all $p < 0.01$; Table 1), on average, soil moisture varied significantly among the treatments (Table 2). Compared with the ambient treatment (C*P0), all other treatments reduced soil moisture by 0.44 Vol% to 1.94 Vol% (Table 2). In addition, warming had a drying effect on soil moisture, reducing it by an average of 1.87 Vol%, 1.13 Vol%, and 0.85 Vol% in the control, reduced, and elevated precipitation plots, respectively (Table 2).

Table 1. Repeated-measure ANOVAs for soil temperature (ST), soil moisture (SM), and soil respiration (R_S) in response to warming, precipitation manipulation, and their interaction.

Effect	df	ST			SM			R_S		
		F	p		F	p		F	p	
Warming (W)	1	6.1900	0.0285	*	12.060	0.0046	**	6.4500	0.0259	*
Precipitation (P)	2	1.5400	0.2535	ns	0.7300	0.5019	ns	1.8800	0.1948	ns
Month (M)	2	2924.7	<0.0001	**	89.360	<0.0001	**	22.620	<0.0001	**
W × P	2	0.3100	0.7397	ns	0.6700	0.5293	ns	5.4300	0.0209	*
W × M	2	2.3800	0.1138	ns	0.3000	0.7451	ns	0.0500	0.9553	ns
P × M	4	1.2200	0.3304	ns	1.7300	0.1763	ns	2.3900	0.0786	ns
W × P × M	4	0.3300	0.8565	ns	0.0800	0.9869	ns	0.1000	0.9831	ns

× = interaction effect; ns = not significant ($p > 0.05$); * = significant ($p < 0.05$); ** = significant ($p < 0.01$).

Table 2. Mean air temperature (AT), soil temperature (ST), soil moisture (SM), and soil respiration (R_S) under different treatments, presented as mean ± standard error.

Variables	C*P0	W*P0	C*P−	W*P−	C*P+	W*P+
AT (°C)	21.02 ± 3.22 a	24.11 ± 3.37 a	21.30 ± 3.34 a	23.64 ± 3.23 a	21.00 ± 3.24 a	23.90 ± 3.34 a
ST (°C)	21.80 ± 2.82 a	22.43 ± 2.90 a	22.36 ± 2.86 a	22.65 ± 2.91 a	22.08 ± 2.80 a	22.52 ± 2.79 a
SM (Vol %)	7.34 ± 0.56 a	5.47 ± 0.50 b	6.53 ± 0.43 ab	5.40 ± 0.39 b	6.90 ± 0.39 a	6.04 ± 0.36 ab
R_S ($\mu mol \cdot CO_2 \cdot m^{-2} \, s^{-1}$)	2.02 ± 0.46 a	3.12 ± 0.49 ab	2.76 ± 0.41 ab	2.23 ± 0.38 a	2.36 ± 0.48 ab	3.76 ± 0.55 b

W = warming; P = precipitation manipulation; C*P0 = atmospheric temperature control and precipitation control; W*P0 = atmospheric warming and precipitation control; C*P− = atmospheric temperature control and reduced precipitation; W*P− = atmospheric warming and reduced precipitation; C*P+ = atmospheric temperature control and elevated precipitation; and W*P+ = atmospheric warming and elevated precipitation. Values followed by a different letter are significantly different to each other ($p < 0.05$).

3.2. Treatment Effects on R_S

Warming and month also significantly affected R_S ($p < 0.05$ and $p < 0.01$, respectively), as did the interactive effect of warming and precipitation manipulation ($p < 0.05$; Table 1). Compared with the ambient treatment, all other treatments significantly enhanced R_S by 16.9% to 86.6% (Table 2). R_S exhibited similar temporal variations under all treatments over the course of the study period, increasing from June to August and subsequently decreasing thereafter (Figure 1). However, the relative R_S between warmed plots and non-warmed plots under different precipitation levels exhibited the opposite temporal variation, with a smaller R_S in August than in June and October (Figure 2a). Moreover, R_S in June and October differed significantly among the three precipitation treatments. Warming had a positive effect on R_S in the control and elevated precipitation plots, but a negative effect on R_S in reduced precipitation plots, since the relative R_S in the latter plots was negative (Figure 2b). Specifically, warming enhanced R_S by an average of 54.6% and 59.7% in the control and elevated precipitation plots, respectively, whereas warming reduced R_S by 19.4% in reduced precipitation plots.

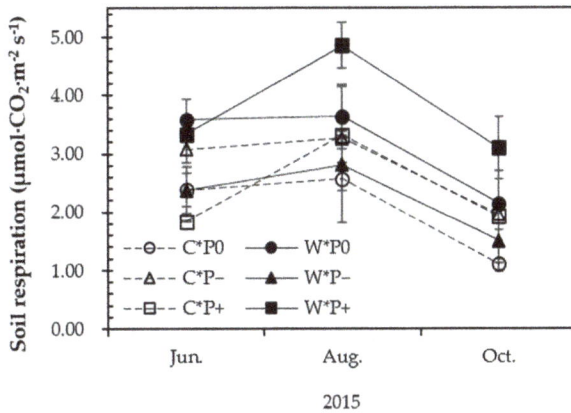

Figure 1. Variations in soil respiration among treatments. Vertical bars indicate the standard error. Treatments: C*P0 = atmospheric temperature control and precipitation control; W*P0 = atmospheric warming and precipitation control; C*P− = atmospheric temperature control and reduced precipitation; W*P− = atmospheric warming and reduced precipitation; C*P+ = atmospheric temperature control and elevated precipitation; and W*P+ = atmospheric warming and elevated precipitation.

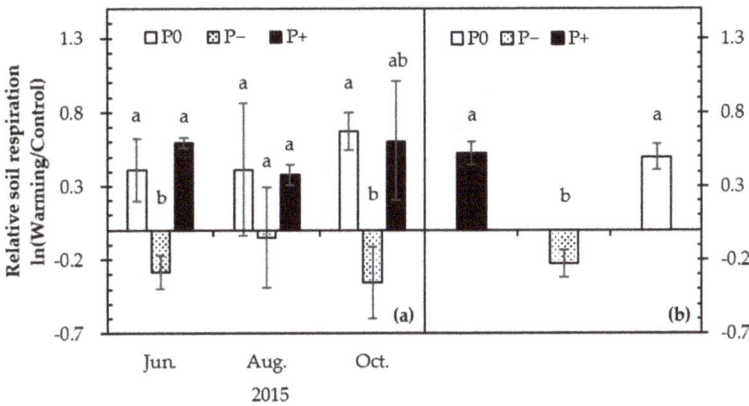

Figure 2. (a) Variations in relative soil respiration between the warmed plots and non-warmed plots subjected to three precipitation manipulations; (b) mean relative soil respiration between the warmed plots and non-warmed plots under different precipitation manipulations. Vertical bars indicate the standard error. Different lower-case letters represent significant differences ($p < 0.05$). Treatments: P0 = precipitation control; P− = reduced precipitation; P+ = elevated precipitation.

3.3. Correlations Between R_S and Soil Moisture

There was a positive correlation between R_S and soil temperature ($r = 0.52$, $p < 0.01$) and moisture ($r = 0.33$, $p < 0.05$) across all plots. Specifically, there were significant positive correlations between R_S and soil moisture in both warmed plots ($r = 0.72$, $p < 0.01$) and non-warmed plots ($r = 0.60$, $p < 0.01$; Figure 3a). The dependency of R_S on soil moisture in warmed plots was higher than that in non-warmed plots. The relative R_S was positively correlated with the relative soil moisture between warmed plots and non-warmed plots ($r = 0.42$, $p < 0.05$; Figure 3b).

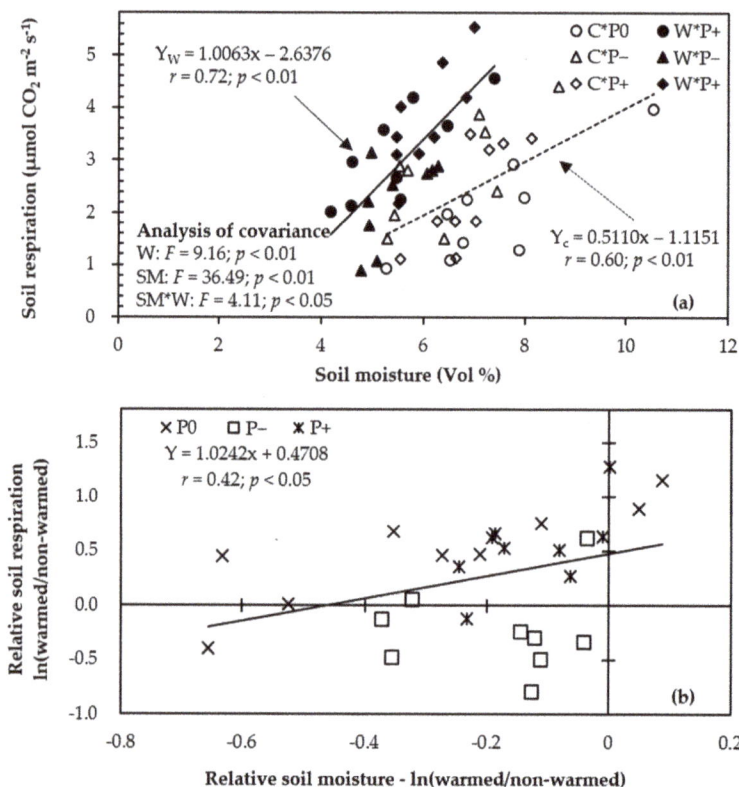

Figure 3. (a) Soil respiration plotted against soil moisture in warmed plots (YW) and non-warmed plots (YC); (b) relative soil respiration between warmed and non-warmed plots plotted against relative soil moisture between the warmed plots and non-warmed plots. Treatments: P0 = precipitation control; P− = reduced precipitation; P+ = elevated precipitation; C*P0 = atmospheric temperature control and precipitation control; W*P0 = atmospheric warming and precipitation control; C*P− = atmospheric temperature control and reduced precipitation; W*P− = atmospheric warming and reduced precipitation; C*P+ = atmospheric temperature control and elevated precipitation; and W*P+ = atmospheric warming and elevated precipitation.

4. Discussion

4.1. Effects of Treatments on R_S

Generally, we found that warming and the interaction of warming and precipitation manipulation had significant effects on R_S, whereas precipitation manipulation alone had no effect on R_S (Table 1). A previous experiment conducted at the same site in 2015 reported that precipitation manipulation alone had no effect on above- and below-ground seedling biomass [15,16]. Given that plant growth has a strong influence on R_S via root respiration, the lack of plant biomass response to precipitation manipulation could explain that the precipitation manipulation did not significantly affect R_S in our study, a result that was consistent with the findings of Wei et al. [18].

Furthermore, the significant positive effects of warming and the interaction of warming and precipitation manipulation on R_S were likely due to the treatment effects on the seedlings, which could directly contribute to R_S, and the treatment effects on soil conditions (i.e., temperature and moisture), which could result in site-specific soil conditions and indirectly affect the R_S. In this study, soil temperature remained largely unaffected by treatment, but the soil dried out, likely due to increased seedling transpiration rates, especially in warmed and reduced precipitation plots. According to previous studies in the same study site, warming not only increased the root collar diameter and above- and below-ground biomass of seedlings [15,16], but also altered the soil microbial activity and community [17]. Thus, the warming effect on R_S might be primarily driven by the seedling and soil microbial responses to the treatments. Hence, combining the results of the present study and other studies, we suggest that the shifts in R_S caused by the treatment-induced modifications to seedling growth and soil conditions might involve a complex of mechanisms that interact to determine root responses to variable soil environments, changes in the soil microbial community, and allocation of assimilated C in the plant–soil–microbe system [5,6,11].

Soil at the study site was a loamy sand [19], and thus, soil moisture at the study site was naturally low. As such, it is possible that soil moisture played a stronger role than did soil temperature in determining R_S over the course of the study period. We found that although warming generally enhanced R_S in the control and elevated precipitation plots, warming reduced R_S in the reduced precipitation plots, despite soil temperatures being increased by warming (Table 2). The drier conditions caused by the combination of warming and reduced precipitation most likely led to this decline in R_S. This finding was consistent with those of other studies showing that soil drying induced by warming can offset the effects of increasing temperature in a water-limited environment [5,6,14].

4.2. Soil Moisture and Warming Effect on R_S

Our results indicated that there might be a warming effect/soil moisture threshold on R_S, which suggested that the warming effect on R_S would be influenced by soil moisture within this water-limited environment. Several lines of evidence support this conclusion. First, it was observed that R_S generally increased with increasing soil moisture in both warmed plots and non-warmed plots; however, covariance analysis revealed a significant interaction between warming and soil moisture ($p < 0.05$). Moreover, warming caused a shift in the R_S-moisture response curve (Figure 3a). These patterns might be the result of complex interactions among the mechanisms involved in the warming effect and soil moisture. Second, we also observed that warming had a positive effect on R_S in the control and elevated precipitation plots but a negative effect on R_S in reduced precipitation plots (Table 2 and Figure 2b). Third, warming effects on R_S were positively correlated with soil moisture among all precipitation treatments (Figure 3b).

Both autotrophic and heterotrophic respiration were correlated with soil moisture. In previous studies at the same site, it was reported that warming enhanced root biomass, as well as soil microbial biomass and activity [15–17]. Hence, we concluded that warming enhanced R_S by increasing both autotrophic and heterotrophic respiration under sufficient soil available water [7,20,21]. Soil moisture can greatly influence the diffusion of soluble nutrients, and consequently reduce available nutrients for soil microbes and uptake by roots [22,23]. In water-limited environments, increasing competition for nutrients between roots and soil microbes induced by water stress might therefore offset, at least somewhat, the positive effect of warming on microbial biomass and activity [24]. On the other hand, when soil moisture is limited, microbes and plant roots have to invest more energy to produce protective secondary compounds, which would hamper their growth and the amount of C allocated to respiration [25]. Therefore, in water-limited environments, the negative warming effect of R_S might be attributed to reductions in heterotrophic and autotrophic respiration associated with drought induced by warming.

5. Conclusions

The results of our study indicated that the direct and/or interactive effects of changes in air temperature and precipitation regimes would likely alter R_S. Both R_S and the warming effect on R_S, reflected by the relative R_S between warmed and non-warmed plots, varied within treatments, suggesting that treatment-induced changes in R_S and the warming effect on R_S were directly and/or indirectly related to soil conditions (temperature and moisture) and the growth of seedlings. Most notably, soil moisture appears to be a key factor controlling C fluxes from soil to atmosphere, and warming-induced soil drying may weaken or offset the atmospheric warming effect on R_S in water-limited environments. Our study demonstrated the preliminary response of R_S to the effects of warming and altered precipitation regimes on *P. densiflora* seedlings. Given that *P. densiflora* is a common temperate coniferous tree species and is widely distributed throughout South Korea [26], future monitoring will provide important parameters for predicting the response of forest belowground C turnover to a warmer climate.

Acknowledgments: This study was supported by the National Research Foundation of Korea (NRF-2013R1A1A 2012242) and Korean Ministry of Environment (2016001300004).

Author Contributions: Guanlin Li, Seongjun Kim, Seung Hyun Han and Hanna Chang were responsible for the collection of data from the study site; Guanlin Li and Seongjun Kim analyzed the data; Yowhan Son conceived and designed the experiments; Guanlin Li wrote the manuscript.

Conflicts of Interest: The authors declare no conflicts of interest. The funding sponsors had no role in the design of this study; in the collection, analyses, or interpretation of data; in the writing of the manuscript; or in the decision to publish the results.

Appendix

Table A1. Mean air temperature (AT), soil temperature (ST), soil moisture (SM), and soil respiration (R_S) measured in June, August and October, 2015 under different treatments, presented as mean ± standard error.

Month	Treatment	AT (°C)	ST (°C)	SM (Vol %)	R_S ($\mu mol \cdot CO_2 \cdot m^{-2} s^{-1}$)
June	C*P0	23.22 ± 0.24	23.91 ± 0.33	7.03 ± 0.39	2.39 ± 0.28
	W*P0	26.16 ± 0.56	25.09 ± 0.39	5.20 ± 0.34	3.58 ± 0.35
	C*P−	23.49 ± 0.19	24.72 ± 0.12	6.14 ± 0.54	3.08 ± 0.23
	W*P−	26.45 ± 0.41	25.18 ± 0.36	4.94 ± 0.02	2.38 ± 0.41
	C*P+	23.17 ± 0.11	24.03 ± 0.29	6.64 ± 0.22	1.84 ± 0.00
	W*P+	26.30 ± 0.01	24.72 ± 0.21	5.87 ± 0.21	3.34 ± 0.11
August	C*P0	25.16 ± 0.20	25.26 ± 0.10	8.43 ± 1.10	2.57 ± 0.75
	W*P0	28.65 ± 1.54	25.58 ± 0.22	6.45 ± 0.56	3.63 ± 0.55
	C*P−	25.67 ± 0.00	25.69 ± 0.14	7.39 ± 0.67	3.26 ± 0.89
	W*P−	27.27 ± 1.04	25.93 ± 0.19	6.18 ± 0.06	2.81 ± 0.04
	C*P+	25.21 ± 0.07	25.65 ± 0.12	7.67 ± 0.25	3.32 ± 0.06
	W*P+	28.08 ± 0.46	25.87 ± 0.05	6.73 ± 0.19	4.86 ± 0.39
October	C*P0	14.68 ± 0.05	16.22 ± 0.42	6.56 ± 0.75	1.10 ± 0.10
	W*P0	17.53 ± 1.32	16.63 ± 0.16	4.78 ± 0.41	2.14 ± 0.07
	C*P−	14.74 ± 0.28	16.68 ± 0.26	6.06 ± 0.70	1.95 ± 0.26
	W*P−	17.20 ± 0.00	16.84 ± 0.59	5.09 ± 0.18	1.50 ± 0.52
	C*P+	14.63 ± 0.27	16.56 ± 0.29	6.37 ± 0.42	1.92 ± 0.79
	W*P+	17.30 ± 0.58	16.97 ± 0.35	5.52 ± 0.02	3.10 ± 0.53

C*P0 = atmospheric temperature control and precipitation control; W*P0 = atmospheric warming and precipitation control; C*P− = atmospheric temperature control and reduced precipitation; W*P− = atmospheric warming and reduced precipitation; C*P+ = atmospheric temperature control and elevated precipitation; and W*P+ = atmospheric warming and elevated precipitation.

Table A2. The code of statistical analysis performed with SAS v.9.3 (SAS Systems, Cary, NC, USA).

Repeated measures ANOVA	data RMANOVA; do P = 'n', 'd', 'I'; do T = 'c', 'w'; do s = 1 to 3; input M1–M3; output; end; end; end; cards; proc glm; class P T; model M1–M3 = P T P*T/nouni; repeated t 3 (6 8 10) contrast (1)/summary printe; run;	/P = precipitation manipulation; n = precipitation control; d = reduced precipitation; i = elevated precipitation/ /T = warming; c = atmospheric temperature control; w = atmospheric warming/ /month/

Fisher's least significant difference (LSD) test	data LSD;
	do a = 1 to 6;
	do i = 1 to 3;
	input x @@; output; end; end; cards;
	proc anova; class a; model x = a; means a/LSD; run;
		/a = Treatment/

Analysis of Linear regression	data regression; input ST SM RS @@; cards; proc corr pearson; var ST SM RS; run;	/ST = soil temperature; SM = soil moisture; RS = CO_2 efflux/
Analysis of covariance	data covariance; input T $ RS ST SM @@; cards; proc sort; by T; run; proc glm; class T; model RS = T ∣ SM; run;	/T = warming; ST = soil temperature; SM = soil moisture; RS = CO_2 efflux/

References

1. IPCC. *Climate Change 2013: The Physical Science Basis: Working Group I Contribution to the Fifth Assessment Report of the Intergovernmental Panel on Climate Change*; Cambridge University Press: Cambridge, UK, 2014.
2. Haghshenas, M.; Mohadjer, M.R.M.; Attarod, P.; Pourtahmasi, K.; Feldhaus, J.; Sadeghi, S.M.M. Climate effect on tree-ring widths of *Fagus orientalis* in the Caspian forests, northern Iran. *For. Sci. Technol.* **2016**, *12*, 176–182. [CrossRef]
3. Lim, S.J.; Park, H.J.; Kim, H.S.; Park, S.I.; Han, S.S.; Kim, H.J.; Lee, S.H. Vulnerability assessment of forest landslide risk using GIS adaptation to climate change. *For. Sci. Technol.* **2016**, *12*, 207–213. [CrossRef]
4. Liu, T.; Xu, Z.Z.; Hou, Y.H.; Zhou, G.S. Effects of warming and changing precipitation rates on soil respiration over two years in a desert steppe of northern China. *Plant Soil* **2016**, *400*, 15–27. [CrossRef]
5. Schindlbacher, A.; Wunderlich, S.; Borken, W.; Kitzler, B.; Zechmeister-Boltenstern, S.; Jandl, R. Soil respiration under climate change: Prolonged summer drought offsets soil warming effects. *Glob. Chang. Biol.* **2012**, *18*, 2270–2279. [CrossRef]
6. Liu, Y.; Liu, S.; Wan, S.; Wang, J.; Luan, J.; Wang, H. Differential responses of soil respiration to soil warming and experimental throughfall reduction in a transitional oak forest in central China. *Agric. For. Meteorol.* **2016**, *226*, 186–198. [CrossRef]
7. Bao, X.; Zhu, X.; Chang, X.; Wang, S.; Xu, B.; Luo, C.; Zhang, Z.; Wang, Q.; Rui, Y.; Cui, X. Effects of soil temperature and moisture on soil respiration on the Tibetan Plateau. *PLoS ONE* **2016**, *11*, e0165212. [CrossRef] [PubMed]
8. Noh, N.J.; Kuribayashi, M.; Saitoh, T.M.; Nakaji, T.; Nakamura, M.; Hiura, T.; Muraoka, H. Responses of soil, heterotrophic, and autotrophic respiration to experimental open-field soil warming in a cool-temperate deciduous forest. *Ecosystems* **2016**, *19*, 504–520. [CrossRef]
9. Noh, N.J.; Lee, S.J.; Jo, W.; Han, S.; Yoon, T.K.; Chung, H.; Son, Y. Effects of experimental warming on soil respiration and biomass in *Quercus variabilis* Blume and *Pinu sdensiflora* Sieb. et Zucc. seedlings. *Ann. For. Sci.* **2016**, *73*, 533–545. [CrossRef]
10. Zhong, Z.M.; Shen, Z.X.; Fu, G. Response of soil respiration to experimental warming in a highland barley of the Tibet. *SpringerPlus* **2016**, *5*, 137. [CrossRef] [PubMed]

11. Wertin, T.M.; Belnap, J.; Reed, S.C. Experimental warming in a dryland community reduced plant photosynthesis and soil CO_2 efflux although the relationship between the fluxes remained unchanged. *Funct. Ecol.* **2016**. [CrossRef]

12. Liu, L.; Wang, X.; Lajeunesse, M.J.; Miao, G.; Piao, S.; Wan, S.; Wu, Y.; Wang, Z.; Yang, S.; Li, P.; et al. A cross-biome synthesis of soil respiration and its determinants under simulated precipitation changes. *Glob. Chang. Biol.* **2016**, *22*, 1394–1405. [CrossRef] [PubMed]

13. Sharkhuu, A.; Plante, A.F.; Enkhmandal, O.; Gonneau, C.; Casper, B.B.; Boldgiv, B.; Petraitis, P.S. Soil and ecosystem respiration responses to grazing, watering and experimental warming chamber treatments across topographical gradients in northern Mongolia. *Geoderma* **2016**, *269*, 91–98. [CrossRef]

14. Feng, J.; Wang, J.; Ding, L.; Yao, P.; Qiao, M.; Yao, S. Meta-analyses of the effects of major global change drivers on soil respiration across China. *Atmos. Environ.* **2017**, *150*, 181–186. [CrossRef]

15. Yun, S.J.; Han, S.; Han, S.H.; Kim, S.; Li, G.L.; Park, M.; Son, Y. Short-term effects of warming treatment and precipitation manipulation on the ecophysiological responses of *Pinus densiflora* seedlings. *Turk. J. Agric. For.* **2016**, *40*, 621–630. [CrossRef]

16. Park, M.J. Effects of Open-Field Artificial Warming and Precipitation Manipulation on Physiological Characteristics and Growth of *Pinus densiflora* Seedlings. Master's Thesis, Korea University, Seoul, Korea, 2016.

17. Li, G.L.; Kim, S.; Park, M.J.; Son, Y. Short-term effects of experimental warming and precipitation manipulation on soil microbial biomass, substrate utilization patterns, and community composition. *Pedosphere* **2017**, in press.

18. Wei, X.; Zhang, Y.; Liu, J.; Gao, H.; Fan, J.; Jia, X.; Cheng, J.; Shao, M.; Zhang, X. Response of soil CO_2 efflux to precipitation manipulation in a semiarid grassland. *J. Environ. Sci.* **2016**, *45*, 207–214. [CrossRef] [PubMed]

19. Yun, S.J.; Han, S.; Han, S.H.; Lee, S.J.; Jung, Y.; Kim, S.; Son, Y. Open-field experimental warming and precipitation manipulation system design to simulate climate change impact. *J. Korean For. Soc.* **2014**, *103*, 159–164. [CrossRef]

20. Akinremi, O.O.; McGinn, S.M.; McLean, H.D.J. Effects of soil temperature and moisture on soil respiration in barley and fallow plots. *Can. J. Soil Sci.* **1999**, *79*, 5–13. [CrossRef]

21. Jarvi, M.P.; Burton, A.J. Acclimation and soil moisture constrain sugar maple respiration in experimentally warmed soil. *Tree Physiol.* **2013**, *33*, 949–959. [CrossRef] [PubMed]

22. Van Meeteren, M.J.M.; Tietema, A.; van Loon, E.E.; Verstraten, J.M. Microbial dynamics and litter decomposition under a changed climate in a Dutch heathland. *Appl. Soil Ecol.* **2008**, *38*, 119–127. [CrossRef]

23. Geng, S.M.; Yan, D.H.; Zhang, T.X.; Weng, B.S.; Zhang, Z.B.; Gang, W. Effects of extreme drought on agriculture soil and sustainability of different drought soil. *Hydrol. Earth Syst. Sci. Discuss.* **2014**, *11*, 1–29. [CrossRef]

24. Williams, M.A. Response of microbial communities to water stress in irrigated and drought-prone tallgrass prairie soils. *Soil Biol. Biochem.* **2007**, *39*, 2750–2757. [CrossRef]

25. Chang, C.T.; Sperlich, D.; Sabaté, S.; Sánchez-Costa, E.; Cotillas, M.; Espelta, J.M.; Gracia, C. Mitigating the stress of drought on soil respiration by selective thinning: Contrasting effects of drought on soil respiration of two oak species in a Mediterranean forest. *Forests* **2016**, *7*, 263. [CrossRef]

26. Kim, D.H.; Kim, J.H.; Park, J.H.; Ewane, E.B.; Lee, D.H. Correlation between above-ground and below-ground biomass of 13-year-old *Pinus densiflora* S. et Z. planted in a post-fire area in Samcheok. *For. Sci. Technol.* **2016**, *12*, 115–124.

forests

MDPI

Article

Elevated CO_2 and Tree Species Affect Microbial Activity and Associated Aggregate Stability in Soil Amended with Litter

Salwan M. J. Al-Maliki [1,2], David L. Jones [3], Douglas L. Godbold [4], Dylan Gwynn-Jones [1] and John Scullion [1,*]

1   Institute of Biological, Environmental & Rural Sciences, Aberystwyth University, Penglais,
    Aberystwyth SY23 3DA, UK; salwan.mohammed@yahoo.com (S.M.J.A.-M.); dyj@aber.ac.uk (D.G.-J.)
2   Soil and Water Science Department, College of Agriculture, AlQasim Green University, Al Qasim 13239, Iraq
3   School of Environment, Natural Resources & Geography, Bangor University, Bangor LL57 2UW, UK;
    d.jones@bangor.ac.uk
4   Institute of Forest Ecology, University of Natural Resources and Life Sciences, Vienna 1190, Austria;
    douglas.godbold@boku.ac.at
*   Correspondence: jos@aber.ac.uk; Tel.: +44-197-062-2304; Fax: +44-197-062-2350

Academic Editors: Robert Jandl and Mirco Rodeghiero
Received: 31 January 2017; Accepted: 28 February 2017; Published: 3 March 2017

Abstract: (1) Elevated atmospheric CO_2 (eCO_2) may affect organic inputs to woodland soils with potential consequences for C dynamics and associated aggregation; (2) The Bangor Free Air Concentration Enrichment experiment compared ambient (330 ppmv) and elevated (550 ppmv) CO_2 regimes over four growing seasons (2005–2008) under *Alnus glutinosa*, *Betula pendula* and *Fagus sylvatica*. Litter from the experiment (autumn 2008) and *Lumbricus terrestris* were added to mesocosm soils. Microbial properties and aggregate stability were investigated in soil and earthworm casts. Soils taken from the field experiment in spring 2009 were also investigated; (3) eCO_2 litter had lower N and higher C:N ratios. *F. sylvatica* and *B. pendula* litter had lower N and P than *A. glutinosa*; *F. sylvatica* had higher cellulose. In mesocosms, eCO_2 litter decreased respiration, mineralization constant (respired C:total organic C) and soluble carbon in soil but not earthworm casts; microbial-C and fungal hyphal length differed by species (*A. glutinosa* = *B. pendula* > *F. sylvatica*) not CO_2 regime. eCO_2 increased respiration in field aggregates but increased stability only under *F. sylvatica*; (4) Lower litter quality under eCO_2 may restrict its initial decomposition, affecting C stabilization in aggregates. Later resistant materials may support microbial activity and increase aggregate stability. In woodland, C and soil aggregation dynamics may alter under eCO_2, but outcomes may be influenced by tree species and earthworm activity.

Keywords: FACE; litter quality; respiration; carbon; microbial biomass; fungal hyphae

---

1. Introduction

Atmospheric CO_2 concentrations have increased significantly over recent decades [1]. Atmospheric CO_2 taken up by plants and incorporated into soil may be partly protected from decomposition within stable aggregates [2], a key mechanism facilitating soil carbon gain under elevated CO_2 [3]. Consequently, changes in soil aggregation may influence soil respiration responses to atmospheric CO_2 increases.

Leaf litter is a major C input to woodland soils and thus plays key roles in nutrient cycling and organic matter dynamics [4]. Litters vary in their nutrient, cellulose and lignin contents, factors which largely determine decomposition rates [5]. Litter quality is affected by intrinsic species characteristics. Thus, the nitrogen concentration of *A. glutinosa* leaves may be twice that of western red cedar and western hemlock [6]. Concentrations of N are low and C:N ratios are high in beech (*Fagus sylvatica*) compared to oak (*Quercus robur*) leaves [7]. Elevated atmospheric CO_2 (eCO_2) may affect litter quality. In litter from three *Populus* spp., eCO_2 caused a general decrease in nitrogen concentrations [8]. From a meta-analysis [9], it was estimated that nitrogen concentrations in leaf litter under eCO_2 were reduced by approximately 7% compared to ambient CO_2, with lignin concentration about 6.5% higher. However, these changes were not linked to any consistent effect on litter decomposition.

Tree species influence soil microbial communities, although these effects are often indirect and complex; leaf litter diversity increased microbial abundance at the soil surface but nutrients were the primary drivers at depth [10]. There are also some inconsistencies in reported responses of soil microbiota to varying CO_2 regimes. Thus, eCO_2 over six growing seasons did not significantly alter soil microbial biomass carbon, metabolic quotient and basal respiration in nutrient-poor grassland [11]; these findings were attributed to nitrogen limitations. However, eCO_2 over five years did not affect microbial biomass, even where nitrogen fertilizer was added [12], suggesting that microbial biomass is insensitive to changes in atmospheric CO_2. In contrast, under chickpea, eCO_2 increased labile soil C, microbial biomass and respiration [13]. In a review based mainly on forest ecosystems [14], evidence of eCO_2 effects on soil microbial composition was considered weak, whereas that for increased physiological activity was more consistent.

Interactions between soil aggregation and atmospheric CO_2 are poorly understood with contrasting findings arising from different studies and limited information on woodland soils. A Californian chaparral ecosystem exposed for six years to CO_2 concentrations ranging from 250 to 750 $\mu L \cdot L^{-1}$ CO_2 exhibited decreased macro-aggregate and micro-aggregate stability at the highest CO_2 levels [15]; the same study also found reduced protection of soil organic matter and faster turnover of carbon at high CO_2 levels. No effect of elevated CO_2 on aggregate stability was found after four years of CO_2 exposure in a grassland mesocosm experiment, confounding expectations that increased labile carbon and root biomass under eCO_2 would promote aggregation [16]. However, increased aggregate stability under eCO_2 was found in grassland soils and attributed to higher glomalin production by fungi [17]. In poplar plantations, eCO_2 did not affect macro-aggregation, but increased micro-aggregation, although these changes were small relative to those between poplar genotypes [18]. Earthworms at the ORNL-FACE temperate woodland site [3] directly contributed to the formation of soil aggregates, stabilizing increased carbon inputs resulting from atmospheric CO_2 enrichment.

In general, aggregate stability increases rapidly following addition of more labile organic matter but effects are transient. Thus, initial increases in aggregate stability were lost four weeks after glucose application but were more persistent with cellulose [19]. Organic materials have been classified into four groups [20]: (i) simple labile compounds which increased aggregate stability in less than month; (ii) more complex residues which maximized aggregate stability between one and three months; (iii) stable compounds which increased aggregate stability only after three months; (iv) ligneous residues which might have some benefit in the longer-term.

The objectives of this study were to evaluate the effects of eCO_2 on C dynamics and associated aggregate stability in response to litter inputs (mainly) from three contrasting tree species. Working with materials from a plantation experiment allowed for the evaluation of species-specific responses to eCO_2 under both mesocosm and field conditions. It was hypothesized that elevated CO_2 would reduce litter quality, thus slowing its decomposition, and that this would reduce initial aggregate stabilization; aggregate formation was expected to mitigate this effect. Elevated CO_2 effects were expected to be more pronounced in species producing higher quality litter and these litters were expected to have more pronounced early effects than those of low quality litter. Ingestion of soil and litter by earthworms was expected to mitigate these effects.

2. Materials and Methods

For practical reasons, the main investigation was based on a mesocosm experiment with a single visit to the experimental site providing the opportunity to corroborate mesocosm findings under field conditions.

2.1. Experimental Set-Up

The investigation was undertaken in field and, in greater detail, greenhouse experiments. The field experiment was the Bangor FACE site, at Henfaes experimental farm (UK 53°14′N; 4°01′W). The soil at Henfaes is a fine loamy brown earth over gravel (Rheidol series) classified as a Dystric Cambisol in the Food Agriculture Organisation system [21]. CO_2 levels were manipulated in eight site rings which included four ambient and four elevated treatments. Within each ring, groups of three tree species (*Alnus glutinosa* L. Gaertn., *Betula pendula* Roth and *Fagus sylvatica* L.) were planted either individually or in mixtures; in this study, only single species plots were investigated to avoid complex responses to heterogenous litter inputs.

The FACE rings operated at CO_2 330 ppmv for ambient and at 550 ppmv for eCO_2. For fuller details of the experimental design and its operation, see [22]. In the field, litter was collected on a weekly basis in baskets located within the plots, then air-dried and stored; amounts collected were highest for *A. glutinosa*, intermediate for *B. pendula*, but markedly lower for *F. sylvatica*. There was a similar species ranking for woody tissue, which increased significantly at eCO_2 [22]. Litters collected in autumn 2008 were used as inputs in the mesocosm experiment.

In spring 2009, field soil samples (24 in total) were taken from the 0–5 cm depth after removal of any litter layer, from each of the CO_2 regime X tree species combinations for subsequent measurement of field soil responses to treatments and as microbial inocula for mesocosm soils. For *A. glutinosa* and *B. pendula*, little litter remained at sampling, but some *F. sylvatica* litter was still apparent. This depth was chosen as being most affected by differences in litter quality. Sub-samples were either stored at 2 °C prior to microbial analyses or their use as inocula, or air-dried over 5 days at room temperature for subsequent organic C and aggregate stability determinations.

The experimental design of the mesocosm trial was based on that of the field experiment, with litters from each plot representing all species and CO_2 regime combinations being tested (four replicates per treatment combination). A soil similar to that on the experimental site (Rheidol series—Dystric Cambisol) was taken from 10–20 cm depth to avoid initial differences arising from experimental treatments. The percentage of sand, silt and clay was 44%, 35% and 20% respectively; initial aggregate stability (see Section 2.2 for method) was 30%. Bulk soil was mixed thoroughly with 5 g of soil, sampled as described above from the Bangor FACE site, as a microbial inoculum for mesocosms (2 liter pots with 10 cm diameter to which 2 kg of soil was added; *n* = 24). Soil inocula were matched to the litters subsequently applied as amendments, to take account of field treatment adaptations in microbial populations. Litters were ground to <1 mm and added to the soil surface (2 g) three times a week. Soil moisture was maintained at approximately 20% *w*/*w* during this experiment by addition of water and the shaded glasshouse had a temperature range of 10–21 °C over the 30-day duration of the experiment. The experiment was terminated at this point because supplies of litter from *F. sylvatica* plots had been exhausted.

Lumbricus terrestris L. was the predominant anecic earthworm on the field site [23] and has been shown to promote aggregation [24]. As supplies of litter were limited, the inclusion of a no earthworm treatment would have compromised the experiment. Thus, in order to assess the effects of earthworm processes on measured parameters, their casts were sampled and considered to represent recently formed potential aggregates. Three adult earthworms were added to each mesocosm (Neptune Ecology, Ipswich).

In the mesocosm experiment, the first collection of surface casts was two weeks after commencement of litter additions; very little casting occurred prior to 2 days before this collection. These materials were collected three times a week and bulked after air-drying for the entire experimental period (ambient temperatures) for total C and stability measurements. The final collection

(28–30 days) was retained fresh for cast microbial measurements, with sub-samples incubated for 22 days (respiration had stabilized by this time) at 20 °C and moisture content similar to that at sampling; this investigation aimed to assess temporal trends in microbial activity and C turnover in fresh cast aggregates. Soil from the mesocosms was sampled at the end of the experiment, 30 days after amendments commenced, for assessment of soil aggregate properties. Sub-samples of these stable aggregates were then incubated for 37 and 67 days at 20% moisture content and 20 °C during which time respiration was measured in order to monitor medium-term trends in aggregate C dynamics.

2.2. Measurements

Litter quality was assessed by measuring lignin, cellulose, nutrients (N and P) and the C:N ratio using standard methods [25]. Briefly, acid detergent fibre was determined, then lignin in residues from this analysis was dissolved using a saturated potassium permanganate/buffer solution; lignin was determined by weight loss and cellulose by weight loss from these residues after ignition at 500 °C. C and N were measured using a LECO CHN analyzer (LECO Corp., St Joseph, MI, USA); P was determined colorimetrically.

Microbial biomass carbon (MBC) was estimated by the fumigation-extraction method [26] using 0.5 M K_2SO_4 to extract organic C from chloroform-fumigated and non-fumigated samples. Extracts from the fumigated and non-fumigated samples were analyzed using a Total Organic Carbon analyzer (Shimadzu TOC-5050, Shimadzu UK Ltd., Milton Keynes, UK). MBC was calculated by subtracting the extracted organic carbon in the non-fumigated samples from that in the fumigated samples and using a standard conversion factor. Fungal hyphal length was measured using the membrane filtration method [27]. Soils (0.5 g) were placed in 30 mL of distilled water and shaken (Griffin & George Ltd., London, UK) for 1 h. The suspensions were allowed to settle, the supernatant was decanted, mixed with Calcofluor white M2R (Sigma-Aldrich, Gillingham, UK; final concentration 0.1%) and then passed through 2 μm filters using a vacuum pump. Hyphal lengths were estimated using a grid intersect method (20 μm grid) at ×100 magnification [28].

Soil respiration (20 °C and moisture content at sampling) was measured by two methods. The first, used for casts, where material available was limited, was based on the MicroResp procedure [29], using a BioTek ELx808 plate reader (BioTek Instruments, Inc, Swindon, UK) to monitor pH change (phenol red indicator) due to increasing dissolved CO_2. The second method for soil aggregates was based on a static alkali trap procedure [30].

Organic carbon in mesocosm aggregates and casts, and in field aggregates, was estimated by loss on ignition at 400 °C for 16 h in a muffle furnace [31] using a conversion factor of 1.724. Extractable (K_2SO_4) carbon (KSE-C) in mesocosm aggregates and casts was estimated [32] from the C concentrations of the non-fumigated extracts used for the microbial biomass C determinations. Mineralization constant [33] was determined as a means of discounting variations in respiration due to differing amounts of organic matter in samples; it was calculated as the ratio of respired C to total organic C.

Aggregate stability was measured on both cast and bulk soil samples using an adaptation of a standard method [34]. Air-dry cast and bulk soil samples (10 and 50 g respectively) were gently passed through an 8 mm sieve and recovered on a 4 mm sieve. Wet macro-aggregate (>2 mm) stability was then measured by wet sieving (Russell Finex, Feltham, UK, model 85521) on a 2-mm sieve for 2 min under a flow rate of 6.8 liters per minute; soil remaining on the sieve was counted as stable aggregates >2 mm (SA). The %SA was then calculated as the proportion of dry (8–4 mm) aggregates stable >2 mm (both weights corrected for moisture and gravel contents).

2.3. Statistical Analyses

Data were analyzed using Minitab 14 (Minitab Inc., State College, PA, USA, two-way ANOVA) following checks for normality and equality of variance. The analyses tested two factors; species including (*A. glutinosa*, *B. pendula* and *F. sylvatica*) and CO_2 regime (ambient and elevated); interaction effects were also tested. Incubation time or differences between soil and casts were not assessed as experimental factors; in the cast incubation experiment, the former was of lesser interest than the differences between treatments on each sampling occasion whilst differences in sampling protocols rendered soil and cast samples non-comparable other than in a general sense. Differences between individual species' means were assessed where appropriate using the TUKEY multiple range test, with a significance level of $p < 0.05$. Relationships between various parameters were assessed by linear correlation analyses.

3. Results

3.1. Chemical Composition of Litter

There were marked differences in chemical composition between litter types (Table 1a). Cellulose, C:N ratios, and lignin:nitrogen ratios were lower ($p = 0.009$) for *B. pendula* and particularly *A. glutinosa* litter compared with *F. sylvatica*. Nitrogen and phosphorous concentrations were higher for *A. glutinosa* and *B. pendula* than for *F. sylvatica*. Litter from the ambient CO_2 regime ($p = 0.037$) had a higher N concentration and lower C:N ratio ($p = 0.041$) than that from the eCO_2 treatment (Table 1b). There were no other significant CO_2 regime effects; although interaction effects were non-significant ($p = 0.072$ and 0.064 respectively), the CO_2 effects described above tended to be most pronounced for *A. glutinosa* and least pronounced for *F. sylvatica*.

Table 1. Effect of (**a**) litter species and (**b**) CO_2 regime on the chemical composition of the litter used in the mesocosm experiment.

	N g·kg^{-1}	C:N Ratio	P g·kg^{-1}	Cellulose g·kg^{-1}	Lignin g·kg^{-1}	Lignin:N Ratio
(a)						
Alnus	27.95 [a]	16.8 [b]	1.95 [a]	71.1 [b]	300.9 [a]	10.76 [c]
Betula	18.83 [b]	24.9 [b]	1.65 [a]	67.6 [b]	281.1 [a]	14.92 [b]
Fagus	9.95 [c]	46.8 [a]	0.81 [b]	139.9 [a]	290.3 [a]	29.17 [a]
(b)						
Ambient	20.2 [a]	27.6 [b]	1.52 [a]	86.0 [a]	283.5 [a]	14.03 [a]
Elevated	17.8 [b]	32.1 [a]	1.41 [a]	99.8 [a]	302.5 [a]	17.00 [a]

Litter species and CO_2 regime means with a common letter superscript do not differ significantly ($p < 0.05$).

3.2. Microbial Indices

In casts (Table 2), *B. pendula* litter produced significantly greater ($p = 0.019$) hyphal length than *A. glutinosa* and *F. sylvatica*. MBC, in contrast, was markedly higher ($p = 0.017$) in *A. glutinosa* compared to the other species; CO_2 regime had no significant effects although MBC was substantially lower for eCO_2. In mesocosm soil aggregates (Table 2), although MBC and fungal hyphal length tended to be higher in *B. pendula* compared to *A. glutinosa* and *F. sylvatica*, and in elevated compared with ambient CO_2, these differences were non-significant.

Table 2. Effect of (**a**) litter species and (**b**) CO_2 regime on microbial biomass carbon and fungal hyphal length in the soil and cast aggregates—mesocosm experiment.

	Microbial Biomass C (mg·kg^{-1})		Fungal Hyphal Length (m·g^{-1})	
	Soil	Cast	Soil	Cast
(a)				
Alnus	133.9 ± 28.4 [a]	2798 ± 265 [a]	2.71 ± 0.59 [a]	12.74 ± 1.86 [a,b]
Betula	157.5 ± 15.1 [a]	1773 ± 210 [b]	3.98 ± 1.59 [a]	19.16 ± 4.01 [a]
Fagus	129.9 ± 19.0 [a]	2198 ± 179 [a,b]	0.67 ± 0.16 [a]	6.75 ± 1.53 [b]
(b)				
Ambient	124.0 ± 14.0 [a]	2428 ± 250 [a]	2.20 ± 0.49 [a]	13.44 ± 2.97 [a]
Elevated	156.9 ± 19.3 [a]	2085 ± 162 [a]	2.71 ± 1.15 [a]	12.33 ± 2.25 [a]

Mean ± standard error: litter species and CO_2 regime means with a common letter superscript do not differ significantly ($p < 0.05$).

Respiration in incubated earthworm casts (Table 3) was generally higher for *A. glutinosa* compared to *B. pendula* and *F. sylvatica* litter. Differences between *A. glutinosa* and *F. sylvatica* were significant ($p < 0.05$ or $p < 0.01$) up to the final incubation period, but those with *B. pendula* were significant ($p = 0.031$) at day 18 only. By 22 days, respiration had stabilized across all treatments at low rates. Initially respiration for ambient and eCO_2 litter did not differ, however in later stages of the incubations, eCO_2 increased respiration ($p < 0.001$). There was a CO_2 litter X interaction ($p = 0.036$) at 18 days (Figure 1). Although respiration under eCO_2 was higher for all species, this effect was more pronounced for *A. glutinosa*.

Table 3. Effect of (**a**) litter species and (**b**) CO_2 regime on respiration in earthworm casts at four incubation times—mesocosm experiment.

	Respiration (mg CO_2-C $kg^{-1} \cdot h^{-1}$)			
	Time (Days)			
	2	6	18	22
(a)				
Alnus	5.2 ± 0.32 [a]	3.6 ± 0.33 [a]	2.5 ± 0.52 [a]	0.16 ± 0.12 [a]
Betula	4.1 ± 0.33 [a,b]	2.7 ± 0.32 [a,b]	1.5 ± 0.25 [b]	0.09 ± 0.04 [a]
Fagus	3.0 ± 0.47 [b]	1.8 ± 0.27 [b]	0.7 ± 0.26 [c]	0.13 ± 0.07 [a]
(b)				
Ambient	4.1 ± 0.34 [a]	2.8 ± 0.29 [a]	0.8 ± 0.17 [b]	0.02 ± 0.03 [b]
Elevated	4.0 ± 0.46 [a]	2.6 ± 0.36 [a]	2.3 ± 0.37 [a]	0.24 ± 0.07 [a]

Mean $\pm$ standard error: litter species and CO_2 regime means with a common letter superscript do not differ significantly ($p < 0.05$).

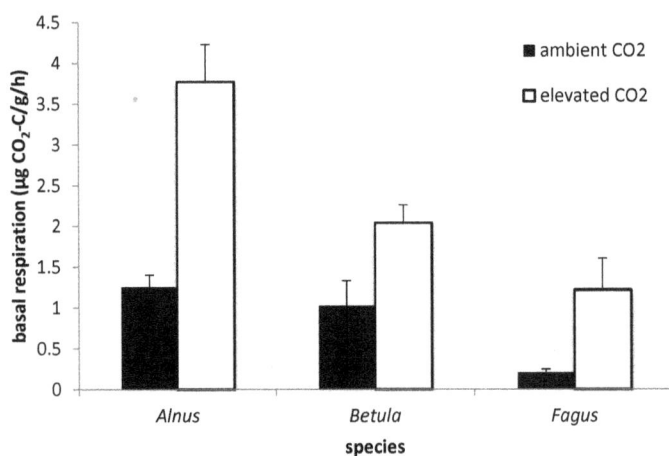

Figure 1. Interaction ($p = 0.036$) plot for CO_2 regime and litter species (*A. glutinosa*, *F. sylvatica* and *B. pendula*) for respiration in earthworm casts at day 18 of incubation. Error bars represent standard errors ($n = 4$).

Respiration in mesocosm soil (30 days) and in field aggregates (Table 4) was not significantly affected by species. In mesocosm soil aggregates, respiration 30 days after inputs commenced was reduced by eCO_2 ($p = 0.019$). At day 30, aggregate stability in mesocosm soil was negatively correlated ($n = 24$; $r = 0.64$; $p < 0.001$) with respiration; no similar relationships were found in mesocosm cast or in field aggregates. In field aggregates, eCO_2 increased respiration significantly compared to ambient CO_2.

Table 4. Effects of (**a**) litter species and (**b**) CO_2 regime on respiration in soil aggregates from mesocosm and field experiments.

	Respiration (mg CO_2-C $kg^{-1} \cdot h^{-1}$)			
		Mesocosm		Field
	Soil Day 30	Soil 30–67 Days	Soil 67–97 Days	Soil
(a)				
Alnus	0.32 ± 0.08 [a]	1.09 ± 0.13 [a,b]	0.62 ± 0.09 [a,b]	1.16 ± 0.16 [a]
Betula	0.45 ± 0.11 [a]	0.61 ± 0.08 [b]	0.41 ± 0.04 [b]	1.42 ± 0.19 [a]
Fagus	0.39 ± 0.09 [a]	3.23 ± 0.42 [a]	0.86 ± 0.17 [a]	1.37 ± 0.15 [a]
(b)				
Ambient	0.52 ± 0.08 [a]	1.65 ± 0.43 [a]	0.63 ± 0.12 [a]	1.08 ± 0.08 [b]
Elevated	0.25 ± 0.05 [b]	1.64 ± 0.36 [a]	0.62 ± 0.09 [a]	1.55 ± 0.15 [a]

Mean ± standard error: litter species and CO_2 regime means with a common letter superscript do not differ significantly ($p < 0.05$).

When mesocosm aggregates were incubated for 37 and 67 days after the end (days 67–97) of the experiment (Table 4), cumulative respiration was significantly higher ($p < 0.001$) in *F. sylvatica* compared to *A. glutinosa* and *B. pendula* (30–67 days), but later (67–97 days) only the difference between *F. sylvatica* and *B. pendula* remained significant ($p = 0.038$). Respiration was not significantly affected by CO_2 regime nor were there any significant interaction effects.

Mineralization constant did not significantly differ between species in mesocosm or field aggregates (Table 5). The mineralization constant in recently formed casts was significantly higher ($p = 0.028$) for *A. glutinosa* compared to *F. sylvatica* but not *B. pendula*. Elevated CO_2 significantly decreased ($p = 0.035$) the mineralization constant in mesocosm aggregates, but increased it significantly ($p = 0.022$) in field aggregates; CO_2 regime had no effect on cast aggregates from the mesocosm experiment.

Table 5. Effect of (**a**) litter species and (**b**) CO_2 regime on mineralization constant in aggregates from the mesocosm and field experiments.

	Mineralization Constant (mg CO_2-C $g^{-1}C$ h^{-1})		
	Mesocosm		Field
	Soil	Cast	Soil
(a)			
Alnus	0.0058 ± 0.0015 [a]	0.06172 ± 0.0033 [a]	0.0246 ± 0.0042 [a]
Betula	0.0099 ± 0.0028 [a]	0.04750 ± 0.0057 [a,b]	0.0297 ± 0.0045 [a]
Fagus	0.0073 ± 0.0018 [a]	0.03687 ± 0.0073 [b]	0.0322 ± 0.0031 [a]
(b)			
Ambient	0.0103 ± 0.0020 [a]	0.04906 ± 0.0049 [a]	0.0235 ± 0.0021 [b]
Elevated	0.0051 ± 0.0010 [b]	0.04833 ± 0.0059 [a]	0.0341 ± 0.0035 [a]

Mean ± standard error: litter species and CO_2 regime means with a common letter superscript do not differ significantly ($p < 0.05$).

3.3. Total Carbon and K_2SO_4 Extractable Carbon (KSE-C)

At the end of the mesocosm experiment (day 30), aggregates had significantly higher organic carbon ($p = 0.004$) and KSE-C ($p < 0.001$) for *B. pendula* than for *F. sylvatica* (Table 6); this effect arose in part from the observed lesser incorporation of *F. sylvatica* litter into soil. In mesocosm casts (day 30), there were no significant species effects. Elevated CO_2 litter decreased ($p = 0.010$) KSE-C in mesocosm soil aggregates. In the field, eCO_2 had no effect on either organic carbon index. Mesocosm soil (day 30) aggregate stability was positively correlated with organic carbon ($n = 24$; $r = 0.58$; $p = 0.002$) and

KSE-C ($n = 24$; $r = 0.57$; $p = 0.003$), but neither of these parameters correlated with the stability of cast aggregates.

Table 6. Effect of (**a**) litter species and (**b**) CO_2 regime on organic carbon and K_2SO_4 extractable carbon (KSE-C) in aggregates from the mesocosm and field experiments.

	Organic Carbon%			KSE-C mg kg^{-1}	
	Mesocosm		Field	Mesocosm	
	Soil	Cast	Soil	Soil	Cast
(a)					
Alnus	4.17 ± 0.04 [a,b]	8.49 ± 0.24 [a]	4.91 ± 0.30 [a]	155.05 ± 5.04 [b]	413.4 ± 21.6 [a]
Betula	4.23 ± 0.05 [a]	8.97 ± 0.44 [a]	4.99 ± 0.30 [a]	174.61 ± 4.15 [a]	358.3 ± 34.0 [a]
Fagus	3.93 ± 0.08 [b]	8.66 ± 0.10 [a]	4.22 ± 0.12 [a]	144.67 ± 3.36 [b]	366.3 ± 24.8 [a]
(b)					
Ambient	4.16 ± 0.07 [a]	8.79 ± 0.32 [a]	4.77 ± 0.23 [a]	163.80 ± 5.82 [a]	358.0 ± 25.9 [a]
Elevated	4.07 ± 0.06 [a]	8.63 ± 0.36 [a]	4.64 ± 0.22 [a]	152.42 ± 3.34 [b]	400.6 ± 17.4 [a]

Mean $\pm$ standard error: litter species and CO_2 regime means with a common letter superscript do not differ significantly ($p < 0.05$).

3.4. Aggregate Stability

In both soils and casts, *B. pendula* litter had the highest aggregate stability, with *F. sylvatica* lowest and *A. glutinosa* intermediate (Table 7). These trends were statistically significant ($p < 0.001$) for the mesocosm soils only. In casts, aggregate stability differences between species were close to significance ($p = 0.068$).

There were no direct CO_2 regime effects on aggregate stability (Table 7). However, there was a significant CO_2 regime X litter species interaction ($p = 0.046$) in the field soils (Figure 2). Elevated CO_2, increased aggregate stability under *F. sylvatica* but decreased stability under the other two species; these effects were small relative to background stability. Although not statistically significant ($p = 0.09$), a broadly similar interaction was found for cast aggregate stability in the mesocosm study, with only *F. sylvatica* litter showing a positive response to eCO_2.

Table 7. Effect of (**a**) litter species and (**b**) CO_2 regime on % stable aggregates in soil and casts from mesocosm and field experiments.

	% Stability		
	Mesocosm		Field
	Soil	Cast	Soil
(a)			
Alnus	35.0 ± 2.16 [b]	86.9 ± 1.49 [a]	90.1 ± 2.20 [a]
Betula	44.9 ± 4.43 [a]	87.5 ± 1.43 [a]	92.2 ± 1.40 [a]
Fagus	25.8 ± 0.89 [c]	79.1 ± 4.48 [a]	88.8 ± 1.98 [a]
(b)			
Ambient	36.9 ± 3.45 [a]	82.4 ± 3.35 [a]	91.3 ± 1.71 [a]
Elevated	33.6 ± 3.04 [a]	86.6 ± 0.99 [a]	89.4 ± 1.34 [a]

Mean $\pm$ standard error: litter species and CO_2 regime means with a common letter superscript do not differ significantly ($p < 0.05$).

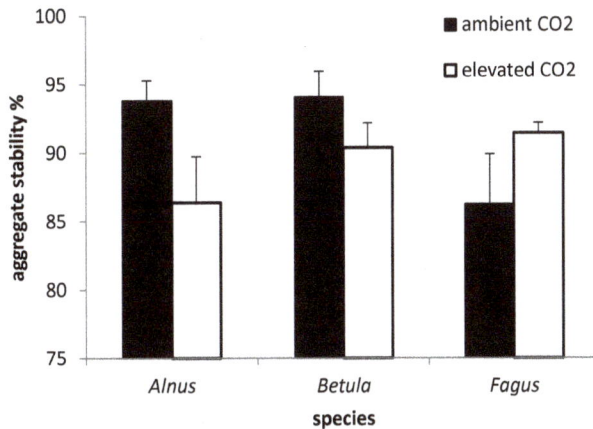

Figure 2. Interaction (p = 0.046) plot for CO_2 regime and litter species *(A. glutinosa, B. pendula* and *F. sylvatica)* for stable aggregation—field experiment. Error bars represent standard errors (n = 4).

4. Discussion

For most indices, litter quality was ranked *A. glutinosa* > *B. pendula* > *F. sylvatica* although in some cases differences between *A. glutinosa* and *B. pendula* were small. Evidence from other studies has indicated similar rankings of litter quality by species [6,7]. Elevated CO_2 tended to reduce litter quality in general, although this effect was marked for N and C:N only. Again, these findings are broadly in line with those of other similar studies [9]. For most quality parameters, species differences were larger than those for CO_2 regime.

Treatment effects on microbial biomass C (MBC) and fungal hyphal length were found in cast aggregates only and then only between species, not CO_2 regimes. Both indices tended to be greater for *B. pendula* and *A. glutinosa* compared to *F. sylvatica* litter; higher nutrient contents in these litters may explain the enhanced microbial responses observed. Litter quality effects on MBC in casts may vary with time; thus, high quality litter (lucerne) gave higher MBC at 14 and 28 days compared to wheat straw, but these amendments did not differ at 7 and 56 days [35]; data here for very recent casts showed higher respiration with better quality litter but in later soil aggregates these differences were inconsistent. Given the effects of eCO_2 on N concentrations, a reduced microbial community might have been expected; other studies [11] have attributed the absence of any eCO_2 effect on MBC in grassland soils to N limitation. If N is a key determinant of MBC responses to eCO_2 [14], in our study, some variation between species in microbial eCO_2 responses might have been expected given their very different litter N contents; only species' differences were observed and then only in casts.

The effects of treatments on respiration were complex and time-dependent. Higher quality litters *(A. glutinosa* and *B. pendula)* had higher respiration in recently formed earthworm casts up to 18 days. The generally low rates of cast respiration at 22 days were consistent with trends in other studies [35], with respiration and microbial biomass decreasing steadily over 100 days. In contrast, for incubated soil aggregates (37 and 67 days after inputs ceased) the quality effect was reversed with respiration being higher in poor quality *F. sylvatica* litter, due probably to delayed mineralization of this litter. As with MBC, litter quality X time interactions have been shown for respiration [36]; here, for casts and incubated aggregates, the ranking of litters by respiration varied with time.

When casts were incubated, respiration was unaffected by CO_2 regime initially, but was significantly higher for eCO_2 litter in the later stages of cast incubation, a trend also seen in soil aggregates; there was also evidence of CO_2 X litter species treatment interactions, with the later stimulatory effect of eCO_2 being more pronounced in *A. glutinosa* litter. Again, eCO_2-induced reductions in litter quality may have delayed decomposition by some days, a response more pronounced in otherwise rapidly decomposing litter. Elevated CO_2 decreased respiration and mineralization constant in soil aggregates from the mesocosm trial, but the reverse was the case in field aggregates sampled some months after litterfall. Other studies [14,37] have found that eCO_2 stimulates respiration in field soils; findings reported here suggest a more complex and time-dependent response in aggregates. Differences in mineralization constant suggest that respiration responses were influenced by the composition of organic materials present in aggregates. Findings concur with the view [38] that litter quality influences aggregate C dynamics but only in the short-term. It is likely that a higher C:N ratio litter at eCO_2 initially restricted microbial decomposition but that this led to increased materials available for mineralization later. Slower litter decay rates have been linked to increases in lignin:N ratios [39].

Organic carbon and KSE-C in earthworm casts were not affected by treatments. In mesocosm soil aggregates, organic carbon and KSE-C were high for *A. glutinosa* and *B. pendula* compared to *F. sylvatica*. Lower soil aggregate KSE-C for *F. sylvatica* and under elevated CO_2 may in part be explained by limited mineralization of this litter; evidence of higher respiration associated with *F. sylvatica* litter occurred at a later stage to that when KSE-C data were obtained. There were no significant treatment effects on these parameters in the field, as found in a previous investigation [22] on the same experimental site.

Elevated CO_2 can affect soil aggregation, a dynamic property that responds to environmental changes, feeding back to other ecosystem functions [39]. Interactions between micro-organisms, organic residues and mineral particles affect the extent and temporal dynamics of aggregation. There is evidence that high quality residues improve aggregation quickly, whereas lower quality residues are more effective in the longer-term [20]. Species litter was ranked as *A. glutinosa* > *B. pendula* > *F. sylvatica* in terms of decomposability; the effects of eCO_2 were less pronounced and more complex.

In the short term, more easily decomposable *A. glutinosa* and *B. pendula* litters produced higher aggregate stability than *F. sylvatica* litter. Compared to low and high C:N ratio litter, intermediate quality *B. pendula* litter may have provided a better balance between stimulation of microbial activity to promote early aggregate stabilization and greater persistence of stabilizing agents. In this context, the markedly higher KSE-C contents in soil aggregates with added *B. pendula* litter at the end of the mesocosm experiment may indicate mineralization of stabilizing compounds; soluble carbon is considered a key source of energy for microorganisms [40]. Additionally, organic carbon, MBC, fungal hyphae and microbial activity tended to be higher in aggregates of *B. pendula* compared to the other litters. Some months after litterfall, *F. sylvatica* had similar aggregate stability to the other species in the field. *F. sylvatica* litter may sustain microbial activity, generating stabilizing compounds over a longer time period after inputs compared with more labile litters. There is some evidence supporting this interpretation in the low respiration rates (Table 3) in casts for *F. sylvatica*, but high rates in aggregates incubated to 67 days (Table 4) after amendments to the mesocosm experiment had ceased. Tree root and associated fungal responses, in addition to litter effects, may have influenced aggregate stability in the field.

Species effects on aggregate stability, at least in the mesocosm experiment, were larger than those of the CO_2 regime; this was expected given that CO_2 differences in litter quality were generally smaller, as found previously on this experimental site [22]. There was, however, evidence of species X CO_2 interactions, with elevated CO_2 having a positive effect on aggregate stability only with *F. sylvatica*. These findings suggest that atmospheric CO_2 effects on aggregation in woodland soils may be species-specific. The aggregation interaction may be explained in part by associated, though non-significant interactions for respiration. Microbial activity enhances aggregate stability soon after aggregate formation or organic inputs [41], but in the mesocosm study was negatively correlated with stability at day 30, suggesting that, by this stage, respiration was associated with decomposition of aggregate stabilizing agents. Field soils sampled some months after litterfall may have entered a microbial 'degradative' phase in aggregates receiving the more easily decomposable litter of *A. glutinosa* and *B. pendula*, whereas delayed microbial decomposition of lower quality *F. sylvatica* litter may not have reached this phase. This interpretation is consistent with evidence that aggregates formed with higher quality litter have a faster turnover than those with lower quality litter [42]. Findings relating to differential eCO_2 effects are broadly similar to other studies [15]; thus, the composition of the vegetation cover and its response to eCO_2 will have a significant influence on outcomes for aggregation and associated parameters.

The effects of earthworm activity were not directly addressed in this study as soil and cast materials in the mesocosm study were not directly comparable. Nevertheless, it is clear that most microbial indices were markedly higher in casts, as was the case for C and aggregate stability indices. For most measured indices, though not respiration, treatment effects observed in soil aggregates were absent or reduced in casts. This may indicate that earthworm comminution of litter and its thorough mixing with mineral fractions mitigated at least some of the differences in C inputs.

5. Conclusions

Exposure to elevated CO_2 may alter microbial populations and their activity, in addition to aggregate stability. Where there are marked differences in litter quality between tree species, species effects are likely to be more pronounced. Earthworms may mitigate some of these CO_2 and species effects. At least for the contrasting species investigated, significant shifts in their relative proportion in response to atmospheric and climatic changes may have more impact on forest C and aggregate dynamics than intra-specific respiration responses to eCO_2; any such changes may have implications for the seasonal dynamics of aggregation and nutrient cycling. Resolving the complex and often species-specific responses to eCO_2 as they affect C inputs to soils, with associated impacts on soil characteristics, remains an important challenge [14].

Acknowledgments: S. Al-Maliki acknowledges financial support from the Ministry of Higher Education and Scientific Research, Iraq. This study was undertaken as part of CIRRE under the Bangor–Aberystwyth University initiative.

Author Contributions: S.M.J.A. and J.S. respectively carried out and supervised the research. D.L.G. and D.L.J. designed the field experiment. D.G.-J. and J.S. had the major role in revising an initial draft by S.M.J.A. D.L.G. and D.L.J. commented on the final draft.

Conflicts of Interest: The authors declare no conflict of interest.

References

1. Intergovernmental Panel on Climate Change; Core Writing Team; Pachauri, R.K.; Meyer, L.A. (Eds.) *Climate Change 2014: Synthesis Report, Contribution of Working Groups I, II and III to the Fifth Assessment Report of the Intergovernmental Panel on Climate Change*; Intergovernmental Panel on Climate Change: Geneva, Switzerland, 2014; 151p.
2. Bronick, C.; Lal, R. Soil structure and management: A review. *Geoderma* **2005**, *124*, 3–22. [CrossRef]
3. Sánchez-de León, Y.; Lugo-Pérez, J.; Wise, D.H.; Jastrow, J.D.; González-Meler, M.A. Aggregate formation and carbon sequestration by earthworms in soil from a temperate forest exposed to elevated atmospheric CO_2: A microcosm experiment. *Soil Biol. Biochem.* **2014**, *68*, 223–230. [CrossRef]
4. Berg, B.; McClaugherty, C. *Plant Litter: Decomposition, Humus Formation, Carbon Sequestration*, 2nd ed.; Springer: Berlin/Heidelberg, Germany, 2008; p. 338.
5. Le Guillou, C.; Angers, D.A.; Maron, P.A.; Leterme, P.; Menasseri-Aubry, S. Linking microbial community to soil water-stable aggregation during crop residue decomposition. *Soil Biol. Biochem.* **2012**, *50*, 126–133. [CrossRef]
6. Richardson, J.S.; Shaughnessy, C.R.; Harrison, P.G. Litter breakdown and invertebrate association with three types of leaves in a temperate rainforest stream. *Arch. Hydrobiol.* **2004**, *159*, 309–325. [CrossRef]
7. Loranger-Merciris, G.; Laossi, K.R.; Bernhard-Reversat, F. Soil aggregation in a laboratory experiment: Interactions between earthworms, woodlice and litter palatability. *Pedobiologia* **2008**, *51*, 439–443. [CrossRef]
8. Cotrufo, M.F.; de Angelis, P.; Polle, A. Leaf litter production and decomposition in a poplar short rotation coppice exposed to free air CO_2 enrichment (POPFACE). *Glob. Chang. Biol.* **2005**, *11*, 971–982. [CrossRef]

9.   Norby, R.J.; Cotrufo, M.F.; Ineson, P.; O'Neill, E.G.; Canadell, J.G. Elevated CO_2, litter chemistry, and decomposition: A synthesis. *Oecologia* **2001**, *127*, 153–165. [CrossRef] [PubMed]
10. Thoms, C.; Gattinger, A.; Jacob, C.; Thomas, F.M.; Gleixner, G. Direct and indirect effects of tree diversity drive soil microbial diversity in temperate deciduous forest. *Soil Biol. Biochem.* **2010**, *42*, 1558–1565. [CrossRef]
11. Niklaus, P.; Alphei, J.; Ebersberger, D.; Kampichler, C.; Kandeler, E.; Tscherko, D. Six years of in situ CO_2 enrichment evoke changes in soil structure and soil biota of nutrient poor grassland. *Glob. Chang. Biol.* **2003**, *9*, 585–600. [CrossRef]
12. Dorodnikov, M.; Blagodatskaya, E.; Blagodatsky, S.; Kuzyakov, Y.; Marhan, S. Stimulation of microbial extracellular enzyme activities by elevated CO_2 depends on soil aggregate size. *Glob. Chang. Biol.* **2009**, *15*, 1603–1614. [CrossRef]
13. Saurav Sahaa, D.; Chakrabortya, V.K.; Sehgal, L.N.; Madan, P. Long-term atmospheric CO_2 enrichment impact on soil biophysical properties and root nodule biophysics in chickpea (*Cicer arietinum* L.). *Eur. J. Agron.* **2016**, *75*, 1–11. [CrossRef]
14. Norby, R.J.; Zak, D.R. Ecological lessons from Free-Air CO_2 Enrichment (FACE) experiments. *Ann. Rev. Ecol. Evol. Syst.* **2011**, *42*, 181–203. [CrossRef]
15. Del Galdo, I.; Oechel, W.C.; Cotrufo, M.F. Effects of past, present and future atmospheric CO_2 concentrations on soil organic matter dynamics in a chaparral ecosystem. *Soil Biol. Biochem.* **2006**, *38*, 3235–3244. [CrossRef]
16. Eviner, V.T.; Stuart Chapin, F. The influence of plant species, fertilization and elevated CO_2 on soil aggregate stability. *Plant Soil* **2002**, *246*, 211–219. [CrossRef]
17. Rillig, M.C.; Wright, S.F.; Allen, M.F.; Field, C.B. Rise in carbon dioxide changes soil structure. *Nature* **1999**, *400*, 628. [CrossRef]
18. Hoosbeek, M.; Vos, J.M.; Bakker, E.J.; Scarascia-Mugnozza, G.E. Effects of free atmospheric CO_2 enrichment (FACE), N fertilization and poplar genotype on the physical protection of carbon in the mineral soil of a polar plantation after five years. *Biogeosciences* **2006**, *3*, 479–487. [CrossRef]
19. Tisdall, J.M.; Oades, J.M. Organic matter and water-stable aggregates in soils. *J. Soil Sci.* **1982**, *33*, 141–163. [CrossRef]
20. Abiven, S.; Menasseri, S.; Chenu, C. The effects of organic inputs over time on soil aggregate stability—A literature analysis. *Soil Biol. Biochem.* **2009**, *41*, 1–12. [CrossRef]
21. Teklehaimanot, Z.; Jones, M.; Sinclair, F. Tree and livestock productivity in relation to tree planting configuration in a silvopastoral system in North Wales, UK. *Agrofor. Syst.* **2002**, *56*, 47–55. [CrossRef]
22. Hoosbeek, M.; Lukac, M.; Velthorst, E.; Smith, A.; Godbold, D. Free atmospheric CO_2 enrichment increased above ground biomass but did not affect symbiotic N_2-fixation and soil carbon dynamics in a mixed deciduous stand in Wales. *Biogeosciences* **2011**, *8*, 353–364. [CrossRef]
23. Scullion, J.; Smith, A.R.; Gwynn-Jones, D.; Jones, D.L.; Godbold, D. Deciduous woodland exposed to elevated atmospheric CO_2 has species-specific impacts on anecic earthworms. *Appl. Soil Ecol.* **2014**, *80*, 84–92. [CrossRef]
24. Scullion, J.; Malik, A. Earthworm activity affecting organic matter, aggregation and microbial activity in soils restored after opencast mining for coal. *Soil Biol. Biochem.* **2000**, *32*, 119–126. [CrossRef]
25. Harborne, J.B.; Harborne, A.J. *Phytochemical Methods: A Guide to Modern Techniques of Plant Analysis*; Taylor and Francis Group: London, UK, 1998; p. 302.
26. Vance, E.; Brookes, P.; Jenkinson, D. Microbial biomass measurements in forest soils: The use of the chloroform fumigation-incubation method in strongly acid soils. *Soil Biol. Biochem.* **1987**, *19*, 697–702. [CrossRef]
27. Hanssen, J.F.; Thingstad, T.F.; Goksøyr, J. Evaluation of hyphal lengths and fungal biomass in soil by a membrane filter technique. *Oikos* **1974**, *25*, 102–107. [CrossRef]
28. Olson, F. Quantitative estimates of filamentous algae. *Trans. Am. Microsc. Soc.* **1950**, *69*, 272–279. [CrossRef]
29. Campbell, C.D.; Chapman, S.J.; Cameron, C.M.; Davidson, M.S.; Potts, J.M. A rapid microtiter plate method to measure carbon dioxide evolved from carbon substrate amendments so as to determine the physiological profiles of soil microbial communities by using whole soil. *Appl. Environ. Microbiol.* **2003**, *69*, 3593–3599. [CrossRef] [PubMed]
30. Rowell, D.L. *Soil Science: Methods and Applications*; Addison Wesley Longman: Harlow, UK, 1994; p. 117.
31. Ball, D. Loss-on-ignition as an estimate of organic matter and organic carbon in non-calcareous soils. *Eur. J. Soil Sci.* **1964**, *15*, 84–92. [CrossRef]

32. Milne, R.; Haynes, R. Soil organic matter, microbial properties, and aggregate stability under annual and perennial pastures. *Biol. Fertil. Soils* **2004**, *39*, 172–178. [CrossRef]

33. Gil-Sotres, F.; Trasar-Cepeda, M.; Ciardi, C.; Ceccanti, B.; Leirós, M. Biochemical characterization of biological activity in very young mine soils. *Biol. Fertil. Soils* **1992**, *13*, 25–30. [CrossRef]

34. Yoder, R. A direct method of aggregate analysis of soils and a study of the physical nature of erosion losses. *J. Am. Soc. Agron.* **1936**, *28*, 337–351. [CrossRef]

35. Haynes, R.; Fraser, P. A comparison of aggregate stability and biological activity in earthworm casts and uningested soil as affected by amendment with wheat or lucerne straw. *Eur. J. Soil Sci.* **1998**, *49*, 629–636. [CrossRef]

36. Tiunov, A.V.; Scheu, S. Microbial biomass, biovolume and respiration in *Lumbricus terrestris* L. cast material of different age. *Soil Biol. Biochem.* **2000**, *32*, 265–275. [CrossRef]

37. Taneva, L.; Gonzalez-Meler, M.A. Decomposition kinetics of soil carbon of different age from a forest exposed to 8 years of elevated atmospheric CO_2 concentration. *Soil Biol. Biochem.* **2008**, *40*, 2670–2677. [CrossRef]

38. Gentile, R.; Vanlauwe, B.; Six, J. Litter quality impacts short- but not long-term soil carbon dynamics in soil aggregate fractions. *Ecol. Appl.* **2011**, *21*, 695–703. [CrossRef] [PubMed]

39. Niklaus, P.A.; Alphei, J.; Kampichler, C.; Kandeler, E.; Korner, C.; Tscherko, D.; Wohlfender, M. Slowed-down soil drying cycles resulted in lower soil aggregation under elevated CO_2. *Ecology* **2007**, *88*, 3153–3163. [CrossRef] [PubMed]

40. Haynes, R. Labile organic matter as an indicator of organic matter quality in arable and pastoral soils in New Zealand. *Soil Biol. Biochem.* **2000**, *32*, 211–219. [CrossRef]

41. Abiven, S.; Menasseri, S.; Angers, D.; Leterme, P. Dynamics of aggregate stability and biological binding agents during decomposition of organic materials. *Eur. J. Soil Sci.* **2007**, *58*, 239–247. [CrossRef]

42. Six, J.; Carpentier, A.; van Kessel, C.; Merckx, R.; Harris, D.; Horwath, W.R.; Lüscher, A. Impact of elevated CO_2 on soil organic matter dynamics as related to changes in aggregate turnover and residue quality. *Plant Soil* **2001**, *234*, 27–36. [CrossRef]

forests

MDPI

Article

Mitigating the Stress of Drought on Soil Respiration by Selective Thinning: Contrasting Effects of Drought on Soil Respiration of Two Oak Species in a Mediterranean Forest

Chao-Ting Chang [1,2,*], Dominik Sperlich [1,2], Santiago Sabaté [1,2], Elisenda Sánchez-Costa [2], Miriam Cotillas [2], Josep Maria Espelta [2] and Carlos Gracia [1,2]

[1] Department of Evolutionary Biology, Ecology and Environmental Sciences, University of Barcelona, Diagonal 643, 08028 Barcelona, Spain; dominik@creaf.uab.cat (D.S.); santi.sabate@ub.edu (S.S.); cgracia@ub.edu (C.G.)

[2] CREAF, Cerdanyola del Vallès, 08193 Barcelona, Spain; e.sanchez@creaf.uab.cat (E.S.-C.); miriam.cotillas@gmail.com (M.C.); Josep.Espelta@uab.cat (J.M.E.)

* Correspondence: chaoting@creaf.uab.cat; Tel.: +34-9358-14850

Academic Editors: Robert Jandl and Mirco Rodeghiero
Received: 7 September 2016; Accepted: 28 October 2016; Published: 4 November 2016

Abstract: Drought has been shown to reduce soil respiration (SR) in previous studies. Meanwhile, studies of the effect of forest management on SR yielded contrasting results. However, little is known about the combined effect of drought and forest management on SR. To investigate if the drought stress on SR can be mitigated by thinning, we implemented plots of selective thinning and 15% reduced rainfall in a mixed forest consisting of the evergreen *Quercus ilex* and deciduous *Quercus cerrioides*; we measured SR seasonally from 2004 to 2007. Our results showed a clear soil moisture threshold of 9%; above this value, SR was strongly dependent on soil temperature, with Q_{10} of 3.0–3.8. Below this threshold, the relationship between SR and soil temperature weakened. We observed contrasting responses of SR of target oak species to drought and thinning. Reduced rainfall had a strong negative impact on SR of *Q. cerrioides*, whereas the effect on SR for *Q. ilex* was marginal or even positive. Meanwhile, selective thinning increased SR of *Q. cerrioides*, but reduced that of *Q. ilex*. Overall, our results showed that the negative effect of drought on SR can be offset through selective thinning, but the effect is attenuated with time.

Keywords: reduced rainfall; selective thinning; soil respiration; *Quercus ilex*; *Quercus cerrioides*

1. Introduction

Forest ecosystems contain one of the largest stocks of carbon and they represent one of the most important potential carbon sinks [1]. Globally, forest ecosystems are estimated to contain 681 ± 66 Pg (1 Pg = 10^{15} g) of carbon, with around 383 ± 28 Pg C (44%) of that total contained in the soil [1]. Therefore, forest soil respiration (SR) plays a crucial role in regulating soil carbon pools and carbon dynamics of terrestrial ecosystems under global warming [2,3]. Climate change scenarios project increases in mean annual temperature, increases in evapotranspiration, and decreases in precipitation [4–6]. Hence, future climate change is expected to have a great impact on SR by altering its main environmental drivers: temperature and moisture [7–10]. Because forest ecosystems may mitigate climate change through carbon sequestration [11], the effects of forest management practices on ecosystem carbon sinks need to be assessed. However, there is still no consensus on how forest management affects the soil's carbon balance; in addition, information on how forest management alters the response of SR to global warming is still limited [12–14].

Selective thinning is a common practice to improve forest health and productivity. Generally, after selective thinning, the remaining trees receive more solar radiation, soil water, soil organic matter, and nutrients, thus enhancing their photosynthetic capacity [15–19]. As a result, SR is expected to increase after forest thinning due to the increase in both soil organic matter and autotrophic respiration caused by the improvement of tree vitality. However, many studies have investigated the effect of forest management on SR with conflicting conclusions. Tang et al. [20] observed a decrease of 13% in total SR after thinning and suggested the decrease may be associated with the decrease in root density. On the contrary, Tian et al. [21] found an increase in SR up to 30% after thinning that slightly declined to 20%–27% in the following four to six years in a Chinese Fir (*Cunninghamia lanceolata* (Lamb.) Hook) plantation. Johnson and Curtis [22] concluded in their review that forest harvesting had little or no effect on soil carbon and nitrogen storage. Overall, the effect of thinning on SR is determined by many interactive factors, such as changes in soil temperature (*Ts*), soil moisture, microbial and root respiration, and decomposition of litter and woody debris. The responses of SR to thinning are the result of the combined effects of a "tug of war" among these factors.

In the Mediterranean region, summer drought has been identified as the main factor that limits plant species distribution and growth [23]. However, studies examining the extent to which drought affects SR have yielded inconsistent results. Some studies have shown that drought conditions will reduce SR due to low root and microbial activities [24–28]. Others report that drought may increase SR through enhancement of root growth [29,30]. Contrasting responses of fine root growth to drought were also found; fine root growth was enhanced in beech [31], but inhibited in spruce [32].

Given its arid and semi-arid climate, the Mediterranean region is a suitable area to study the effects of drought on forest productivity. While being exposed to re-occurring summer droughts, Mediterranean forests are particularly vulnerable to further reductions in water supply under climate change scenarios. Intergovernmental Panel on Climate Change [33], for instance, calls for a 15%–20% reduction of soil water availability over the next three decades in Mediterranean- type ecosystems. However, soil processes in Mediterranean ecosystems have received relatively little attention [7,8,34], and are currently under-represented as priorities for research networks [35,36]. This study may provide a better understanding of responses of SR to soil water deficits and the interaction with selective thinning. Selective thinning is a general practice to recover the structure of oak forests after wildfires, but it is also a potential drought mitigation practice.

The specific objectives of this study were: (i) to examine the time-course of the effects of selective thinning on the pattern of SR under two dominant tree species, *Quercus ilex* L. and *Quercus cerrioides* Willk & Costa in a Mediterranean forest; (ii) to evaluate the possible responses of SR under these two species subjected to experimental drought, and finally; (iii) to investigate whether selective thinning reduces the negative effect of drought on SR.

We expected that: (1) thinning would increase SR due to the deposition of the thinning material on the ground and the increase in nutrient availability; (2) reduced rainfall would decrease SR, especially during the growing season, as a result of decreased soil moisture; (3) due to the combined effect of thinning and reduced rainfall, thinning would compensate for the decrease in SR under drought conditions.

2. Materials and Methods

2.1. Site Description

The experiment was conducted in the region of Bages, Catalonia, NE Spain (41°44' N, 1°39' E, 800 m above sea level). Climate is dry, sub-humid Mediterranean, with a pronounced summer drought from July to September. Mean annual temperature and precipitation are 12 °C and 600 ± 135 mm, respectively (1980–2000) [37]. Soils are developed above calcareous substrate, surface rockiness is high, and the soil is moderately well drained with a mean depth ca. 25–50 cm. Additional information on the site is provided in Cotillas et al. [38].

2.2. Stand History and Tree Species Composition

Our study site is a mixed oak forest dominated by *Q. ilex* (Holm oak) and *Q. cerrioides* that regenerated by resprouting after a large wildfire in 1998. *Q. ilex* is a sclerophyllous evergreen tree species that is distributed widely over the Iberian Peninsula. *Q. cerrioides* is a winter semi-deciduous (marcescent) species. Both tree species have the ability to resprout from stumps and roots after disturbances [39]. When starting the experiment in 2004, the post-fire regeneration was six years old. The stem basal area and height of *Q. cerrioides* and *Q. ilex* from the study site were significantly different. *Q. cerrioides* individuals had a larger mean stem basal area (12.4 ± 0.8 cm^2) and height (177 ± 4 cm) than those of *Q. ilex* (9.7 ± 0.8 cm^2 and 144 ± 4 cm) [38].

2.3. Experimental Design

Our experiment was designed to test the effects of thinning and experimental drought in a Mediterranean oak forest. A total of 12 plots were installed with three replicates each for (1) control, (2) 15% rainfall exclusion, (3) selective thinning, and (4) combined (thinning with 15% rainfall exclusion). The plots (15 m × 20 m) were distributed randomly in the sampling area with a minimum buffer of 10 m surrounding every plot. To intercept runoff water, a ditch of ca. 50 cm depth was excavated along the entire top edge of the rainfall exclusion plots and covered with Poly Vinyl Chloride (PVC) strips. Due to instrumental limitations, SR rates were measured only in one replicate of each treatment. Tree height, basal area, and density were measured before starting the experiment and no significant differences were found in structural characteristics among plots [38]. Selective thinning was done in spring 2004. Traditional criteria of selective low-thinning for young oak coppices were applied [40,41]: 20%–30% of total stump basal area per plot was reduced, the weakest stems were eliminated, and from one to three dominant stems per stump were left. After selective thinning, mean stem basal area and height in thinning and combined treatments were 14.3 ± 0.8 cm^3 and 180 ± 4 cm, respectively, and in the unthinned plots, those same characteristics were 7.7 ± 0.8 cm^3 and 146 ± 4 cm, respectively. In the reduced rainfall and combined treatment plots, parallel drainage channels were installed at ca. 50 cm height above the soil and covered 15% of the ground surface. The channels were installed after the measurement of autumn 2004.

2.4. Field Measurements

SR and Ts under *Q. ilex* and *Q. cerrioides* individuals were measured seasonally from 2004 to 2007 during three-day periods for each treatment. In each plot, four stainless-steel rings were inserted permanently at a soil depth of 3 cm. The rings were weeded regularly. CO_2 concentration was measured in situ with an automatic changeover open system. The system consisted of an infrared gas analyzer (IRGA, LiCor 6262, LiCor, Inc., Lincoln, NE, USA), a data logger (CR10 Data logger, Campbell Scientific Inc., Logan, UT, USA), 12 pairs of channels, 12 chambers, 12 pairs of rotameters, six pumps, and two flowmeters. Four pairs of channels were connected with the soil chambers. Each pair of channels consisted of two tubes, one attached to the top of the chamber (reference CO_2 concentration) and another attached to the base for calculating the increment in CO_2 concentration (sample CO_2 concentration). The other eight pairs of channels were connected to leaf and stem chambers, which were measured in parallel, but are not presented in this work. The stainless steel soil chambers were closed cylindrical chambers 28 cm in diameter and 15 cm high. Air was pumped through all chambers continuously at 1 L· min^{-1}, but only one chamber at a time was directed to the gas analyzer for 1 min. Meanwhile, air through the other chamber was exhausted to the atmosphere. When air was directed to the gas analyzer, only the last 40 seconds of recordings from the gas analyzer were averaged and recorded by the data logger. A complete measurement cycle took 60 min, including four rounds of measurements of absolute, ambient air, and CO_2 concentration (ppm) from all chambers and one additional zero calibration cycle.

Soil chambers were shaded by placing a 50×50 cm green fine mesh on top to avoid possible heating by direct sunlight during the measurements. Soil temperatures in the upper 5 cm of soil were measured continuously with Pt100 temperature sensors ($n = 4$) and recorded in parallel with the CO_2 concentration analysis. Soil moisture (cm^3/cm^3) in the upper 20 cm of soil was recorded manually once per day during the three-day measurement of each plot using 10 Time Domain Reflectometry Probes (Tektronix, 1520C Beaverton, OR, USA), which were installed randomly within each plot. Due to instrument failure, no SR data were recorded during winter 2007. Starting from summer 2005, seasonal litter fall per tree species was collected from each treatment. After collecting the litter, its fresh weight was determined. Samples were oven-dried at 65 °C for 48 h and then the dry weight was determined.

2.5. Data Analysis

We used analysis of variance (ANOVA) with treatment (thinning, reduced rainfall, both thinning and reduced rainfall combined, and control), season (winter, spring, summer, autumn) and year (2004, 2005, 2006, and 2007) as main factors to examine their effects on SR, Ts, and soil moisture. The daily or seasonal averages were used in these analyses. The relationship between SR and Ts in different treatments was based on daily average data using regression analysis, where a univariate exponential model was fitted [42]:

$$R = R_0 \left(e^{KT} \right) \tag{1}$$

where R is the measured soil respiration rate (μmol C m^{-2}·s^{-1}), R_0 is the basal respiration at temperature of 0 °C, T is the measured soil temperature (°C), and K is the fitted parameter. Thereafter, the temperature sensitivity of soil respiration can be derived as:

$$Q_{10} = e^{10K} \tag{2}$$

where Q_{10} is the apparent field-observed proportional increase in SR related to a 10 °C increase in temperature. We also used recursive partitioning analysis to separate the relationship between SR and Ts by soil moisture regime. As models based on partitioning can only handle linear models, the equation above was transformed by linearizing with logarithms:

$$\text{Ln } R = \ln R_0 + KT \tag{3}$$

Logarithmic transformed SR values were used as the dependent variable. Once the soil moisture thresholds were obtained, nonlinear regression analyses (model 1) were used to determine the relationship between SR and Ts in each soil moisture interval. All statistical analyses were performed with PASW statistics 18 (SPSS Inc., 2009, Chicago, IL, USA), except the recursive partitioning analysis, which was conducted with R statistical software version 2.15.3 (R Development Core Team, 2013) using the *party* package [43]. For all statistical tests, significance was accepted at $P < 0.05$. Values are given as mean ± standard error (SE).

3. Results

3.1. Temporal Variation in Ts and Soil Moisture

The average temperature showed no significant difference between treatments (Table 1). The seasonal course of soil temperature was pronounced in our study site. The highest recorded Ts was 32.2 °C in summer 2005 and the lowest was −0.3 °C in winter 2005. Soil moisture varied largely over the study period, ranging from 2.3% to 18.4% (Figure 1). Mean annual precipitation was lowest in 2006 (400 mm) and highest in 2007 (830 mm). The highest soil moisture occurred in winter and spring, but then dropped sharply in summer. The lowest soil moisture (2.3%) was recorded during the thinning treatment in summer 2005. Soil moisture was correlated negatively with Ts; the peak of Ts in summer coincided with the lowest soil moisture values. Throughout the four monitored years, the mean seasonal soil moisture in the control treatment was consistently higher than in the other treatments. Despite the reduced rainfall treatment, we did not find lower soil moisture in the plots subjected to reduced rainfall during most of the measurement campaigns.

Table 1. Treatment effects on soil temperature (Ts) and soil respiration (SR) of *Q. ilex* and *Q. cerrioides*.

Variable	Treatment	*Q. ilex*	*Q. cerrioides*	Average
Ts (°C)	Natural rainfall	14.88 a	14.98 a	14.93 a
	Reduced rainfall	16.77 a	15.99 a	16.38 a
	No Thinning	16.31 a	15.67 a	15.99 a
	Thinning	15.30 a	15.28 a	15.29 a
SR (μmol C m^{-2}·s^{-1})	Natural rainfall	0.45 a	0.47 a	0.46 a
	Reduced rainfall	0.38 a	0.30 b	0.34 b
	No Thinning	0.47 a	0.33 a	0.40 a
	Thinning	0.36 b	0.44 b	0.40 a

The different letters indicate the significant differences between treatments ($p < 0.05$).

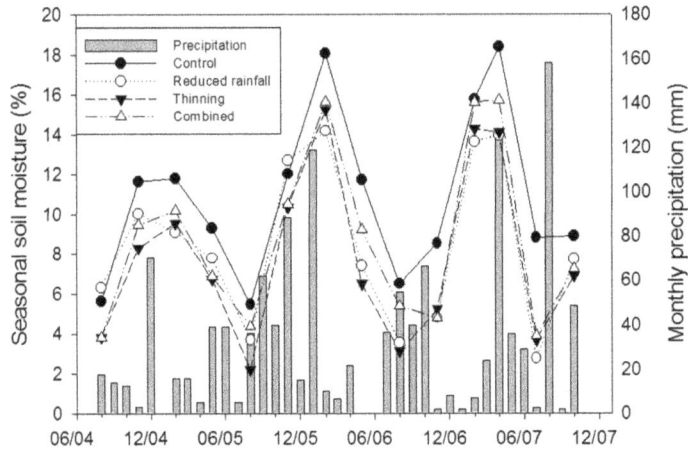

Figure 1. Seasonal variation in soil moisture (lines) and monthly variation in precipitation (bars) for each treatment during the study period. Different symbols represent different treatments. Labels on the x-axis represent time in month/year format.

3.2. Treatment Effect on SR

Within the four treatments, SR was between 0.00 and 1.82 μmol C m^{-2}·s^{-1}, with an overall mean ($\pm$SD) of 0.43 $\pm$ 0.28 μmol C m^{-2}·s^{-1}. Reduced rainfall treatment significanly depressed SR, with around 26% lower in comparison to natural rainfal (Table 1). Selective thinning showed no effect on overall SR (Table 1). SR under *Q. ilex* (0.44 $\pm$ 0.28 μmol C m^{-2}·s^{-1}) was significantly higher than SR under *Q. cerrioides* (0.41 $\pm$ 0.28 μmol C m^{-2}·s^{-1}, P < 0.001). Meanwhile, SR under *Q. ilex* showed no significant difference in subjected to reduced rainfall while SR under *Q. cerrioides* showed a pronounced decrease. Selective thinning, however, had different effects on SR under *Q. ilex* and *Q. cerrioides*; thinning enhanced SR under *Q. cerrioides*, but it reduced SR under *Q. ilex*.

Figure 2 shows the mean seasonal variations of SR under *Q. ilex* and *Q. cerrioides* in the four treatments. Generally, SR was higher during the growing season and lower in winter. Due to high precipitation in spring 2007, the SR in the control, thinning, and combined treatments showed the highest peak during this period. In the control treatment, SR under *Q. ilex* was significantly higher than under *Q. cerrioides*, except in autumn 2005 and spring 2006. In the reduced rainfall treatment, SR under *Q. ilex* showed a significantly higher rate compared to SR under *Q. cerrioides*, especially in spring and summer. Besides, there was almost no seasonality of SR under *Q. cerrioides*. SR under *Q. ilex* even showed higher values in comparison to the SR in the control treatment in the first year after treatment installation. In the thinning treatment, SR under *Q. cerrioides* was significantly higher than under *Q. ilex*, especially in spring. In the combined treatment, the seasonal patterns of SR under both tree species were very similar in the first 2 years. In the following years, SR under *Q. cerrioides* showed a higher value, which was very similar to the pattern of SR in the thinning treatment.

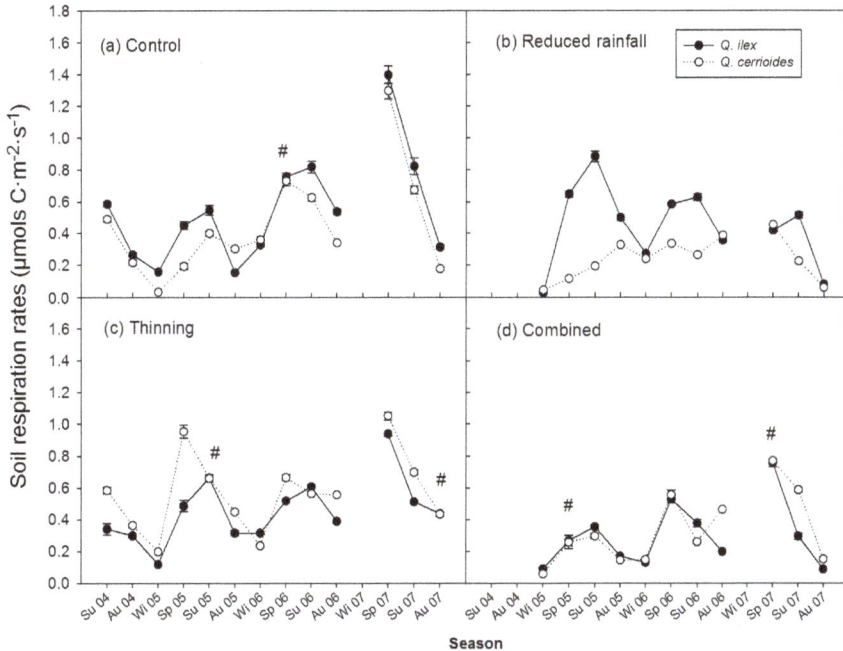

Figure 2. Seasonal variation in soil respiration of *Q. ilex* and *Q. cerrioides* for each treatment: (**a**) control; (**b**) reduced rainfall; (**c**) thinning; (**d**) combined treatment. Reduced rainfall treatment was installed at the end of 2004, therefore, the data for reduced rainfall and the combined treatments started in 2005. Data represent seasonal means with SE. Differences in SR between species were statistically significant except when marked with # ($p > 0.05$).

We also compared the diurnal variation in SR under the two tree species during spring and summer campaigns (Figures 3 and 4). During the spring campaigns, SR under both tree species in the control treatment showed a clear diurnal pattern, except for SR under *Q. cerrioides* in spring 2005. Meanwhile, in the reduced rainfall treatment, the diurnal changes of SR almost diminished. In the thinning treatment, SR under *Q. ilex* in 2005 showed a reversed diurnal pattern, but in the following two years the patterns turned back to be flat. The diurnal patterns of SR under *Q. cerrioides* in the thinning treatment were similar to the patterns in the control treatment, but with limited range and a clear depressed SR at noon. In the combined treatment, SR under both *Q. ilex* and *Q. cerrioides* showed a significant reduction during the day in 2005, but the reduction decreased in the following years. The diurnal variation of SR during summer campaigns was slightly different compared to spring. In the control treatment, although SR under the two tree species showed similar daily patterns, the variation of SR under *Q. ilex* was much higher than SR under *Q. cerrioides*. In the reduced rainfall treatment, SR under *Q. ilex* still exhibited a clear diurnal change, while SR under *Q. cerrioides* was almost steady.

In both thinning and combined treatments, SR under two tree species showed a pronounced reduction during the day.

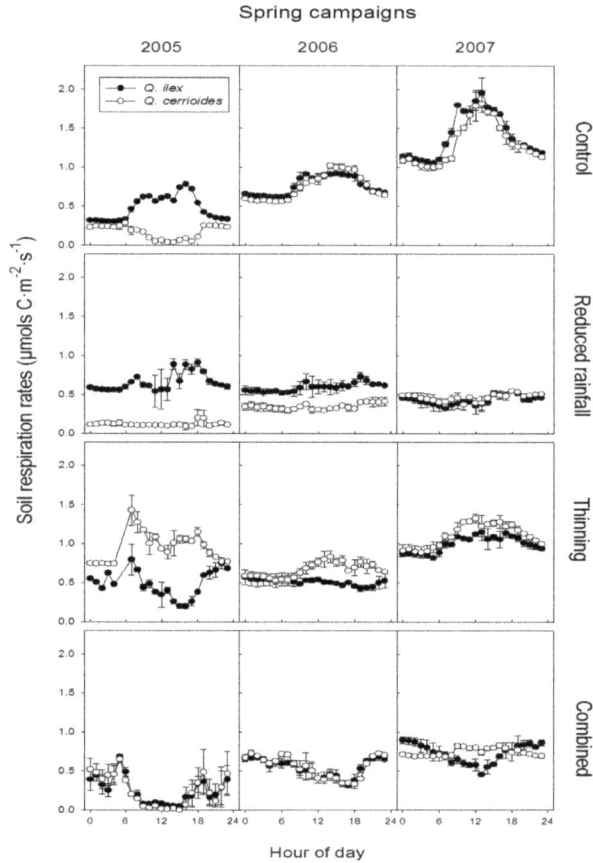

Figure 3. Diurnal variation of soil respiration rates (SR) with standard errors under *Q. ilex* and *Q. cerrioides* during spring in 2005, 2006, and 2007 (from left to right) and for each treatment: control, reduced rainfall, thinning, and combined treatment (from up to down). Shown are hourly rates of SR averaged over each campaign.

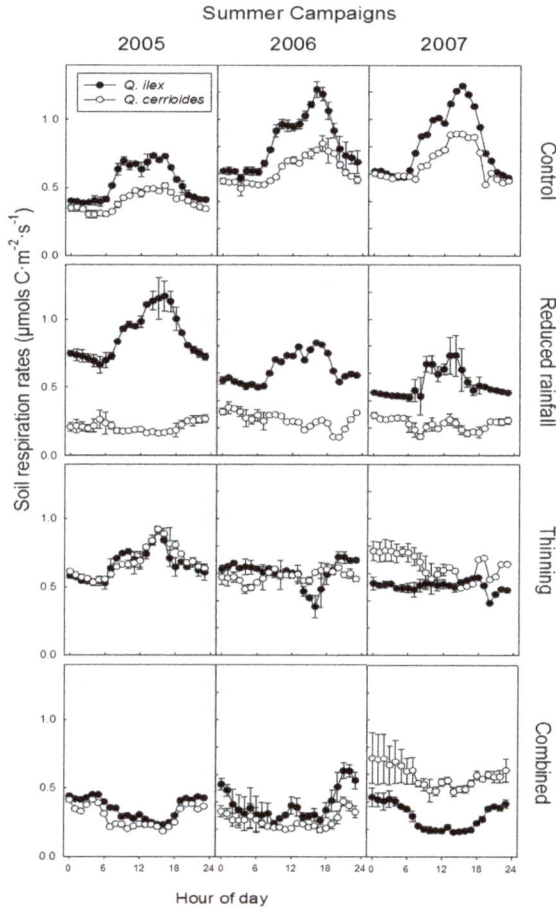

Figure 4. Diurnal variation of soil respiration rates (SR) with standard errors under *Q. ilex* and *Q. cerrioides* during summer in 2005, 2006, and 2007 (from left to right) and for each treatment: control, reduced rainfall, thinning, and combined treatment (from up to down). Shown are hourly rates of SR averaged over each campaign.

3.3. Relationship Between SR and Ts

By using recursive partitioning, we identified a soil moisture threshold around 8%–9%; when soil moisture was higher than 8%, SR and Ts were highly correlated, with apparent Q_{10} values from 2.99 to 3.83, and Ts explained 91%–96% of the variation in SR. When soil moisture was lower than 8%, apparent Q_{10} values declined to 1.23–1.44. Figure 5 shows the daily average SR of each treatment as a function of Ts separated by soil moisture regimes. In the control treatment, apparent Q_{10} was 3.0 when soil moisture was higher than 9%, and declined to 1.37 when soil moisture was lower than 9%. Thinning and combined treatments showed a similar pattern, except that the soil moisture threshold was slightly lower than the threshold of the control. In the reduced rainfall treatment, we could not identify the soil moisture threshold by using recursive partitioning, although the recorded soil moisture ranged from 2.8% to 14.2%. The overall apparent Q_{10} in the reduced rainfall treatment was 1.36. When we separated the SR under different species and compared its relationship with Ts, similar relationships between SR and Ts were found in all treatments except in the reduced rainfall treatment (inset in Figure 5b and Figures S1–S3); SR under *Q. ilex* showed a positive correlation with *Ts* with a Q_{10} of 1.53, whereas the SR under *Q. cerrioides* showed no relationship with *Ts*.

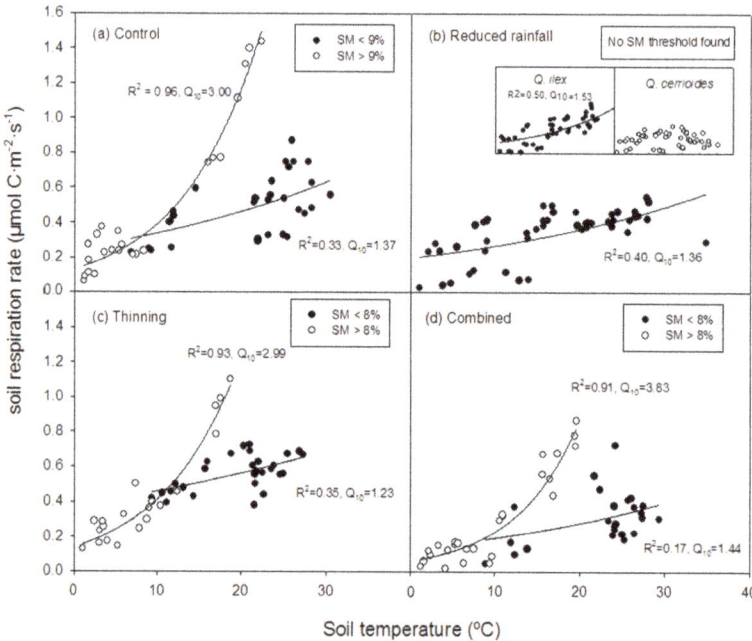

Figure 5. Relationship between daily SR and *Ts* (5cm) separated by soil moisture regime in each treatment: (**a**) control; (**b**) reduced rainfall; (**c**) thinning; (**d**) combined treatment. Closed circles indicate the lower soil moisture regime, and open circles indicate the higher soil moisture regime. Lines show fit to Equation (1) for SR and *Ts* within the same soil moisture regime. R^2 and Q_{10} values are given for each panel. In the reduced rainfall treatment, the relationship between SR and *Ts* cannot be separated by soil moisture regime by using recursive partitioning; therefore, the closed circles represent all soil moisture regimes. Inset in (**b**) shows the relationship between daily SR and *Ts* under two tree species (n = 49–53).

3.4. Temporal Variation in Litterfall

The peak of litterfall differed between the two tree species; in the control, *Q. ilex* mainly dropped leaves during spring and summer, while *Q. cerrioides* dropped leaves all year except during summer (Figure 6). In the reduced rainfall treatment, the peak of litterfall from *Q. ilex* was in spring, while *Q. cerrioides* remained the same throughout the year. In the thinning and combined treatments, the peak of litterfall from *Q. ilex* occurred in summer. Moreover, the total litterfall amount from *Q. cerrioides* was less in the thinning treatment and showed a peak of litterfall in spring. Although *Q. ilex* is an evergreen species, the amount of litterfall from *Q. ilex* was larger than from *Q. cerrioides*, especially during the driest summer of 2006.

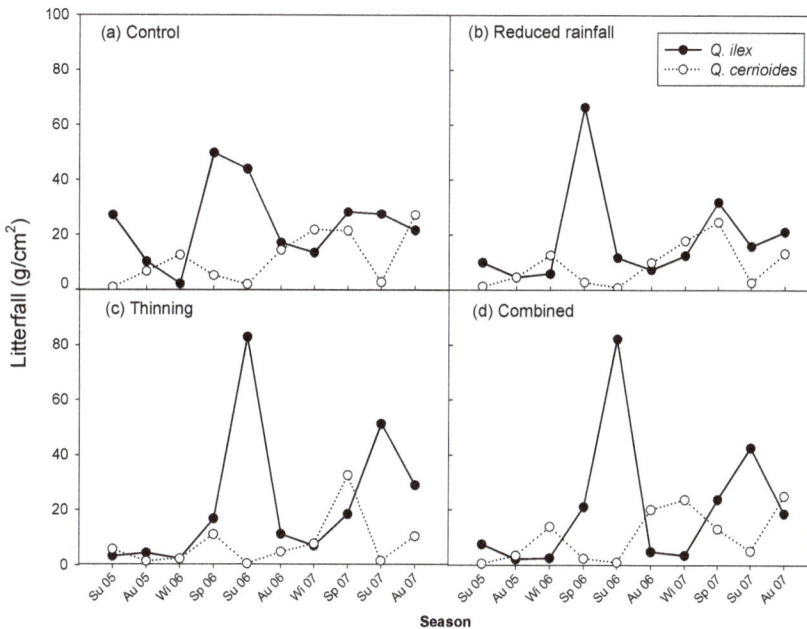

Figure 6. Seasonal variations in litter fall of *Q. ilex* and *Q. cerrioides* for each treatment: (**a**) control; (**b**) reduced rainfall; (**c**) thinning; (**d**) combined treatment. Reduced rainfall treatment was installed at the end of 2004, therefore, the data for reduced rainfall and combined treatments started in 2005.

4. Discussion

We expected to find the lowest soil moisture in the reduced rainfall treatment. However, the observed soil moisture data suggested that the channels installed in the reduced rainfall treatment only had partially or no effect. This may be due to the low precipitation during this period which probably diminished the treatment effect of reduced rainfall. We also suspect that the channels installed to reduce rainfall may have created some shadow and somewhat prevented the direct top-soil water evaporation. Despite the reduced rainfall treatment, we observed a tendency for soil moisture to be lower in the selective thinning treatments, especially during the summers of 2005 and 2006. Many studies have shown that thinning influences site-specific microclimatic conditions [14,44]. The removal of aboveground vegetation is known to increase *Ts* [45] and soil moisture as a consequence of reduced root and canopy interception and, hence, reduced evapotranspiration [46]. The observed lower soil moisture in the selective thinning treatment may be due to the way that selective thinning retained the roots, but increased the opening of the canopy. Moreover, thinning has been shown to increase transpiration rate through enhancement of tree growth, and this may consequently reduce soil moisture [46,47].

The observed decrease in overall SR from our study is similar to other research. Studies have shown how drought stress depressed SR from several aspects. First, the low water content of the soil created an environment that slowed the diffusion of solutes and, thus, suppressed microbial respiration by limiting the supply of substrate [48]. Additionally, microbes and plant roots have to invest more energy to produce protective molecules and this reduces their growth and respiration [49]. From hourly to daily scales, drought has been shown to decrease the recently assimilated C allocation to roots ca. 33%–50% [50,51]. The decrease in plant substrate and photosynthetic activity caused by drought may explain the reduction in SR [52,53]. With the prolongation of reduced rainfall over time, annual SR, especially root respiration, would have decreased followed by the depression of forest productivity and growth. For example, Brando et al. [54] found a decline in net primary productivity of 13% in the first year and up to 62% in the following four years in a throughfall reduction experiment.

Interestingly, despite the effect of drought on SR, we observed an increase in SR under *Q. ilex* in the reduced rainfall treatment in the first year after the reduced rainfall treatment. A similar pattern was observed in South Catalonia, where Asensio et al. [30] found significantly higher SR in the drought treatment compared to the control treatment during summer. First, they argued, that the prolonged low availability of soil water compelled roots to uptake deeper soil water; second, they also argued that moderate drought enhanced photosynthetic rates [55] to support roots with the majority of the photosynthetic assimilates. In our study site, Miguel [56] measured the treatment effects on mineral soil nutrients, and root density and distribution during the summers of 2007 and 2008, which is right after our measurement, and found a significant increase of fine roots of *Q. ilex* only in the reduced rainfall treatment. The high C/N ratio and low soil water content found in our study site [56] also implied a very low microbial respiration. Hinko-Najera et al. [57] also found that a reduction in throughfall mainly decreased autotrophic respiration, but not heterotrophic respiration, in a Mediterranean to cool temperate forest. As a result, we conjecture that the increase in SR under *Q. ilex* observed in our reduced rainfall treatment was caused by the increase of fine roots while the decrease in SR under *Q. cerrioides* may have been caused mainly by the decrease in root respiration. Miguel [56] also found that the fine and small roots of *Q. cerrioides* were distributed mainly in the 0–30 cm depth layer, but the roots of *Q. ilex* were found to be deeper. In other words, the different responses of SR under *Q. ilex* and *Q. cerrioides* may have been due to different rooting systems.

Previous studies have shown contradictory results of how thinning affects SR: SR has been found to increase, decrease, or even remain unchanged after thinning [18,44,58–63]. The different responses likely are due to thinning intensity, timing, and duration of the measurement campaigns after thinning. In our study, we observed an increase in SR in the selective thinning treatment during the first two years after selective thinning. We also found a significant reduction in SR during the daytime in the first summer campaign. We explain the possible reasons how thinning affects SR from a different temporal scale. Over the hourly to daily scales, selective thinning increased water and nutrient availability and, therefore, increased both microbial and root respiration. In the meantime, the woody debris and dead roots produced during thinning stimulated heterotrophic respiration [21,64]. Additionally, Sohlenius [65] found that slash produced by logging promoted productivity of soil microflora due to the increase in moisture and microbial biomass, which increased SR. However, selective thinning may also decrease SR because of the lower soil moisture caused by more solar radiation and higher transpiration in the initial phase after selective thinning [47]. From daily to seasonal scales, the enhancement of tree growth and photosynthesis due to selective thinning may promote more root respiration [66–68]. Cotillas et al. [38] investigated tree growth in the same study site and observed a remarkable improvement in residual stem growth (ca. 50%) and a reduction in stem mortality after selective thinning. However, they also found that the positive effects of thinning declined rapidly during the three-year experiment. López et al. [69] found an increase of more than 100% in root biomass and 76% in root production in a *Q. ilex* forest after thinning, especially during winter and autumn. We also found higher soil organic matter and soil phosphorous in the selective thinning treatments [56], which may also enhance SR. From seasonal to annual scales, selective thinning increased annual SR as a result of a longer growing period due to the absence of drought [70]. Supported by our litterfall data, the total amount of litterfall from *Q. cerrioides* was less in the thinning treatment; during the same time, we also observed a stronger effect of thinning on SR under *Q. cerrioides*. Overall, the effect of selective thinning on SR over time is likely to be reduced with the recovery of stands.

The apparent soil Q_{10} was affected significantly by soil moisture. However, this soil moisture threshold is not applicable to the relationship between SR and Ts in the reduced rainfall treatment. In the reduced rainfall treatment, we observed some campaigns with soil moisture higher than 8%, but SR of these campaigns were still lower than the SR in the control treatment of the same campaigns. The reduction of Q_{10} due to drought has been found in many studies [71–74]. As the apparent Q_{10}

Forests **2016**, *7*, 263

in this study was calculated as annual Q_{10}, the low Q_{10} in the reduced rainfall treatment could be attributed by the diminished seasonal amplitude of SR, especially SR under *Q. cerrioides*. We found relatively few studies on the response of Q_{10} to forest management. At our study site, we found Q_{10} did not vary in response to thinning, which is similar to the finding of Tang et al. [20]. Our result is also consistent with Pang et al. [62], who showed that thinning increased the seasonal Q_{10} significantly, but not the yearly Q_{10}. Overall, the different SR-Ts relationship between the reduced rainfall treatment and combined treatment indicated that selective thinning treatment had at least partially mitigated the drought stress by improving the SR in response to environmental change.

Our study demonstrates that evergreen and deciduous trees growing in the same environmental conditions can emit different quantities of CO_2 from the soil. We found that thinning and reduced rainfall treatments have different effects on SR and litterfall of the two investigated tree species. This may be explained by the plant functional type (i.e., evergreen and deciduous species). *Q. ilex* is an evergreen species, which is well adapted to poor environments, and has low resource-loss ratios [75,76]. Therefore, the SR under *Q. ilex* was less affected by selective thinning. In contrast, deciduous species, such as *Q. cerrioides*, have a shorter period of active photosynthesis and a higher sensitivity to drought [77]. Therefore, deciduous species may require higher levels of nutrients and water to support higher rates of foliar net CO_2 assimilation to compensate for the shorter active period [78].

5. Conclusion

In conclusion, we examined the effects of drought and thinning on SR in a Mediterranean oak forest and observed a significant change in SR due to thinning and reduced rainfall. Both treatments influenced SR over different time scales. The main conclusions drawn from this study are as follows:

- Q_{10} of SR was clearly modulated by soil moisture, with a threshold value around 8%–9%. Reduced rainfall decreased both SR and Q_{10}, unlike selective thinning;
- Selective thinning had less effect on SR under *Q. ilex*, but increased the SR rate under *Q. cerrioides* in the first two years;
- Reduced rainfall significantly depressed SR rate under *Q. cerrioides* by 50%, especially during the growing season, and the drought effect accumulated over years. Reduced rainfall increased SR rate under *Q. ilex* during the growing season by 50%;
- Selective thinning mitigated the negative effect of drought on SR rate under *Q. cerrioides*, although the mitigation was only significant during spring and during the last year of the experiment.

Supplementary Materials: The following are available online at http://www.mdpi.com/1999-4907/7/11/263/s1, Figure S1: Relationship between daily SR and Ts under *Q. ilex* and *Q. cerrioides* separated by soil moisture regime in the control treatment, Figure S2: Relationship between daily SR and Ts under *Q. ilex* and *Q. cerrioides* separated by soil moisture regime in the selective thinning treatment, Figure S3: Relationship between daily SR and Ts under *Q. ilex* and *Q. cerrioides* separated by soil moisture regime in the combined treatment.

Acknowledgments: We gratefully acknowledge the help from Altug Ekici, Laura Albaladejo, Belén Sánchez-Humanes, Roger Sallas, Josep Barba, Carme Barba, and Jaume Casadesús. We would like to thank three anonymous reviewers for their insightful comments and Thomas A. Gavin, Professor Emeritus, Cornell University, for help with editing the English in this paper. The research leading to these results received funding from INIA (RTA04-028-02 and RTA2005-00100-CO2) and the European Community's Seventh Framework Programme GREENCYCLEII (FP7 2007–2013) under grant agreement n 238366. Chao-Ting Chang, Dominik Sperlich, Santiago Sabaté, and Carlos Gracia are members of the research group ForeStream (AGAUR, Catalonia 2014SGR949).

Author Contributions: Conceived and designed the experiments: Santiago Sabaté, Josep Maria Espelta, and Carlos Gracia; Performed the experiments: Elisenda Sánchez-Costa, Santiago Sabaté, and Miriam Cotillas; Analyzed the data: Chao-Ting Chang, Elisenda Sánchez-Costa, and Dominik Sperlich; Wrote the paper: Chao-Ting Chang.

Conflicts of Interest: The authors declare no conflict of interest.

References

1. Pan, Y.; Birdsey, R.A.; Fang, J.; Houghton, R.; Kauppi, P.E.; Kurz, W.A.; Phillips, O.L.; Shvidenko, A.; Lewis, S.L.; Canadell, J.G.; et al. A large and persistent carbon sink in the world's forests. *Science* **2011**, *333*, 988–993. [CrossRef] [PubMed]
2. Schlesinger, W.H.; Andrews, J.A. Soil respiration and the global carbon cycle. *Biogeochemistry* **2000**, *48*, 7–20. [CrossRef]
3. Melillo, J.M.; Steudler, P.A.; Aber, J.D.; Newkirk, K.; Lux, H.; Bowles, F.P.; Catricala, C.; Magill, A.; Ahrens, T.; Morrisseau, S.; et al. Soil warming and carbon-cycle feedbacks to the climate system. *Science* **2002**, *298*, 2173–2176. [CrossRef] [PubMed]
4. Giorgi, F. Climate change hot-spots. *Geophys. Res. Lett.* **2006**, *33*, L08707. [CrossRef]
5. Rowell, D.P.; Jones, R.G. Causes and uncertainty of future summer drying over Europe. *Clim. Dyn.* **2006**, *27*, 281–299. [CrossRef]
6. Le Quéré, C.; Peters, G.P.; Andres, R.J.; Andrew, R.M.; Boden, T.; Ciais, P.; Friedlingstein, P.; Houghton, R.A.; Marland, G.; Moriarty, R.; et al. Global carbon budget 2013. *Earth Syst. Sci. Data Discuss.* **2013**, *6*, 689–760. [CrossRef]
7. Raich, J.W.; Schlesinger, W.H. The global carbon dioxide flux in soil respiration and its relationship to vegetation and climate. *Tellus B* **1992**, *44*, 81–99. [CrossRef]
8. Raich, J.W.; Potter, C.S. Global patterns of carbon dioxide emissions from soils. *Global Biogeochem. Cycles* **1995**, *9*, 23–36. [CrossRef]
9. Gärdenäs, A.I. Soil respiration fluxes measured along a hydrological gradient in a Norway spruce stand in south Sweden (Skogaby). *Plant Soil* **2000**, *221*, 273–280. [CrossRef]
10. Fang, C.; Moncrieff, J.B. The dependence of soil CO_2 efflux on temperature. *Soil Biol. Biochem.* **2001**, *33*, 155–165. [CrossRef]
11. Sullivan, B.W.; Kolb, T.E.; Hart, S.C.; Kaye, J.P.; Dore, S.; Montes-Helu, M. Thinning reduces soil carbon dioxide but not methane flux from southwestern USA ponderosa pine forests. *For. Ecol. Manag.* **2008**, *255*, 4047–4055. [CrossRef]
12. Cheng, X.; Han, H.; Kang, F.; Liu, K.; Song, Y.; Zhou, B.; Li, Y. Short-term effects of thinning on soil respiration in a pine (*Pinus tabulaeformis*) plantation. *Biol. Fertil. Soils* **2013**, *50*, 357–367. [CrossRef]
13. Sullivan, P.F.; Arens, S.J.T.; Chimner, R.A.; Welker, J.M. Temperature and Microtopography Interact to Control Carbon Cycling in a High Arctic Fen. *Ecosystems* **2008**, *11*, 61–76. [CrossRef]
14. Wang, H.; Liu, W.; Wang, W.; Zu, Y. Influence of long-term thinning on the biomass carbon and soil respiration in a larch (*Larix gmelinii*) forest in Northeastern China. *Sci. World J.* **2013**, *2013*. [CrossRef] [PubMed]
15. Ginn, S.E.; Seiler, J.R.; Cazell, B.H.; Kreh, R.E. Physiological and growth responses of eight-year-old loblolly pine stands to thinning. *For. Sci.* **1991**, *37*, 1030–1040.
16. Peterson, J.A.; Seiler, J.R.; Nowak, J.; Ginn, S.E.; Kreh, R.E. Growth and physiological responses of young loblolly pine stands to thinning. *For. Sci.* **1997**, *43*, 529–534.
17. Tang, Z.; Chambers, J.L.; Guddanti, S.; Barmett, J.P. Thinning, fertilization, and crown position interact to control physiological responses of loblolly pine. *Tree Physiol.* **1999**, *19*, 87–94. [CrossRef] [PubMed]
18. Tang, J.; Qi, Y.; Xu, M.; Misson, L.; Goldstein, A.H. Forest thinning and soil respiration in a ponderosa pine plantation in the Sierra Nevada. *Tree Physiol.* **2005**, *25*, 57–66. [CrossRef] [PubMed]
19. Selig, M.F.; Seiler, J.R.; Tyree, M.C. Soil carbon and CO_2 efflux as influenced by the thinning of loblolly pine (*Pinus taeda* L.) plantations on the Piedmont of Virginia. *For. Sci.* **2008**, *54*, 58–66.
20. Tang, J.; Baldocchi, D.D. Spatial–temporal variation in soil respiration in an oak–grass savanna ecosystem in California and its partitioning into autotrophic and heterotrophic components. *Biogeochemistry* **2005**, *73*, 183–207. [CrossRef]
21. Tian, D.-L.; Yan, W.-D.; Fang, X.; Kang, W.-X.; Deng, X.-W.; Wang, G.-J. Influence of Thinning on Soil CO_2 Efflux in Chinemse Fir Plantations. *Pedosphere* **2009**, *19*, 273–280. [CrossRef]
22. Johnson, D.W.; Curtis, P.S. Effects of forest management on soil C and N storage: Meta analysis. *For. Ecol. Manag.* **2001**, *140*, 227–238. [CrossRef]
23. Mooney, H.A. Carbon-gaining capacity and allocation patterns of mediterranean-climate plants. In *Mediterranean-Type Ecosystems*; Springer: Berlin, Germany, 1983; pp. 103–119.

24. Davidson, E.; Belk, E.; Boone, R.D. Soil water content and temperature as independent or confounded factors controlling soil respiration in a temperate mixed hardwood forest. *Glob. Chang. Biol.* **1998**, *4*, 217–227. [CrossRef]

25. Kaye, J.P.; Hart, S.C. Restoration and canopy-type effects on soil respiration in a ponderosa pine-bunchgrass ecosystem. *Soil Sci. Soc. Am. J.* **1998**, *62*, 1062–1072. [CrossRef]

26. Borken, W.; Savage, K.; Davidson, E.A.; Trumbore, S.E. Effects of experimental drought on soil respiration and radiocarbon efflux from a temperate forest soil. *Glob. Chang. Biol.* **2006**, *12*, 177–193. [CrossRef]

27. Savage, K.E.; Davidson, E.A. Interannual variation of soil respiration in two New England forests. *Glob. Biogeochem. Cycles* **2001**, *15*, 337–350. [CrossRef]

28. Borken, W.; Davidson, E.A.; Savage, K.; Gaudinski, J.; Trumbore, S.E. Drying and Wetting Effects on Carbon Dioxide Release from Organic Horizons. *Soil Sci. Soc. Am. J.* **2003**, *67*, 1888. [CrossRef]

29. Liu, H.; Li, F.; Jia, Y. Effects of shoot removal and soil water content on root respiration of spring wheat and soybean. *Environ. Exp. Bot.* **2006**, *56*, 28–35. [CrossRef]

30. Asensio, D.; Penuelas, J.; Ogaya, R.; Llusià, J. Seasonal soil and leaf CO_2 exchange rates in a Mediterranean holm oak forest and their responses to drought conditions. *Atmos. Environ.* **2007**, *41*, 2447–2455. [CrossRef]

31. Leuschner, C.; Hertel, D.; Coners, H.; Büttner, V. Root competition between beech and oak: A hypothesis. *Oecologia* **2001**, *126*, 276–284. [CrossRef]

32. Puhe, J. Growth and development of the root system of Norway spruce *Picea abies* in forest stands—A review. *For. Ecol. Manag.* **2003**, *175*, 253–273. [CrossRef]

33. Stocker, T.F.; Qin, D.; Plattner, G.-K.; Tignor, M.; Allen, S.K.; Boschung, J.; Nauels, A.; Xia, Y.; Bex, V.; Midgley, P.M. Climate change 2013: The physical science basis. In *The Working Group I contribution to the Fifth Assessment Report of the Intergovernmental Panel on Climate Change*; Cambridge University Press: New York, NY, USA, 2013.

34. Rey, A.N.A.; Pegoraro, E.; Tedeschi, V.; Parri, I.D.E.; Jarvis, G.; Valentini, R. Annual variation in soil respiration and its components in a coppice oak forest in Central Italy. *Glob. Chang. Biol.* **2002**, *8*, 851–866. [CrossRef]

35. Reichstein, M.; Rey, A.; Freibauer, A.; Tenhunen, J.; Valentini, R.; Banza, J.; Casals, P.; Cheng, Y.F.; Grünzweig, M.J.; Irvine, J.; Joffre, R.; et al. Modeling temporal and large-scale spatial variability of soil respiration from soil water availability, temperature and vegetation productivity indices. *Glob. Biogeochem. Cycles* **2003**, *17*, 1104. [CrossRef]

36. Valentini, R. EUROFLUX: An integrated network for studying the long-term responses of biospheric exchanges of carbon, water, and energy of European forests. In *Fluxes of Carbon, Water and Energy of European Forests*; Springer: Berlin/Heidelberg, Germany, 2003; pp. 1–8.

37. Ninyerola, M.; Pons, X.; Roure, J.M. A methodological approach of climatological modelling of air temperature and precipitation through GIS techniques. *Int. J. Climatol.* **2000**, *20*, 1823–1841. [CrossRef]

38. Cotillas, M.; Sabaté, S.; Gracia, C.; Espelta, J.M. Growth response of mixed mediterranean oak coppices to rainfall reduction: Could selective thinning have any influence on it? *For. Ecol. Manag.* **2009**, *258*, 1677–1683. [CrossRef]

39. Retana, J.; Espelta, J.M.; Gracia, M.; Riba, M. *Ecology of Mediterranean Evergreen Oak Forests*; Rodà, F., Retana, J., Gracia, C.A., Bellot, J., Eds.; Ecological Studies; Springer: Berlin/Heidelberg, Germany, 1999; Volume 137.

40. Espelta, J.M.; Rodrigo, A.; Habrouk, A.; Meghelli, N.; Ordónez, J.L.; Retana, J. Land use changes, natural regeneration patterns, and restoration practices after a large wildfire in NE Spa. In *Challenges for Fire Ecology, Landscape, Restoration*; Trabaud, L., Prodon, R., Eds.; Fire and Biological Processes; Backhuys Publications: Leiden, The Netherland, 2002; pp. 315–324.

41. Espelta, J.M.; Retana, J.; Habrouk, A. Resprouting patterns after fire and response to stool cleaning of two coexisting Mediterranean oaks with contrasting leaf habits on two different sites. *For. Ecol. Manag.* **2003**, *179*, 401–414. [CrossRef]

42. Van't Hoff, J.H. *Lectures on Theoretical and Physical Chemistry*; Edward Arnold: London, UK, 1898.

43. Zeileis, A.; Hothorn, T.; Hornik, K. Model-Based Recursive Partitioning. *J. Comput. Graph. Stat.* **2008**, *17*, 492–514. [CrossRef]

44. Masyagina, O.V.; Prokushkin, S.G.; Koike, T. The influence of thinning on the ecological conditions and soil respiration in a larch forest on Hokkaido Island. *Eurasian Soil Sci.* **2010**, *43*, 693–700. [CrossRef]

45. Köster, K.; Püttsepp, Ü.; Pumpanen, J. Comparison of soil CO_2 flux between uncleared and cleared windthrow areas in Estonia and Latvia. *For. Ecol. Manag.* **2011**, *262*, 65–70. [CrossRef]

46. Bréda, N.; Granier, A.; Aussenac, G. Effects of thinning on soil and tree water relations, transpiration and growth in an oak forest (*Quercus petraea* (Matt.) Liebl.). *Tree Physiol.* **1995**, *15*, 295–306. [CrossRef] [PubMed]

47. Olivar, J.; Bogino, S.; Rathgeber, C.; Bonnesoeur, V.; Bravo, F. Thinning has a positive effect on growth dynamics and growth-climate relationships in Aleppo pine (*Pinus halepensis*) trees of different crown classes. *Ann. For. Sci.* **2014**, *71*, 395–404. [CrossRef]

48. Skopp, J.; Jawson, M.D.; Doran, J.W. Steady-State Aerobic Microbial Activity as a Function of Soil Water Content. *Soil Sci. Soc. Am. J.* **1990**, *54*, 1619–1625. [CrossRef]

49. Schimel, J.; Balser, T.C.; Wallenstein, M. Microbial stress, response physiology and its implications for ecosystem function. *Ecology* **2007**, *88*, 1386–1394. [CrossRef] [PubMed]

50. Ruehr, N.K.; Offermann, C.A.; Gessler, A.; Winkler, J.B.; Ferrio, J.P.; Buchmann, N.; Barnard, R.L. Drought effects on allocation of recent carbon: From beech leaves to soil CO_2 efflux. *New Phytol.* **2009**, *184*, 950–961. [CrossRef] [PubMed]

51. Hasibeder, R.; Fuchslueger, L.; Richter, A.; Bahn, M. Summer drought alters carbon allocation to roots and root respiration in mountain grassland. *New Phytol.* **2015**, *205*, 1117–1127. [CrossRef] [PubMed]

52. Burton, A.J.; Pregitzer, K.S.; Zogg, G.P.; Zak, D.R. Drought reduces root respiration in sugar maple forests. *Ecol. Appl.* **1998**, *8*, 771–778. [CrossRef]

53. Yan, L.; Chen, S.; Huang, J.; Lin, G. Water regulated effects of photosynthetic substrate supply on soil respiration in a semiarid steppe. *Glob. Chang. Biol.* **2011**, *17*, 1990–2001. [CrossRef]

54. Brando, P.M.; Nepstad, D.C.; Davidson, E.A.; Trumbore, S.E.; Ray, D.; Camargo, P. Drought effects on litterfall, wood production and belowground carbon cycling in an Amazon forest: Results of a throughfall reduction experiment. *Philos. Trans. R. Soc. Lond. B. Biol. Sci.* **2008**, *363*, 1839–1848. [CrossRef] [PubMed]

55. Lu, C.; Zhang, J. Effects of water stress on photosynthesis, chlorophyll fluorescence and photoinhibition in wheat plants. *Funct. Plant Biol.* **1998**, *25*, 883–892. [CrossRef]

56. Miguel, P.C. Respuestas Ecofisiológicas Y Estructurales a la Recurrencia, Duración E Intensidad de la Sequía en Plantaciones Y Bosques Mixtos de Quercus ilex, Quercus pubescens Y Quercus cerrioides. Doctoral Dissertation, Institut de Recerca i Tecnologia Agroalimentaries IRTA, Universitat Autonoma de Barcelona UAB, Catalonia, Spain, 2010. (In English)

57. Hinko-Najera, N.; Fest, B.; Livesley, S.J.; Arndt, S.K. Reduced throughfall decreases autotrophic respiration, but not heterotrophic respiration in a dry temperate broadleaved evergreen forest. *Agric. For. Meteorol.* **2015**, *200*, 66–77. [CrossRef]

58. Londo, A.J.; Messina, M.G.; Schoenholtz, S.H. Forest Harvesting Effects on Soil Temperature, Moisture, and Respiration in a Bottomland Hardwood Forest. *Soil Sci. Soc. Am. J.* **1999**, *63*, 637. [CrossRef]

59. Ma, S.; Chen, J.; North, M.; Erickson, H.E.; Bresee, M.; Le Moine, J. Short-term effects of experimental burning and thinning on soil respiration in an old-growth, mixed-conifer forest. *Environ. Manag.* **2004**, *33*, S148–S159. [CrossRef]

60. Jonsson, J.A.; Sigurdsson, B.D. Effects of early thinning and fertilization on soil temperature and soil respiration in a poplar plantation. *Icel. Agric. Sci.* **2010**, *23*, 97–109.

61. Olajuyigbe, S.; Tobin, B.; Saunders, M.; Nieuwenhuis, M. Forest thinning and soil respiration in a Sitka spruce forest in Ireland. *Agric. For. Meteorol.* **2012**, *157*, 86–95. [CrossRef]

62. Pang, X.; Bao, W.; Zhu, B.; Cheng, W. Responses of soil respiration and its temperature sensitivity to thinning in a pine plantation. *Agric. For. Meteorol.* **2013**, *171–172*, 57–64. [CrossRef]

63. Dai, Z.; Birdsey, R.A.; Johnson, K.D.; Dupuy, J.M.; Hernandez-Stefanoni, J.L.; Richardson, K. Modeling Carbon Stocks in a Secondary Tropical Dry Forest in the Yucatan Peninsula, Mexico. *Water Air Soil Pollut.* **2014**, *225*, 1925. [CrossRef]

64. Rustad, L.E.; Huntington, T.G.; Boone, R.D. Controls on soil respiration: Implications for climate change. *Biogeochemistry* **2000**, *48*, 1–6. [CrossRef]

65. Sohlenius, B. Short-Term Influence of Clear-Cutting on Abundance of Soil-Microfauna (Nematoda, Rotatoria and Tardigrada) in a Swedish Pine Forest Soil. *J. Appl. Ecol.* **1982**, *19*, 349. [CrossRef]

66. Janssens, I.A.; Lankreijer, H.; Matteucci, G.; Kowalski, A.S.; Buchmann, N.; Epron, D.; Pilegaard, K.; Kutsch, W.; Longdoz, B.; Grünwald, T.; et al. Productivity overshadows temperature in determining soil and ecosystem respiration across European forests. *Glob. Chang. Biol.* **2001**, *7*, 269–278. [CrossRef]

67. Kuzyakov, Y.; Cheng, W. Photosynthesis controls of rhizosphere respiration and organic matter decomposition. *Soil Biol. Biochem.* **2001**, *33*, 1915–1925. [CrossRef]

68. Högberg, P.; Nordgren, A.; Buchmann, N.; Taylor, A.F.S.; Ekblad, A.; Högberg, M.N.; Nyberg, G.; Ottosson-Löfvenius, M.; Read, D.J. Large-scale forest girdling shows that current photosynthesis drives soil respiration. *Nature* **2001**, *411*, 789–792. [CrossRef] [PubMed]

69. López, B.C.; Sabate, S.; Gracia, C.A. Thinning effects on carbon allocation to fine roots in a *Quercus ilex* forest. *Tree Physiol.* **2003**, *23*, 1217–1224. [CrossRef] [PubMed]

70. Aussenac, G.; Granier, A. Effects of thinning on water stress and growth in Douglas-fir. *Can. J. For. Res.* **1988**, *18*, 100–105. [CrossRef]

71. Jassal, R.S.; Black, T.A.; Novak, M.D.; Gaumont-Guay, D.; Nesic, Z. Effect of soil water stress on soil respiration and its temperature sensitivity in an 18-year-old temperate Douglas-fir stand. *Glob. Chang. Biol.* **2008**, *14*, 1305–1318. [CrossRef]

72. Suseela, V.; Conant, R.T.; Wallenstein, M.D.; Dukes, J.S. Effects of soil moisture on the temperature sensitivity of heterotrophic respiration vary seasonally in an old-field climate change experiment. *Glob. Chang. Biol.* **2012**, *18*, 336–348. [CrossRef]

73. Suseela, V.; Dukes, J.S. The responses of soil and rhizosphere respiration to simulated climatic changes vary by season. *Ecology* **2013**, *94*, 403–413. [CrossRef] [PubMed]

74. Wang, B.; Zha, T.S.; Jia, X.; Wu, B.; Zhang, Y.Q.; Qin, S.G. Soil moisture modifies the response of soil respiration to temperature in a desert shrub ecosystem. *Biogeosciences* **2014**, *11*, 259–268. [CrossRef]

75. Aerts, R.; der Peijl, M.J. A simple model to explain the dominance of low-productive perennials in nutrient-poor habitats. *Oikos* **1993**, *66*, 144–147. [CrossRef]

76. Berendse, F. Competition between plant populations at low and high nutrient supplies. *Oikos* **1994**, *71*, 253–260. [CrossRef]

77. Thomas, D.S.; Eamus, D. Seasonal patterns of xylem sap pH, xylem abscisic acid concentration, leaf water potential and stomatal conductance of six evergreen and deciduous Australian savanna tree species. *Aust. J. Bot.* **2002**, *50*, 229–236. [CrossRef]

78. Eamus, D.; Myers, B.; Duff, G.; Williams, D. Seasonal changes in photosynthesis of eight savanna tree species. *Tree Physiol.* **1999**, *19*, 665–671. [CrossRef] [PubMed]

forests

MDPI

Article

Optimization Forest Thinning Measures for Carbon Budget in a Mixed Pine-Oak Stand of the Qingling Mountains, China: A Case Study

Lin Hou *, Zhe Li, Chunlin Luo, Longlong Bai and Ningning Dong

College of Forestry, Northwest A & F University, Yangling 712100, Shaanxi, China; nwsuaflizhe@163.com (Z.L.); clluo@139.com (C.L.); blyrly@163.com (L.B.); 18829781178@163.com (N.D.)
* Correspondence: houlin_1969@nwsuaf.edu.cn; Tel./Fax: +86-29-8708-2124

Academic Editors: Robert Jandl and Mirco Rodeghiero
Received: 18 September 2016; Accepted: 3 November 2016; Published: 12 November 2016

Abstract: Forest thinning is a silviculture treatment for sustainable forest management. It may promote growth of the remaining individuals by decreasing stand density, reducing competition, and increasing light and nutrient availability to increase carbon sequestration in the forest ecosystem. However, the action also increases carbon loss simultaneously by reducing carbon and other nutrient inputs as well as exacerbating soil CO_2 efflux. To achieve a maximum forest carbon budget, the central composite design with two independent variables (thinning intensity and thinning residual removal rate) was explored in a natural pine-oak mixed stand in the Qinling Mountains, China. The net primary productivity of living trees was estimated and soil CO_2 efflux was stimulated by the Yasso07 model. Based on two years observation, the preliminary results indicated the following. Evidently chemical compounds of the litter of the tree species affected soil CO_2 efflux stimulation. The thinning residual removal rate had a larger effect than thinning intensity on the net ecosystem productivity. When the selective thinning intensity and residual removal rate was 12.59% and 66.62% concurrently, the net ecosystem productivity reached its maximum 53.93 $t \cdot ha^{-1} \cdot year^{-1}$. The lower thinning intensity and higher thinning residual removal rated benefited the net ecosystem productivity.

Keywords: selective thinning; thinning residual removal; carbon budget; optimization; Yasso07; Qinling Mountains

1. Introduction

Forest thinning is one of the most efficient tending measures and it is also an effective management technique [1,2]. Selective thinning is one type of high thinning, which removes dominant and co-dominant trees [3]. It improves the vigor of residual trees as they benefit from the water, nutrient, and light resources no longer exploited by the felled trees [4]. Although the objectives of thinning vary, they typically include increasing the share of large diameter trees, improving the quality of timbers, increasing yield value, improving stand stability and influencing tree species composition [3] by decreasing stand density, reducing competition, and increasing the light and nutrient availability for the remaining trees primarily to promote growth of the remaining individuals [5]. Efforts to optimize carbon sequestration in forest ecosystems have mainly focused on enhancing stand biomass productivity and density by adapting thinning intensity and tree species composition [6]. However, timbers and thinning residuals (branches and foliage) are usually removed from stands after thinning which reduces carbon and other nutrient inputs [7–9] and the soil microclimate [10] is changed. For this reason, forest thinning may reduce carbon stocks in forest soil and vegetation [11] due to the increase of soil CO_2 efflux [7]. Thinning intensity and residual removal rate are important for carbon budget in forest systems. Litter decomposition is an important process in the global carbon cycle [12].

Its decomposition affects the soil carbon content, carbon dioxide emissions and is closely related to the chemical quality of litter types and climatic conditions [13] in forest ecosystems, and in particular for litter.

A considerable amount of literature has addressed the effects of forest thinning on the forest carbon cycle. Based on these observations, some studies examined the effects of forest thinning on soil CO_2 efflux with equivocal results ranging from positive effect on soil carbon release [14], negative effect [5,15] and no effect [16–18]. Others reported that intensive biomass harvesting may negatively impact carbon stocks in forest soil and vegetation [11], forest thinning did not have a significant impact on carbon stocks or fluxes [19], and remaining residues after harvesting increased carbon storage [20]. Unfortunately, few studies have examined the concurrent effects of two or more factors simultaneously, e.g., thinning and residual removal (branches, needle/leaf and twigs removed with thinning) on the forest net ecosystem productivity (NEP). In addition, how to balance the tree biomass increase and carbon loss as a consequence of forest thinning is still uncertain.

To attain a preliminary combination of thinning intensity and thinning residual rate for the largest carbon budget in thinned stands, a central composite design with two independent variables (thinning intensity and thinning residual removal rate) was explored in a natural pine-oak mixed stand in the Qinling Mountains, China.

The objectives of our study were (i) to examine the effect of a chemical compound of litter on the stimulation of soil CO_2 efflux; (ii) to estimate the effects of selective thinning and residual removal on the carbon budget of the pine-oak mixed stand; and (iii) to optimize the combined thinning intensity and residual removal rate to achieve the highest total carbon budget.

2. Material and Methods

2.1. Site Description

The Qinling Mountains (32°30′–34°45′ N, 104°30′–112°45′ E) in central China constitute a substantial physical obstacle for northward and southward movement of air masses, because of their high elevation and east-to-west arrangement. Therefore, these mountains are critical in stabilizing the distribution of climate and life zones in eastern China [21]. The pine-oak mixed stand is extensive in the middle of the altitudinal gradient of the Qinling Mountains and it plays important ecological roles in water purification and carbon sequestration.

Experiments were conducted at the Qinling National Forest Ecosystem Research Station (QNFERS), located on the southern slope of the Qinling Mountains, in Huoditang, Ningshan County, Shaanxi Province (32°18′ N, 108°20′ E). The altitude of the study area is from 1500–2500 m above sea level. The area experiences a subtropical climate, with annual mean temperatures around 8–10 °C, annual mean precipitation around 900–1200 mm, and annual mean evaporation around 800–950 mm. The main soil type is mountain brown soil, developed from granite material, with depths ranging from 30 to 50 cm. The total forest area is 2037 hectares in the station. Natural forest occupies 93% of the total forest area in QNFERS, with various vegetation types distributed in this region along the altitudinal gradient, such as evergreen deciduous mixed forest (pine-oak mixed forest), deciduous broad-leaved forest (oak, red birch), temperate coniferous forest (Chinese red pine, Armand pine) and cold temperate coniferous forest (spruce, fir). The most dominant forest type is pine-oak mixed forest with an average stand age of the stands of 42 years and an average height of 9.2 m. Common tree species include *Pinus tabulaeformis*, *Pinus armandii*, and *Quercusaliena* var. *acuteserrata*. The understory species are abundant.

2.2. Experimental Design

Our experimental plots were located on steep slopes (average slope gradient 30°) with thin soil depth (<50 cm), and in fragmented terrain. These characteristics make it extremely difficult to obtain the amount of replications for a randomized block design or orthogonal experiment design.

The Central Composite design (CCD) is the most popular of the many response surface methodology (RSM) classes, and is widely used for estimating second-order response surfaces [22]. The application of these statistical techniques in experiments has the advantages of requiring fewer resources (time, numbers of duplication, and amount of experimentation), but can also reduce process variability [23]. RSM is a collection of statistical and mathematical techniques useful for developing, improving, and optimizing processes [24]. These techniques relate a response variable to predictors that have multiple levels. The coded levels and the actual levels of the independent variables were calculated according to Equations (1)–(3):

$$X_{0j} = \frac{X_{1j} + X_{2j}}{2} \tag{1}$$

$$\Delta_j = \frac{X_{2j} - X_{0j}}{\alpha} \tag{2}$$

$$x_{\alpha j} = \frac{X_{\alpha j} - X_{0j}}{\Delta_j} \tag{3}$$

where X_j is the real value of the independent variable, X_0 is the real value of an independent variable at the center point, Δ_j is the step change value, and x_j is the coded value of an independent variable [25].

The CCD consists of 2^k full or 2^{k-1} half-replicate (k = number of independent variables) factorial points ($\pm 1, \pm 1, \ldots, \pm 1$); $2k$ axial or star points of the form ($\pm \alpha, 0, \ldots, 0$), ($0, \pm \alpha, \ldots, 0$); and a center point ($0, 0, \ldots, 0$). The axial points are replicated one and two times, and allow for the efficient estimation of pure quadratic terms. The center points are replicated one and three times, and provide information about the existence of curvature. The number of center runs can be altered, providing flexibility to improve error estimates and power. Finally, the factorial points allow estimation of the first-order and interaction terms. The CCD can be summarized with the following equation (Equation (4)):

$$N = f + (2k)\,\alpha + n_0 \tag{4}$$

where N is the total number of experimental runs, f is the number of factorial points, $2k$ is the axial point, α is the number of times the axial point is replicated, and n_0 is the center point. The axial distance, α is chosen based on the region of interest. Selecting the appropriate values of α specifies the CCD type, with $\alpha = \sqrt{k}$ being a spherical CCD [22].

The relationship between response and predictor levels can be approximated with a second-order response surface mode [22] (see Equation (5)):

$$y = \beta_0 + \sum_{i=1}^{k} \beta_i x_i + \sum_{i=1}^{k} \beta_{ii} x_{ii}^2 \sum_{j=i+1}^{k} \sum_{i=1}^{k-1} \beta_{ij} x_i x_j + \varepsilon_{ij} \tag{5}$$

where y is the measured response; β_0, β_i, β_j, β_{ii}, and β_{ij} are parameter coefficients; x_i, x_j are the input variables; and ε_{ij} is an error term.

Analysis of Variance (ANOVA) is used to optimize Equation (5) and analyze the interaction effect of the input variables on measured response and the single effect of input variables on measured response by differentiating variable j on variable i and vice versa [25].

The contributions of the controlled variables to the dependent variable, as $F > 1$ were estimated following the method of Tang [25] (see Equations (6) and (7)).

$$\Delta_j = S_j + \frac{1}{2} \sum S_{ij} + S_{jj} \tag{6}$$

$$S = 1 - \frac{1}{F} \tag{7}$$

where Δ_j is the contribution of controlled variable j to the dependent variable; S_j is the linear term for the controlled variable j; S_{ij} is the interaction term for the controlled variables i and j; S_{jj} is the quadratic term for the controlled variable j; F is the F-value in the ANOVA.

Our experiment was based on CCD, generated using Data Processing System (DPS) version 14.50 [25].

In a preliminary investigation conducted over the 15–25 August 2012, we selected 13 plots (20 m × 20 m) with similar slope gradients, canopy cover, tree species composition, and soil depths (Table 1) for the current experiment. For each plot, we surveyed the soil depth, as well as the height (m), diameter at breast height (DBH, cm), and canopy cover (%) of the tree species. The intensity of selective thinning (ST, %) and thinning residual removal rate (TRR, %) were calculated using Equations (8) and (9), respectively:

$$ST = \frac{A_f}{A_T} \tag{8}$$

$$TRR = \frac{Q_i}{Q_t} \tag{9}$$

where A_f, A_T, Q_i, and Q_t are the basal area of logged trees, total basal area of trees, fresh weight of removed residue, and fresh weight of total residue in all plots.

Table 1. General information of plots. Composition of tree species was calculated by their basal area.

Plot No.	Gradient (°)	Canopy Density		Tree Species Composition			Soil Depth (cm)
		2012	2013	2012	2013	2014	
1	35	0.75	0.72	7 Pt 2 Pa 1 Q	9 Pa 1 Pt	9 Pa 1 Pt	30
2	30	0.8	0.73	5 Pt 3 Pa 2 Q	5 Pa 1 Pt 4 Q	5 Pa 1 Pt 4 Q	45
3	30	0.8	0.72	7 Pt 2 Pa 1 Q	7 Pa 2 Pt 1 Q	7 Pa 2 Pt 1 Q	40
4	30	0.8	0.73	6 Pt 2 Pa 2 Q	2 Pa 4 Pt 4 Q	2 Pa 4 Pt 4 Q	50
5	25	0.85	0.81	4 Pt 3 Pa 3 Q	1 Pa 5 Pt 4 Q	1 Pa 5 Pt 4Q	46
6	30	0.8	0.7	1 Pt 7 Pa 2 Q	1 Pa 8 Pt 1 Q	1 Pa 8 Pt 1 Q	35
7	30	0.75	0.73	2 Pt 7 Pa 1 Q	8 Pt 2 Q	8 Pt 2 Q	40
8	30	0.9	0.81	2 Pt 5 Pa 3 Q	2 Pa 5 Pt 3 Q	2 Pa 5 Pt 3 Q	44
9	25	0.9	0.84	2 Pt 7 Pa 1 Q	1 Pa 7 Pt 2 Q	1 Pa 7 Pt 2 Q	33
10	30	0.8	0.72	1 Pa 8 Pt 1 Q	5 Pa 1 Pt 4 Q	6 Pa 4 Q	38
11	30	0.9	0.83	6 Pt 2 Pa 2 Q	1 Pa 3 Pt 6 Q	1 Pa 3 Pt 6 Q	41
12	25	0.8	0.72	6 Pt 3 Pa 1 Q	7 Pa 2 Pt 1 Q	8 Pa 2 Q	45
13	28	0.8	0.73	8 Pt 1 Pa 1 Q	9 Pa 1 Q	8 Pa 1 Pt 1 Q	45

Pa, Pt, and *Q* in the table is *Pinus armandi, Pinus tabulaeformis,* and *Quercus aliena* var. *acutesserata* respectively.

The design consisted of two independent variables (X_1 = thinning and X_2 = thinning residual removal), each with five intensity/rate gradients (Table 2). For the controlled factor (independent variable) in the current study, the value α was $\sqrt{2}$ = 1.414. We set the $+\alpha$ and $-\alpha$ level thinning intensity to 25% and 5% respectively according to the Regulation for Tending of Forest [26], and 100% and 0% for the residual removal rate. To explore the effects of the thinning operation on NEP, zero-treatment of thinning intensity was excluded.

For a central composite design with two independent, five-level variables, 13 experimental runs are required, with four factorial points from treatment I to treatment IV, four axial points from treatment to treatment VIII, and five center points treatment IX (Table 3). The factorial points were a combination of controlled variables at ±1 levels (a thinning intensity at ±1 level represent 22.07% and 7.93% respectively; a thinning residual removal rate at ±1 levels represent 85.36% and 14.64% respectively) in our study. Similarly, the star points were a combination of controlled variables at ±α and 0 levels (a thinning intensity at ±α and 0 levels represented 25%, 5%, and 15% respectively; a thinning residual removal rate at ±α and 0 levels represented 100%, 0%, and 50% respectively). The center point was a combination of controlled variables at 0 levels. Different thinning factors were applied in each plot, except for the plots categorized as center points (Table 3).

Table 2. Experiment design runs in DPS v14.50 (Data Processing System).

Variables	Levels				
	$(-\alpha) - 1.414$	-1	0	1	$(+\alpha) + 1.414$
X_1 (%)	5.00	7.93	15.00	22.07	25.00
X_2 (%)	0.00	14.64	50.00	85.36	100.00

X_1 and X_2 in the table represents actual thinning intensity and thinning residual removal rate respectively. Levels $(-\alpha, -1, 0, +1,$ and $+\alpha)$ in the table are coded values of variables (thinning intensity and thinning residual removal rate) generated by the software Data Processing System (version 14.50).

Table 3. Experiment design runs in DPS v14.50.

Plot No.	Treatment	Design Code		Thinning Factors		Block
		x_1	x_2	X_1 (%)	X_2 (%)	
1	I	1	1	22.07	85.36	
2	II	1	-1	22.07	14.64	Factorial points
3	III	-1	1	7.93	85.36	
4	IV	-1	-1	7.93	14.64	
5	V	$-\alpha$	0	5	50	
6	VI	$+\alpha$	0	25	50	Axial points
7	VII	0	$-\alpha$	15	0	
8	VIII	0	$+\alpha$	15	100	
9	IX	0	0	15	50	
10	IX	0	0	15	50	
11	IX	0	0	15	50	Center points
12	IX	0	0	15	50	
13	IX	0	0	15	50	

The dependent variable in this study was the average of NEP in 2013 and in 2014 in post-treatments. The experimental results were fitted to a second-order polynomial model, and the regression coefficients were determined. The quadratic model for predicting the optimal combination of thinning intensity and removal rate to reach the highest value of NEP (Y_k) is described by Equation (10):

$$Y_k = b_{k0} + \sum_{i=1}^{2} b_{ki}x_i + \sum_{i=1}^{2} b_{kii}x_i^2 + \sum_{i=1}^{2} b_{kij}x_ix_j \tag{10}$$

where b_{k0}, b_{ki}, b_{kii}, and b_{kij} are the constant regression coefficients of the model, and x_i, x_j are codes of the independent variables (x_i = thinning intensity and x_j = thinning residual removal rate).

2.3. Thinning, Residual Removal, and Dynamics of Tree Growth and Litterfall

All trees in the 13 plots with DBH > 10 cm were numbered and tagged between 28 and 30 August 2012. Crop trees in plots were selected. Stem quality, crown size, vitality, spatial distribution of potential crop trees, diameter, and tree damages were taken into account when selecting the crop trees. Competing trees were marked and cut. The intensity of thinning in each plot was determined as shown in Table 2. After log harvesting, the total fresh weight of residuals (branches, needle/leaf and twigs) in each plot was measured. Then, the branches and twigs were cut into pieces of length 60–80 cm and mixed with needle/leaf. According to the experiment design (Table 2), a part of the residuals was removed and the rest was thrown on the forest ground with a similar thickness in each plot. The actions of thinning and residual removal were done manually for steep slopes of plots in September 2012.

The height and DBH of the remaining trees were monitored between 20 and 28 September 2013 and 2014.

Along the slope from bottom to top, each plot was mechanically partitioned in three sections. Nine circular litter traps (diameter 30 cm) were set equidistantly in each plot to monitor the dynamics of litterfall since 20 September 2012. At the end of December 2013 and 2014, littler in each plot was mixed separately in the field and then was brought to the laboratory to dry and measure dry weight.

2.4. Soil CO$_2$ efflux

For the high spatial variability of the soil carbon stocks and the high uncertainty in their changes [27], it is difficult, laborious and expensive to measure soil CO$_2$ efflux directly [13]. Models are needed to estimate the dynamics of carbon in forest soils [28]. Comparing existing soil carbon models Century [29], Q-model [30], ROMUL [31], RothC [32] and DECOMP [33] to Yasso07, the significant advantage of Yasso07 is that the parameters to operate the model are easily accessible [34]. Therefore the Yasso07 model was applied in the current study to estimate CO$_2$ efflux in post-treatments.

2.5. Chemical Analysis

A mixed litter sample 0.50 kg was collected from nine litter traps in each plot. Samples were separated by conjunction of litter types (twig, needle, leaf) and tree species. Chemical compound groups, ESC (ethanol soluble compound), WSC (water soluble compound), ASC (acid soluble compound) and NSC (non-soluble compound) of litter [28] were analyzed by Bai et al. [35]. The analysis processing was the following. (1) ESC: Dried and ground litter sample (diameter is 0.074 mm) of 1.00 gram was put into a cylinder made by filter paper with diameter 4 cm. Then, the cylinder with the sample was removed to a reflux line of a Soxhlet extractor (Yuming Instrument Co. Ltd., Shanghai, China) (BSXT-02-250). Next, 100 mL mixed benzene-alcohol (1:1) solution was added into a fat-wax bottle connected with a reflux line and a condenser pipe and the bottle was placed in a 80 °C thermostat water bath for 20 h until the color of the solution in the reflux line disappeared completely. Afterwards, the cylinder was taken out and put into a draught cupboard until the benzene-alcohol solution volatilized entirely and the first residue was attained. Finally, the residue was dried in a 50 °C oven for 4 h until its weight remained unchanged. The weight of ESC was calculated from the difference between the weight of the sample and the first dried residue; (2) WSC: The first dried residue was moved to a 250 mL beaker and 150 mL distilled water was injected in, and the residue was stirred and broken into pieces with a glass rod. Then, the beaker was covered and placed in a 100 h hydrolysis pot for 3 h. Next, a sand core funnel (type G3) was used to suck filter the contents and the second residue was attained. Afterwards, the second residue was washed in 30 °C pure water until the eluate was without color. After that, the colatuie and the eluate were injected into a new 250 mL beaker and its volume maintained at 200 mL. Finally, the second residue was removed from the sand core funnel to another beaker and dried in a 50 °C oven for 4 h. The weight of WSC was the difference between the weight of dried residues obtained from the first and the second time; (3) ASC: The 150 mL and 2% hydrochloride resolution was injected in the beaker filled with the second dried residue. Then, the residue was stirred and broken into pieces with a glass rod. Next, the beaker was put into a 100 h hydrolysis pot for 5 h and the solution was suck filtered with a G3 sand core funnel and the third residue was attained. After this, the third residue was placed in a porcelain crucible and was washed in 30 °C pure water until sulfate radical could not be detected by 5% barium chloride solution. Finally, the third residue was dried in a 105 °C oven for 4 h. The weight of ASC was the difference between the weight of the dried residues obtained from the second and third time; (4) NSC: The third dried residue was placed in a porcelain crucible and the crucible was put on a 45 °C electric stove to carbonize the third residue for 3 h. Then, the crucible was placed on a 450 °C electric stove to burn the third residue for 8 h. Later, the residue was cooled naturally until room temperature. Finally, the cooled residue was weighed to obtain the fourth residue. The weight of NSC was the difference between the weight of the third dried residue and the fourth.

All these parameters were inputs to run the Yasso07 model.

2.6. Data Processing and Analysis

Ratio of tree species (R_i) was calculated as Equation (11).

$$R_i = \frac{b_i}{B} \times 100\% \tag{11}$$

where b_i and B is basal area of the tree species i and total basal area of tree species in an identical treatment.

Composition of tree species is the proportion R_i: R_m: R_n.

Where R_i, R_m and R_n is ratio of tree species i, m and n respectively.

Diameter classes were used to describe the DBH dynamics of tree species (Table 4). DBH ratio of tree species (D_{ij}) was calculated as follows:

$$D_{ij} = \frac{n_{ij}}{N_i} \times 100\% \tag{12}$$

where n_{ij} is number of tree species i in diameter calss j and N_i is number of total tree species in all diameter classes of 13 plots.

Table 4. Classification of diameter classes.

Diameter Class (cm)	DBH (cm)
4	≤ 6
8	$6.1 \leq \text{DBH} \leq 10$
12	$10.1 \leq \text{DBH} \leq 14$
16	$14.1 \leq \text{DBH} \leq 18$
20	$18.1 \leq \text{DBH} \leq 22$
24	$22.1 \leq \text{DBH} \leq 26$
28	$26.1 \leq \text{DBH} \leq 30$
32	$30.1 \leq \text{DBH} \leq 34$
36	$34.1 \leq \text{DBH} \leq 38$
40	$38.1 \leq \text{DBH} \leq 42$
44	$42.1 \leq \text{DBH} \leq 46$
48	$46.1 \leq \text{DBH} \leq 50$

Chemical compound groups (ESC, WSC, ASC, and NSC) of litterfall were calculated by mass weighted average of tree species. We inputted the quality of litterfall from each post-treatment, the measured chemical compound groups of tree species, and the data of annual precipitation and air temperature from QNFERS into the Yasso07 model to estimate soil CO_2 efflux (Rs) [28].

Based on monitoring DBH and the height of remaining trees in the thirteen plots in 2013 and 2014, living biomass (Mg·ha^{-1}) of whole remaining trees was estimated as Equations (13)–(15) [36].

P. tabulaeformis:

$$Y = 15.525 + 0.6269x \tag{13}$$

P. armandi:

$$Y = 54.280 + 0.4048x \tag{14}$$

Q. aliena var. acutesserata:

$$Y = 13.394 + 1.0564x \tag{15}$$

where Y is the living biomass of trees and x is the stand growing stock (m^3·ha^{-1}).

The stand growing stock was calculated by the stems and volume of each tree. The volume of a single tree was calculated as follows [37] (see Equations (16)–(18)):

P. armandi:

$$\ln V = 0.95697 \ln \left(D^2 H \right) - 9.95738 \tag{16}$$

P. tabulaeformis:

$$\ln V = 0.99138 \ln \left(D^2 H \right) - 10.20211 \tag{17}$$

Q. aliena var. *acutessera*:

$$\ln V = 0.96884 \ln \left(D^2 H - 10.07352 \right) \tag{18}$$

where V (m^3), D (cm), and H (m) are volume, DBH, and height of a tree respectively.

Net primary productivity (NPP) of the current forest was the living tree biomass increment in two consecutive years multiplied by the carbon ration in plants (0.50 in this study).

NEP was calculated as following.

$$NEP = NPP - R_s \tag{19}$$

Figures were plotted using Origin8.0 (OriginLab Corporation, Northampton, MA, USA) software. DPS v14.50 software (Zhejiang University, Hangzhou, China) was used to fit models, analyze the data, determine the effects of a single independent variable and the interaction of independent variables on NEP and optimize the combination of thinning intensity and residual removal rate for the highest NEP.

3. Results

3.1. Dynamics of Tree Species Composition

Selective thinning decreased the canopy density and changed the composition of tree species (Table 1). The proportion of *Quercus aliena* var. *acuteserrata* of the total tree species in each plot was not more than 30% before thinning in 2012 and increased in most plots after thinning in 2013 and 2014 (Table 1).

3.2. DBH Dynamics of Tree Species

Selective thinning and thinning residual removal promoted DBH of the remaining tree species increase in post-treatments (Figure 1).

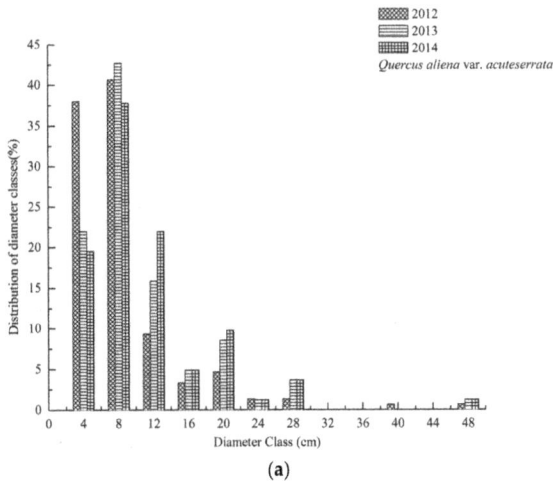

(a)

Figure 1. *Cont.*

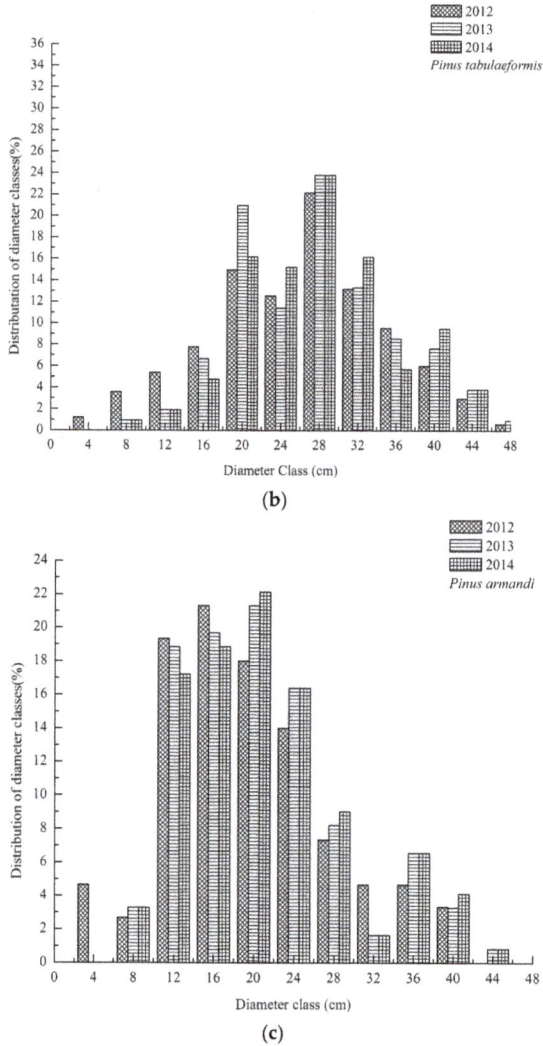

Figure 1. Distribution of diameter classes (%) between pre-treatments and post-treatments. Each plot refers to a different species. Figure (**a–c**) demonstrates distribution of diameter classes of tree species *Quercus aliena* var. *acuteserrata*, *Pinus tabulaeformis* and *Pinus armandi* in all plots before and after thinning.

3.3. Dynamics of Net Primary Productivity of Living Trees

Based on height and DBH of the tree species monitored, the net primary productivity of living trees was estimated. The net primary productivity of living trees decreased between pre-treatments and post-treatments (Figure 2). Comparing to the start point in 2012, the average increment of biomass carbon of living trees decreased respectively 20.46 t·ha^{-1}·year^{-1}·and·15. 64 t·ha^{-1}·year^{-1}·after the first year thinning in 2013 and after the second year thinning in 2014 (Figure 2). With tree growing after thinning, the net primary productivity of the living trees gradually increased after thinning for two years in 2014 (Figure 2).

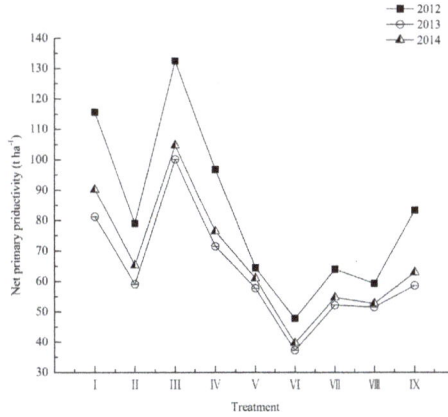

Figure 2. Dynamics of net primary productivity of living trees.

3.4. Soil CO₂ Efflux Stimulation

Based on the chemical compound groups of the litter of the tree species we measured and data of annual precipitation and air temperature in 2013 and 2014 from QNFERS, soil CO_2 efflux was estimated. The results stimulated by Yasso07 demonstrated that selective thinning and residual removal had a hysteretic effect on soil CO_2 efflux. Soil CO_2 efflux in post-treatments in 2013 was lower than that in 2014, except for treatment III (Figure 3).

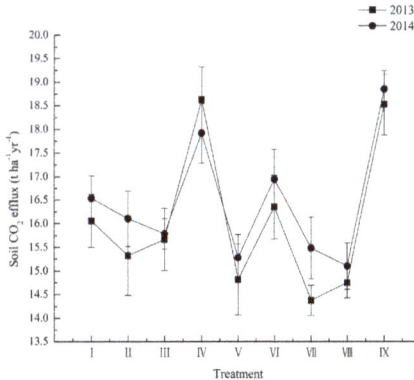

Figure 3. Soil CO_2 efflux in post-treatments.

3.5. Model Fitting

Effects of selective thinning intensity and thinning residual removal rate on net ecosystem productivity were analyzed by fitting a quadratic model. The final quadratic model was obtained for each response and was expressed by the following second-order polynomial equation after optimization.

$$\text{NEP} = 50.10 - 1.87x_1 + 2.85x_2 - 2.00x_1^2 - 2.54x_2^2 + 2.48x_1x_2 \tag{20}$$

Regression coefficient, standard error, and ANOVA for the regression model of NEP are presented (Table 5). Under the condition $p < 0.10$, the quadratic model was a better fit relationship of NEP in conjunction with the selective thinning rate and thinning residual removal rate.

Table 5. Regression coefficient, standard error, and ANOVA (Analysis of Variance) for the regression model of net ecosystem productivity (NEP).

Parameter	Degrees of Freedom	Sum of Squares	Mean Square	F-Value	Pr > F
x_1	1	27.9913	27.9913	5.5179	0.0512
x_2	1	65.1148	65.1148	12.8360	0.0089
x_1^2	1	27.9793	27.9793	5.5155	0.0512
x_2^2	1	44.9868	44.9868	8.8682	0.0206
x_1x_2	1	24.5520	24.5520	4.8399	0.0637
Model	5	182.4717	36.4943	7.1941	0.0226
Lack of fit	3	13.3448	4.4483	0.8028	0.5308
Residual	7	35.5097	5.0728		
Error	4	22.1649	5.5412		
Total	12	217.9815			
		$R^2 = 0.837$			
		Adjusted $R^2 = 0.849$			

3.6. Effects of Forest Management on Carbon Budget

The model Equation (16) demonstrated that both variable x_1 and x_2 affected NEP and they also had an interaction effect on NEP. According to Equations (6) and (7), and the data shown in Table 5, the thinning removal rate ($\Delta x_2 = 2.21$) had a larger effect than selective thinning intensity ($\Delta x_1 = 2.03$) on NEP.

Effects of single factor and their interaction on NEP were analyzed by software DPS v14.50 respectively. The results indicated that selective thinning intensity and thinning residual removal rate were positively related to NEP when the range of the independent variable was $x_1 \in [-1.414, -0.5]$ and $x_2 \in [-1.414, 0.5]$ respectively (Figure 4). In contrast, as independent variables ranged in $x_1 \in [-0.5, 1.414]$ and $x_2 \in [0.5, 1.414]$, both selective thinning intensity and thinning residual removal rate were negatively related to NEP (Figure 4). The modeled effects of thinning and residual removal on NEP are illustrated in a 3D-contour plot (Figure 5). The effect of increasing residual removal rate on NEP was conspicuous when selective thinning intensity was at a low level $x_1 \in [-1.414, -0.25]$ (Figure 5). Thereafter, with an increase of selective thinning intensity, NEP showed a slow increase with increased residual removal rate. Thinning intensity and residual removal rate thus interacted negatively in their effects on NEP. The outcomes suggested that lower selective thinning intensity and higher thinning residual removal rate benefited NEP.

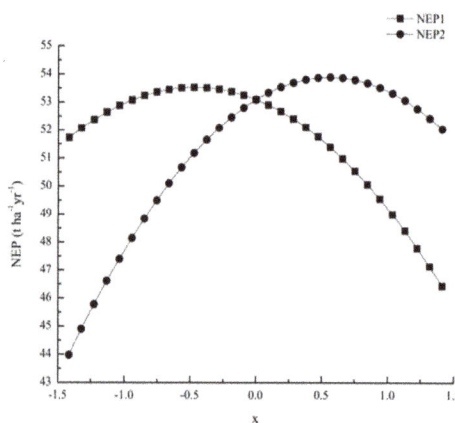

Figure 4. Single effect of selective thinning intensity or thinning residual removal rate on net ecosystem productivity (NEP). x is controlled factor selective thinning intensity or thinning residual removal rate. NEP1 and NEP 2 are effects of selective thinning intensity and thinning residual removal rate on NEP respectively.

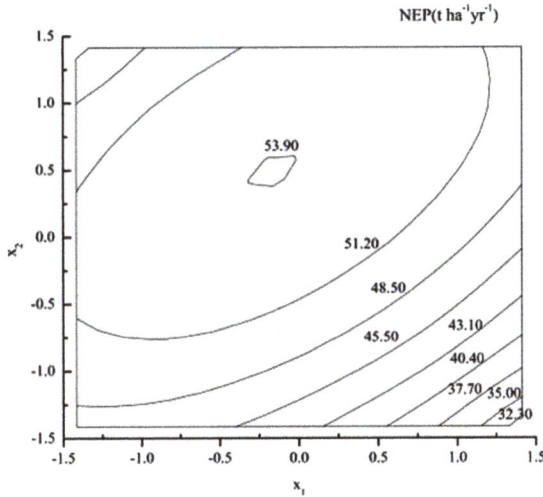

Figure 5. Contour plot for NEP from selective thinning (x_1) and residual removal (x_2).

3.7. Optimization Forest Management Measures for NEP

Each independent variable (thinning intensity, residual removal rate) was individually increased or decreased in an attempt to find the maximum response in NEP. Once the optimal value was found separately for thinning intensity and residual removal rate, the value was selected as the condition for obtaining the overall maximum NEP. Our analysis demonstrated that the maximum NEP (53.93 t·ha^{-1}·year^{-1}) was achieved when the independent variables were $x_1 = -0.17$ and $x_2 = 0.48$. To verify these index values, the codes for the independent variables were incorporated into Equations (5)–(7). For NEP, these codes yielded selective thinning intensity and residual removal rate of 12.59% and 66.62%, resulting in the predicted maximum NEP of 53.93 t·ha^{-1}·year^{-1}.

4. Discussion

4.1. Effect of Chemical Compound Groups of Litterfall on Soil CO_2 Efflux

Chemical compound groups of litter on Euro-American tree species are provided in the Yasso07 manual which are more convenient for the model users in those countries [12]. Whether the parameters of tree species in China affect soil CO_2 efflux stimulation is uncertain. We analyzed chemical compound groups of litter types (leaf/needle, fine root, twig, and coarse root) of three tree species (Table 6). The results indicated that content of ESC, WSC, ASC, and NSC among different tree species with the identical litter type varied significantly (Table 6). A similar trend was also found among litter types with the same tree species (Table 6). To examine the effect of chemical compound groups of litterfall on soil CO_2 efflux, the values we measured (scenario 1) and the global from Yasso07 manual (scenario 2) [13] with litterfall quality, precipitation and air temperature in 2013 in each treatment were adopted in running the Yasso07. The stimulated soil respiration was conspicuously underestimated in scenario 2 than that in scenario 1 with identical treatments (Figure 6). The average underestimated soil CO_2 was 2.55 t·ha^{-1}·year^{-1} and the maximum even reached 3.50 t·ha^{-1}·year^{-1} (treatment IX) (Figure 6).

Table 6. Chemical compound groups of litter among tree species.

Tree Species	Ethanol	Ethanol Std	Water	Water Std	Acid	Acid Std	Non Soluble	Non Soluble Std	Litter Types
Pt	0.125a	0.004	0.153b	0.006	0.433w	0.003	0.289c	0.006	Leaf/needle
Pa	0.122a	0.005	0.151b	0.006	0.418w	0.005	0.309c	0.011	Leaf/needle
Q	0.199b	0.004	0.252c	0.004	0.445w	0.005	0.104d	0.002	Leaf/needle
Pt	0.120a	0.005	0.142d	0.002	0.477w	0.003	0.261e	0.007	Fine root
Pa	0.119a	0.002	0.130ad	0.002	0.476w	0.002	0.276e	0.003	Fine root
Q	0.154c	0.003	0.187e	0.009	0.429w	0.009	0.229f	0.015	Fine root
Pt	0.086k	0.004	0.088kf	0.004	0.525m	0.027	0.301c	0.005	Twig
Pa	0.061e	0.006	0.097f	0.005	0.523m	0.005	0.321c	0.005	Twig
Q	0.038u	0.002	0.081f	0.008	0.562m	0.005	0.319c	0.32	Twig
Pt	0.073g	0.002	0.091f	0.003	0.530m	0.002	0.306c	0.004	Coarse root
Pa	0.072g	0.004	0.085f	0.004	0.524m	0.004	0.319c	0.006	Coarse root
Q	0.111h	0.003	0.139g	0.013	0.473w	0.007	0.278f	0.007	Coarse root

Ethanol, water, acid and non-soluble in the table represents content of ESC, WSC, ASC, and NSC in the litter respectively. The acronym std is standard error. Different lowercase letters indicate differences among tree species and litter types (leaf/needle, fine root, twig, and coarse root) at $p < 0.05$ level by LSD. The acronym of tree species in Table 6 is as same as in Table 1.

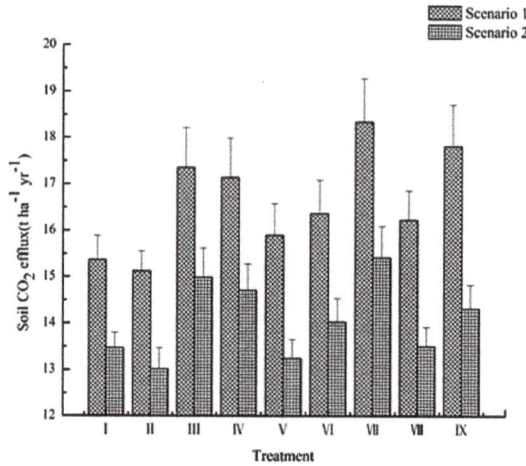

Figure 6. Differences of soil CO_2 efflux stimulation.

Biological characteristics of tree species might lead to varieties of chemical compound groups of litter [38]. Soil CO_2 efflux results from chemical compound groups of litter transforming into different soil carbon compartments [13]. Chemical compound groups of litter differed from each other between tree species even in the same genera. ESC in needles of *Pinus sylvestris* was 10.79 times of that in *Pinus pinaster*, 8.56 times of that in *Pinus pinea* and 7.24 times of that in *Pinus banksiana* respectively [12]. In addition, NSC in leaf of *Quercus robur* was 5.41 times of thath in *Quercus garryana* and ASC in leaf of *Quercus garryana* was 1.54 times of that in *Quercus robur* [38,39]. Cellulose in litter is usually shielded by lignin and the litter decomposition rate decreases with its NSC for high nitrogen and lignin content [40]. ASC is composed of cellulose and hemicellulose, and its decomposition requires cellulase and other special conditions [12]. High NSC and ASC in litter may decrease the litter decomposition rate [12]. In contrast, high ESC and WSC in litter will promote litter decomposition resulting in the main components of dissolved fat, pigment and oil in ESC, and sugars in WSC, with low nitrogen [12].

The current study suggested that the chemical groups of litterfall apparently affected the result of soil CO_2 efflux. We recommended that the chemical compound groups of the litter should be measured before applying the Yasso07 model to stimulate soil CO_2 efflux.

4.2. The Response of Soil CO$_2$ Efflux to Management Measures

We observed that selective thinning and thinning residual removal increased soil CO$_2$ efflux with a hysteretic effect. Soil CO$_2$ efflux in 2014 was generally higher than that in 2013 with identical treatments (Figure 3). Most studies have indicated that temperature and moisture are the main factors positively influencing soil respiration over various climate regions [41–43]. With timber harvesting, thinning residual removal, canopy openness, and residual decomposition, more light with rain having reached the forest ground, increases soil temperature and moisture, and activates soil microorganisms. Thinning also increased soil respiration [14]. Others reported that the prescribed thinning had a negligible effect on soil respiration [16–18]. The likely reasons were that thinning induced increases in shrub abundance might be responsible for commensurate increases in fine root turnover which contributed substantially to the increased light use efficiency of thinned plots [16]. Sullivan et al. indicated that thinning may reduce soil respiration by killing trees, by altering the soil environment, or by changing the amounts and sources of below ground carbon for microbial metabolism [15]. Based on a short-term data, this study reported a preliminary effect of forest thinning on soil CO$_2$ efflux. Future monitoring will assist in clarifying the relationship between soil CO$_2$ efflux and forest thinning.

5. Conclusions

Our two years observation preliminary demonstrated that the chemical compound groups of the litter of tree species should not be ignored in analyzing soil CO$_2$ efflux. Thinning intensity and thinning residual removal rate have different effects on NEP. The thinning removal rate had a larger effect than selective thinning intensity on NEP. When selective thinning intensity and residual removal rate was 12.59% and 66.62% concurrently, the NEP reached its maximum 53.93 t·ha^{-1}·year^{-1}. A lower thinning intensity and higher thinning residual removal rate benefited NEP.

Acknowledgments: This research is a part of project No. 201304307 financed by State Forestry Administration of the People's Republic China.

Author Contributions: L.H. conceived and designed the experiments, and wrote the paper; Z.L. calculated NPP; L.B. analyzed chemical compound groups; C.L. Ran Yasso07 model; N.D. collated the manuscript and evaluated the results.

Conflicts of Interest: The authors declare no conflict of interest.

References

1. Magruder, M.; Chhin, S.; Palik, B.; Bradford, J.B. Thinning increases climatic resilience of red pine. *Can. J. For. Res.* **2013**, *43*, 878–889. [CrossRef]
2. Verschuyl, J.; Riffell, S.; Miller, D.; Wigley, T.B. Biodiversity response to intensive biomass production from forest thinning in north American forests—A meta-analysis. *For. Ecol. Manag.* **2011**, *261*, 221–232. [CrossRef]
3. Boncina, A.; Kadunc, A.; Robic, D. Effects of selective thinning on growth and development of beech (*Fagus sylvatica* L.) forest stands in south-eastern Slovenia. *Ann. For. Sci.* **2007**, *64*, 47–57. [CrossRef]
4. Rodríguez-Calcerrada, J.; Pérez-Ramos, I.M.; Ourcival, J.-M.; Limousin, J.-M.; Joffre, R.; Rambal, S. Is selective thinning an adequate practice for adapting *Quercus ilex* coppices to climate change? *Ann. For. Sci.* **2011**, *68*, 575–582. [CrossRef]
5. Fernandez, I.; Älvarez-Gonzalez, J.G.; Carrasco, B.; Ruíz-González, A.D.; Cabaneiro, A. Post-thinning soil organic matter evolution and soil CO$_2$ effluces in temperate radiata pine plantations: Impacts of moderate thinning regimes on the forest C cycle. *Can. J. For. Res.* **2012**, *42*, 1953–1964. [CrossRef]
6. Dobarco, M.R.; Miegroet, H.V. Soil organic carbon storage and stability in the aspen-conifer ecotone in montane forests in Utah State, USA. *Forests* **2014**, *5*, 666–688. [CrossRef]
7. Houghton, R.A. Revised estimates of the annual net flux of carbon to the atmosphere from changes in land use and land management 1850–2000. *Tellus B* **2003**, *55*, 378–390. [CrossRef]
8. Chatterjee, A.; Vance, G.F.; Pendall, E.; Stahl, P.D. Timber harvesting alters soil carbon mineralization and microbial community structure in coniferous forests. *Soil Biol. Biochem.* **2008**, *40*, 1901–1907. [CrossRef]

9. Diochon, A.; Kellman, L.; Beltrami, H. Looking deeper: An investigation of soil carbon losses following harvesting from a managed northeastern red spruce (*Picea rubens* Sarg.) forest chronosequence. *For. Ecol. Manag.* **2009**, *257*, 413–420. [CrossRef]

10. Jassal, R.S.; Black, T.A.; Cai, T.; Morgenstern, K.; Li, Z.; Gaumont-Guay, D.; Nesic, Z. Components of ecosystem respiration and an estimate of net primary productivity of an intermediate-aged Douglas-fir stand. *Agricu. For. Meteorol.* **2007**, *144*, 44–57. [CrossRef]

11. Mäkipää, R.; Linkosalo, T.; Komarov, A.; Mäkelä, A. Mitigation of climate change with biomass harvesting in Norway spruce stands: Are harvesting practices carbon neutral? *Can. J. For. Res.* **2014**, *45*, 1–9. [CrossRef]

12. User-Interface Software of Yasso07. Available online: http://www.syke.fi/en-US/Research__Development/Research_and_development_projects/Projects/Soil_carbon_model_Yasso/Download (accessed on 5 November 2016).

13. Tuomi, M.; Rasinmäki, J.; Repo, A.; Vanhala, P.; Liski, J. Soil carbon model Yasso07 graphical user interface. *Environ. Model. Softw.* **2011**, *26*, 1358–1362. [CrossRef]

14. Ryu, S.R.; Concilio, A.; Chen, J.; North, M.; Ma, S. Prescribed burning and mechanical thinning effects on belowground conditions and soil respiration in a mixed-conifer forest, California. *For. Ecol. Manag.* **2009**, *257*, 1324–1332. [CrossRef]

15. Sullivan, B.W.; Kolb, T.E.; Hart, S.C.; Kaye, J.P.; Dore, S.; Montes-Helu, M. Thinning reduces soil carbon dioxide but not methane flux from southwestern USA Ponderosa pine forests. *For. Ecol. Manag.* **2008**, *255*, 4047–4055. [CrossRef]

16. Campbell, J.; Alberti, G.; Martin, J.; Law, B.E. Carbon dynamics of a ponderosa pine plantation following a thinning treatment in the northern Sierra Nevada. *For. Ecol. Manag.* **2009**, *257*, 453–463. [CrossRef]

17. Kobziar, L.N.; Stephens, S.L. The effects of fuels treatments on soil carbon respiration in a Sierra Nevada pine plantation. *Agricu. For. Meteorol.* **2006**, *141*, 161–178. [CrossRef]

18. Kobziar, L.N. The role of environmental factors and tree injuries in soil carbon respiration response to fire and fuels treatments in pine plantations. *Biogeochemistry* **2007**, *84*, 191–206. [CrossRef]

19. Saunders, M.; Tobin, B.; Black, K.; Gioria, M.; Nieuwenhuis, M.; Osborne, B.A. Thinning effects on the net ecosystem carbon exchange of a Sitka spruce forest are temperature-dependent. *Agric. For. Meteorol.* **2012**, *157*, 1–10. [CrossRef]

20. Hopmans, P.; Elms, S.R. Changes in total carbon and nutrients in soil profiles and accumulation in biomass after a 30-year rotation of *Pinus radiata* on podzolized sands: Impacts of intensive harvesting on soil resources. *For. Ecol. Manag.* **2009**, *258*, 2183–2193. [CrossRef]

21. Tang, Z.; Fang, J. Temperature variation along the northern and southern slopes of Mt. Taibai, China. *Agric. For. Meteorol.* **2006**, *139*, 200–207. [CrossRef]

22. Oyejola, B.A.; Nwanya, J.C. Selecting the right central composite design. *Int. J. Stat. Appl.* **2015**, *5*, 21–30.

23. Khataee, A.R.; Fathinia, M.; Aber, S.; Zarei, M. Optimization of photocatalytic treatment of dye solution on supported tio$_2$ nanoparticles by central composite design: Intermediates identification. *J. Hazard. Mater.* **2010**, *181*, 886–897. [CrossRef] [PubMed]

24. Kasiri, M.B.; Aleboyeh, H.; Aleboyeh, A. Modeling and optimization of heterogeneous photo-fenton process with response surface methodology and artificial neural networks. *Environ. Sci. Technol.* **2008**, *42*, 7970–7975. [CrossRef] [PubMed]

25. Tang, Q.Y. *Data Processing System-Experimental Design, Statistical Analysis and Data Mining*, 2nd ed.; Science Press: Beijing, China, 2010.

26. Regulation for tending of forest. GB/T 15781 2009. Available online: http://www.doc88.com/p-94457076273.html (accessed on 5 November 2016).

27. Post, W.M.; Emanuel, W.R.; Zinke, P.J.; Stangenberger, A.G. Soil carbon pools and world life zones. *Nature* **1982**, *298*, 156–159. [CrossRef]

28. Tuomi, M.; Laiho, R.; Repo, A.; Liski, J. Wood decomposition model for boreal forests. *Ecol. Model.* **2011**, *222*, 709–718. [CrossRef]

29. Parton, W.J.; Schimel, D.S.; Cole, C.V.; Ojima, D.S. Analysis of factors controlling soil organic matter levels in great plains grasslands. *Soil Sci. Soc. Am. J.* **1987**, *51*, 1173–1179. [CrossRef]

30. Rolff, C.; Ågren, G.I. Predicting effects of different harvesting intensities with a model of nitrogen limited forest growth. *Ecol. Model.* **1999**, *118*, 193–211. [CrossRef]

31. Chertov, O.G.; Komarov, A.S.; Nadporozhskaya, M.; Bykhovets, S.S.; Zudin, S.L. Romul—A model of forest soil organic matter dynamics as a substantial tool for forest ecosystem modeling. *Ecol. Model.* **2001**, *138*, 289–308. [CrossRef]
32. Coleman, K.; Jenkinson, D.S. *Rothc-26.3—A Model for the Turnover of Carbon in Soil*; Springer: Heidelberge, Germany, 1996; pp. 237–246.
33. Wallman, P.; Belyazid, S.; Svensson, M.G.E.; Sverdrup, H. Decomp—A semi-mechanistic model of litter decomposition. *Environ. Model. Softw.* **2006**, *21*, 33–44. [CrossRef]
34. Tuomi, M.; Thum, T.; Järvinen, H.; Fronzek, S.; Berg, B.; Harmon, M.; Trofymow, J.A.; Sevanto, S.; Liski, J. Leaf litter decomposition—Estimates of global variability based on Yasso07 model. *Ecol. Model.* **2009**, *220*, 3362–3371. [CrossRef]
35. Bai, L.L.; Li, Y.; Hou, L.; Luo, C.L.; Geng, Z.C.; Cheng, H.F. Chemical composition of litters from main forests in qinling mountains. *J. Northwest A F Univ. (Nat. Sci. Ed.)* **2016**, *44*, 89–96.
36. Pan, Y.; Luo, T.; Birdsey, R.; Hom, J.; Melillo, J. New estimates of carbon storage and sequestration in China's forests: Effects of age-class and method on inventory-based carbon estimation. *Clim. Chang.* **2004**, *67*, 211–236. [CrossRef]
37. Chen, C.; Peng, H. Standing crops and productivity of the major forest-types at the Huoditang forest region of the Qinling Mountains. *J. Northwest For. Coll.* **1996**, *11*, 92–102.
38. Valachovic, Y.S.; Caldwell, B.A.; Cromack, K.; Griffiths, R.P. Leaf litter chemistry controls on decomposition of Pacific northwest trees and woody shrubs. *Can. J. For. Res.* **2004**, *34*, 2131–2147. [CrossRef]
39. Sariyildiz, T.; Anderson, J.M.; Kucuk, M. Effects of tree species and topography on soil chemistry, litter quality, and decomposition in northeast Turkey. *Soil Biol. Biochem.* **2005**, *37*, 1695–1706. [CrossRef]
40. Berg, B. Litter decomposition and organic matter turnover in northern forest soils. *For. Ecol. Manag.* **2000**, *133*, 13–22. [CrossRef]
41. Davidson, E.A.; Elizabeth, B.; Boone, R.D. Soil water content and temperature as independent or confounded factors controlling soil respiration in a temperate mixed hardwood forest. *Glob. Chang. Biol.* **1998**, *4*, 217–227. [CrossRef]
42. Epron, D.; Farque, L.; Lucot, É.; Badot, P.M. Soil CO_2 efflux in a beech forest: Dependence on soil temperature and soil water content. *Ann. For. Sci.* **1999**, *56*, 221–226. [CrossRef]
43. Burton, A.J.; Pregitzer, K.S. Field measurements of root respiration indicate little to no seasonal temperature acclimation for sugar maple and red pine. *Tree Physiol.* **2003**, *23*, 273–280. [CrossRef] [PubMed]

forests

MDPI

Article

Seasonal Variation in Soil Greenhouse Gas Emissions at Three Age-Stages of Dawn Redwood (*Metasequoia glyptostroboides*) Stands in an Alluvial Island, Eastern China

Shan Yin [1,2,3,†], Xianxian Zhang [1,2,3,†], Jukka Pumpanen [4], Guangrong Shen [1,3], Feng Xiong [1] and Chunjiang Liu [1,2,3,*]

1   School of Agriculture and Biology, Research Centre for Low Carbon Agriculture, Shanghai Jiao Tong University, 800 Dongchuan Rd., Shanghai 200240, China; yinshan@sjtu.edu.cn (S.Y.); xixizi01090@163.com (X.Z.); sgrong@sjtu.edu.cn (G.S.); xiongfeng478@163.com (F.X.)
2   Shanghai Urban Forest Ecosystem Research Station, State Forestry Administration, 800 Dongchuan Rd., Shanghai 200240, China
3   Key Laboratory for Urban Agriculture (South), Ministry of Agriculture, 800 Dongchuan Rd., Shanghai 200240, China
4   Department of Environmental and Biological Sciences, University of Eastern Finland, P.O. Box 1627, Kuopio 70211, Finland; jukka.pumpanen@uef.fi
*   Correspondence: chjliu@sjtu.edu.cn; Tel./Fax: +86-21-3420-6603
†   These authors contributed equally to this work.

Academic Editors: Robert Jandl and Mirco Rodeghiero
Received: 26 August 2016; Accepted: 21 October 2016; Published: 4 November 2016

Abstract: Greenhouse gas (GHG) emissions are an important part of the carbon (C) and nitrogen (N) cycle in forest soil. However, soil greenhouse gas emissions in dawn redwood (*Metasequoia glyptostroboides*) stands of different ages are poorly understood. To elucidate the effect of plantation age and environmental factors on soil GHG emissions, we used static chamber/gas chromatography (GC) system to measure soil GHG emissions in an alluvial island in eastern China for two consecutive years. The soil was a source of CO_2 and N_2O and a sink of CH_4 with annual emissions of 5.5–7.1 Mg C ha^{-1} year^{-1}, 0.15–0.36 kg N ha^{-1} year^{-1}, and 1.7–4.5 kg C ha^{-1} year^{-1}, respectively. A clear exponential correlation was found between soil temperature and CO_2 emission, but a negative linear correlation was found between soil water content and CO_2 emission. Soil temperature had a significantly positive effect on CH_4 uptake and N_2O emission, whereas no significant correlation was found between CH_4 uptake and soil water content, and N_2O emission and soil water content. These results implied that older forest stands might cause more GHG emissions from the soil into the atmosphere because of higher litter/root biomass and soil carbon/nitrogen content compared with younger stands.

Keywords: greenhouse gas; seasonal variation; subtropical; soil temperature; soil moisture

1. Introduction

Establishment and management of forest plantations play an increasingly important role in sequestrating carbon from the atmosphere as one of the major strategies for mitigating global warming. The emissions of greenhouse gases (GHGs) are mostly related to the carbon (C) and nitrogen (N) cycle from forest soils. Forest soils are the sink of carbon in the world and contain about 704 Pg C, with varying C densities under different environmental conditions [1]. On the contrary, they are also the source of N_2O [1,2]. In some countries (e.g., China, India, Russian Fedration, US, Japan, etc.), plantations represent an important part of the national forested areas, and are increasing at the rate of 3–4.5 million hectare per year [3]. China accounts for 24% of the global forest plantations [3]. In China, the plantation area increased by 5.1 million ha per year during the period from 2004–2008 [4]; it is expected that 40 million hectares plantation will be established within the period from 2005 to 2020 [5]. To further our understanding of the patterns of C and N cycles and influential factors, we need to study the soil GHG emissions and their ability to mitigate global warming.

A large number of studies have been conducted about tropical forest soil GHG emissions. For instance, soil CO_2 emissions ranged from 1.45 t C ha^{-1} $year^{-1}$ to 13.74 t C ha^{-1} $year^{-1}$ in subtropical forests of China [6–8], to 10.80 t C ha^{-1} $year^{-1}$ to 11.75 t C ha^{-1} $year^{-1}$ in subtropical Australian rainforests [9], and 25.60 t C ha^{-1} $year^{-1}$ in tropical Thailand forests [10]. Average soil N_2O emissions varied from 1.5 kg N ha^{-1} $year^{-1}$ to 6.07 kg N ha^{-1} $year^{-1}$ in tropical forests [11–13]. Mean annual CH_4 uptake in tropical forest ecosystems ranged from 3.33 kg C ha^{-1} $year^{-1}$ to 57.49 kg C ha^{-1} $year^{-1}$ [14,15], and net CH_4 sinks in tropical Montane tree forests ranged from 0.6 kg C ha^{-1} $year^{-1}$ to 5.9 kg C ha^{-1} $year^{-1}$ in southern Ecuador [16]. These results show that there are drastic variations in GHG emissions in specific sites across different regional biomes, thereby suggesting that the pattern of GHG emissions and influential factors will need to be elucidated at specific sites in the context of considering the management of plantations as a strategy of sequestrating atmospheric CO_2.

The dynamics of soil GHG emissions in forests are influenced by key factors such as soil properties, soil temperature, soil moisture, and vegetation [15,17,18]. In previous reports, seasonal changes in soil GHG emissions were found [19,20]. Soil CO_2 and N_2O emissions both displayed an increasing trend with the progression of succession in natural forests, but no difference in CH_4 emission was observed at different succession stages [2,12]. Few reports had examined GHG emissions at differently aged stages of plantations. Dawn redwood (*Metasequoia glyptostroboides*), as a living fossil tree, is widely distributed as plantations throughout the middle and high latitudes in Eurasian and North American continents [21]. It had high natural durability under the attack of basidomycetes infection and high resistance against soft-rot fungi [22–24]. As a fast-growing species, Dawn redwood plays an important role in carbon stocks and other ecosystem services. To further understand the pattern of GHG emissions in different aged plantations and associated influential factors, soil GHG emissions were measured at 10, 1, 7 and 32 year old dawn redwood stands for two consecutive years in this study.

These are the following objectives of this study: (1) reveal the seasonal variation of soil GHG emissions at different age-stages of plantations; (2) show the relationship between the GHG emissions and soil temperature, and GHG emissions and moisture; (3) determine the relative importance of biomass, soil C and N content, soil temperature, and soil moisture on GHG emissions; and (4) understand the role of dawn redwood stand soil as the source or sink for CO_2, CH_4, and N_2O at different age stages. We hypothesized that different patterns of GHG emissions could exist in differently aged forests. This is partially due to consideration of the different assimilated products of photosynthesis, some of which are allocated into the roots within a short time period after photosynthesis, for example. As such, GHG emissions are not only affected by soil temperature but are also affected by plant photosynthesis via below-ground carbon allocation.

2. Materials and Methods

2.1. Site Description

The experimental stands are located in Dongping National Forest Park (41.68° N, 121.48° E), Chongming Island, Shanghai, China. Chongming Island, the largest alluvial island in the world, is located in the Yangtze River Estuary, which covers an area of 1267 km^2 and which currently increases at the rate of 500 ha $year^{-1}$ through Yangtze River-derived sediment [25]. During the period of 2009–2013, the mean annual temperature and precipitation of this area was 16.6 °C and 1072.3 mm, respectively [26]. Rainfall is concentrated mostly on May–September (Figure 1).

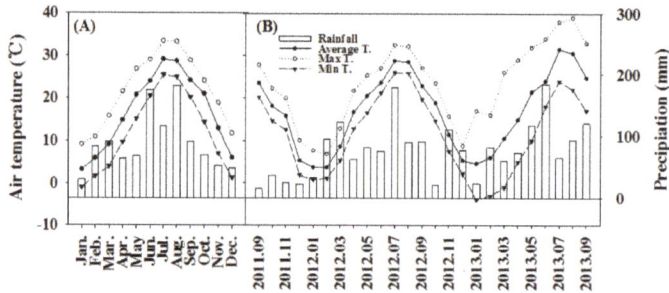

Figure 1. Monthly mean air temperature and precipitation during 2009–2013 (**A**); the monthly mean temperature and precipitation from September 2011 to September 2013 (**B**).

Dongping National Forest Park is the largest forest farm in eastern China, with 70% of the total area covered by dawn redwood plantations. Since the 1960s, plantations have been established to form different aged stands. In order to examine the effects of stand age on soil GHG emission, three different aged stands of 10, 17, and 32 years old were selected. In each stand, three plots (20 m × 20 m) were set up in August 2011 (Table 1).

Biomass carbon storage. In 2011, all trees were counted at all sites. The height of every single tree was determined by using a Haglöf Vertex III Ultrasonic Hypsometer. The diameter at breast height (1.3 m above the ground) (DBH) was measured using a measuring tape. The whole tree dry biomass was calculated by Becuwe's allometric functions ($M = 0.06291$ DBH$^{2.4841}$), and carbon stock in the stands was estimated by considering the carbon contents of tree dry biomasses (around 50%) [27].

Soil properties. To determine the bulk density, pH, total carbon (C), and nitrogen (N) concentrations of the soil in the stand, three soil samples were collected from each plot. Soil bulk density was obtained by the volumetric ring method [28]. Soil pH was measured by 1:5 dry soil: CaCl$_2$ solution (0.01 M) [29]. The total soil C and N concentrations were determined by using an elemental analysis-stable isotope ratio mass spectrometer (Vario ELIII Elementar, Hessen Langenselbold, Germany).

Table 1. Selected sites and soil characteristics for three stands in Dongping National Park, Chongming Island.

	10-Year-Old Stand	17-Year-Old Stand	32-Year-Old Stand
Tree Growth			
Tree density (stems/ha)	1050	725	550
Average height (m)	8.1 ± 1.5	16.2 ± 2.2	28.3 ± 3.4
Average DBH (cm)	10.5 ± 3.1	17.5 ± 3.4	27.2 ± 3.0
Biomass carbon stock (t ha^{-1})	13.96	29.76	64.93
Litter [a]			
Litter amount (t ha^{-1})	1.04 ± 0.008	2.67 ± 0.012	3.87 ± 0.027
Fallen leaf C (%)	47.35 ± 0.61	48.11 ± 0.32	47.67 ± 0.40
Fallen branch C (%)	44.40 ± 0.33	44.86 ± 0.32	46.07 ± 0.37
Fallen leaf N (%)	1.60 ± 0.09	1.84 ± 0.05	1.69 ± 0.08
Fallen branch N (%)	0.75 ± 0.04	0.63 ± 0.04	0.64 ± 0.05
Fallen leaf C:N ratio	29.6	26.1	28.2
Fallen branch C:N ratio	59.2	71.2	72.0
Soil Properties			
Bulk density	1.55 ± 0.01	1.62 ± 0.01	1.62 ± 0.01
pH	8.18 ± 0.068	8.19 ± 0.097	8.12 ± 0.063
Total N (%)	0.11 ± 0.014	0.19 ± 0.038	0.22 ± 0.002
SOC (%)	0.71	1.78	1.94
Total C (%)	1.40 ± 0.014	1.85 ± 0.036	2.11 ± 0.054
C:N ratio	13	10	10
Soil carbon storage (t ha^{-1})	31.87 ± 2.20	37.68 ± 1.07	40.01 ± 2.49

Note: [a] The source of litter data was Xiao's dissertation [30]. DBH, diameter at breast height; SOC, soil organic carbon.

2.2. Measurements

2.2.1. Soil Gas Emissions

Gas emission measurements were based on Forestry Standards "Observation Methodology for Long-term Forest Ecosystem Research" of PR China (LY/T 1952–2011). Because the forest sites were relatively homogeneous, three observation points were systematically arranged in each stand. The static chamber method was employed to measure soil CO_2, CH_4, and N_2O emissions. Gas emissions were measured every two weeks (September 2011–September 2013).

The static chamber consisted of two parts. First, the stainless steel based part (0.5 m × 0.5 m × 0.2 m) was permanently inserted at a 10 cm depth in the soil for each observation point of the plots, and the second upper part was made of a polyvinyl chloride plate with a size of 0.5 m × 0.5 m × 0.5 m. A fan was installed in each upper chamber for air mixing. Next, 30 min after closing the chamber, gas samples were collected with a gastight syringe (100 mL) every 10 min for the next 40 min (0, 10 min, 20 min, 30 min, and 40 min). Five gas samples at each observation point were taken between 9:00 a.m. and 12:00 a.m. and analyzed by gas chromatography (6890N, Agilent, Santa Clara, CA, USA) with an Electron Capture Detector (ECD) for N_2O detection and an Flame Ionization Detector (FID) for CH_4 and CO_2 detection [31,32]. The minimum detectable limit of CO_2, CH_4, and N_2O fluxes were 0.3 mg C m^{-2} h^{-1}, 4.4 µg C m^{-2} h^{-1}, and 0.3 µg N m^{-2} h^{-1}, respectively [33]. The gas emissions were calculated by the rate of gas concentration change during sampling. The calculation details were as follows.

$$F = \frac{dC}{dt} \times \frac{mPV}{ART} = H \times \frac{dC}{dt} \times \frac{mP}{RT} \tag{1}$$

where F is the gas emissions (mg m^{-2} h^{-1} for CO_2 and CH_4, and μg m^{-2} h^{-1} for N_2O), and $\frac{dC}{dt}$ (μL L^{-1} min^{-1} for CO_2 and CH_4, and nL L^{-1} min^{-1} for N_2O) is the emission rate of CO_2, CH_4, or N_2O concentration in the chamber. A linear regression is used to calculate the emission rate. The m (g mol^{-1}) is the molecular weight of trace gas. P indicates the atmospheric pressure ($P = 1.013 \times 10^5$ Pa). R is the gas constant ($R = 8.314$ J mol^{-1} K^{-1}). T (K) is the air temperature in the chamber. V (cm^3), H (cm), and A (cm^2) are the volume, height, and area of the static chamber, respectively.

2.2.2. Soil Temperature and Soil Water Content

The probe of digital thermometer JM 624 (Jinming Insturment Co., LTD, Tianjin, China) was inserted at 5 and 10 cm soil depth to detect the soil temperature on the outside of each chamber when we collected the gas samples. Soil samples were taken by soil auger from 0 cm to 10 cm and 10 cm to 20 cm depths to determine soil water contents gravimetrically by measuring the fresh and dry weights after drying in an oven at 105 °C for two days.

2.3. Data Analysis

Generally, the growing season of dawn redwood in Shanghai is from May to November, and the non-growing season is from December to April. We split our observed data into two parts according to the growing or non-growing season to determine whether soil respiration increases simultaneously with increasing photosynthesis.

2.3.1. Q_{10} Values

The temperature sensitivity of the soil respiration rate at the three stands was calculated by a non-linear regression model with the van't Hoff function, as follows:

$$R_S = \alpha e^{\beta T}, \tag{2}$$

where R_S is the soil respiration (mg CO_2 m^{-2} h^{-1}), α and β are fitted constants, and T is soil temperature, which was measured at 5 cm and 10 cm depths in the soil [34,35]. Q_{10} is the factor explaining the temperature sensitivity of soil respiration, and it is calculated as follows: $Q_{10} = e^{10\beta}$ [36,37].

2.3.2. The Relationship between GHG Emissions and Environmental Factors

One-sample Kolmogorov-Smirnov testing was used to determine whether the GHG emissions, soil temperature, and soil moisture were normally distributed. Soil temperature and soil moisture were normally distributed. Data variation among the sites was tested for significance by using the Duncan test following ANOVA. Pearson correlation analyses were used to analyze the relationship between greenhouse gas and the environment factors. Statistical analysis was conducted using IBM SPSS Statistics 21 software.

Canonical correspondence analysis (CCA) was conducting by using the CCA procedure in PAST 3 to detect the relationship between soil GHG emissions and environmental factors, such as soil temperature, soil water content, soil C and N concentration, and foliage C and N concentrations. A plot of the first two canonical variables (Can 1 and Can 2) was made to visually show the correlation among gases and environmental variables.

3. Results

3.1. Soil Respiration Rate

During the experimental period of 2011 to 2013, the mean CO_2 emission rate was 228.30 ± 142.40 mg m^{-2} h^{-1}, 238.14 ± 142.20 mg m^{-2} h^{-1}, and 297.71 ± 218.09 mg m^{-2} h^{-1} in the 10, 17, and 32-year-old stands, respectively (Table 2). Maximum soil CO_2 emissions were observed in May and August in every year, and the smallest emissions in January and February (Figure 2). The mean soil CO_2 emissions were 346.47 ± 164.23 mg m^{-2} h^{-1} and 117.09 ± 52.34 mg m^{-2} h^{-1} in the growing season and non-growing season, respectively (Figure 3).

Table 2. Average forest soil CO_2, CH_4, and N_2O emissions measured in the 10, 17, and 32-year-old stands during the period from 2011–2013.

	Stand Age	2011–2012	2012–2013	2011–2013
CH_4 (mg m^{-2} h^{-1})	10	-0.030 ± 0.029 b	-0.021 ± 0.016 b	-0.026 ± 0.024 b
	17	-0.035 ± 0.059 b	-0.030 ± 0.025 b	-0.032 ± 0.045 b
	32	-0.081 ± 0.093 a	-0.056 ± 0.049 a	-0.069 ± 0.075 a
CO_2 (mg m^{-2} h^{-1})	10	233.35 ± 152.28 a	223.25 ± 134.76 a	228.30 ± 142.40 a
	17	250.42 ± 146.93 a	225.86 ± 139.22 a	238.14 ± 142.20 a
	32	322.40 ± 241.16 a	273.01 ± 194.12 a	297.71 ± 218.09 a
N_2O (μg m^{-2} h^{-1})	10	7.17 ± 16.12 a	3.40 ± 6.05 a	5.29 ± 12.20 a
	17	15.79 ± 29.95 a	4.38 ± 6.68 a	10.09 ± 22.23 a
	32	15.46 ± 19.23 a	9.04 ± 7.56 b	12.25 ± 14.82 a

Note: The periods of 2011–2012 and 2012–2013 are 15 September 2011–1 September 2012 and 14 September 2012–2 September 2013, respectively. The contents in this table refer to mean average greenhouse gas emissions ± standard deviation. Different lower case letters after these contents indicate significant differences between the treatments, each with $p < 0.05$.

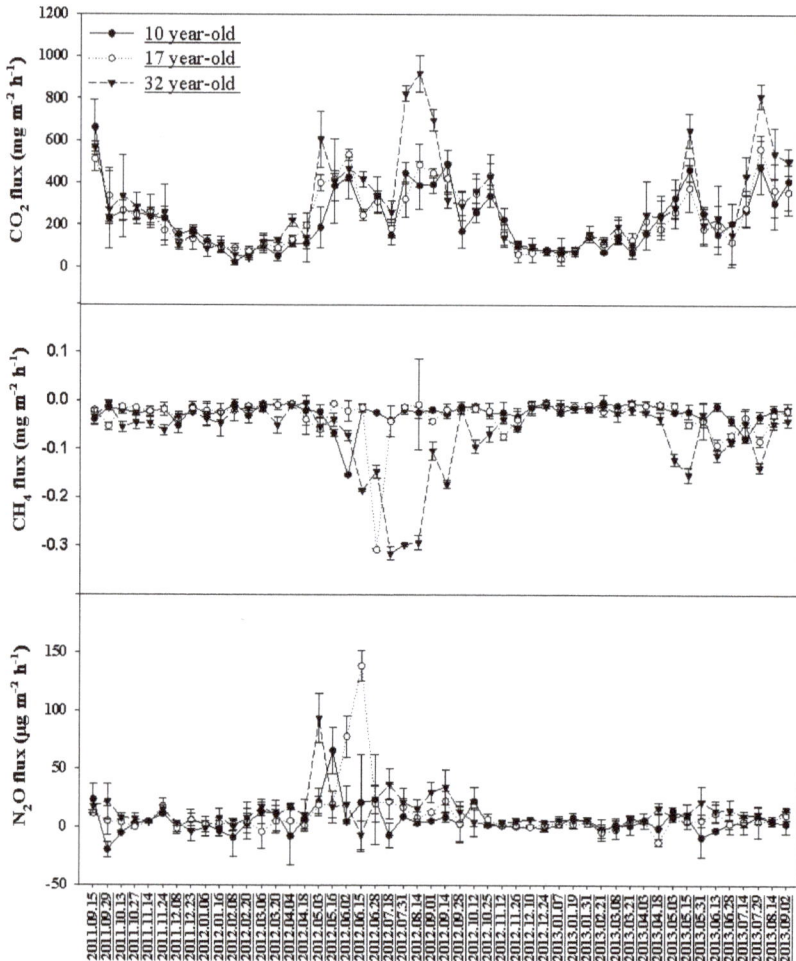

Figure 2. Soil CO_2, CH_4, and N_2O emissions measured in 10, 17, and 32-year-old stands during 2011–2013. The error bars shown in the figure are standard deviations.

3.2. Soil CH$_4$ Uptake

The soil was a sink of CH_4 in all three stands, with the highest uptake of CH_4 occurring in the summer (Figure 2). During 2011–2013, the mean soil CH_4 uptake rates were 0.026 mg m^{-2} h^{-1}, 0.032 mg m^{-2} h^{-1}, and 0.069 mg m^{-2} h^{-1} in the 10, 17, and 32-year-old stands, respectively (Table 2). The CH_4 uptake rates were significantly higher in the older stand compared to the younger stands ($p < 0.05$). The highest CH_4 uptakes were measured in the growing season (Figure 3).

3.3. Soil N$_2$O Emission

There were large differences in N_2O emissions among the three stands, ranging from −19.78 µg m^{-2} h^{-1} to 65.39 µg m^{-2} h^{-1}, −13.02 µg m^{-2} h^{-1} to 138.00 µg m^{-2} h^{-1}, and −6.98 µg m^{-2} h^{-1} to 93.45 µg m^{-2} h^{-1} in the 10, 17, and 32-year-old stands, respectively (Figure 2). The mean N_2O emissions were 5.29, 10.09, and 12.25 µg m^{-2} h^{-1}, respectively (Table 2), thereby showing that the older stand had larger N_2O emissions compared with the younger stands, but it was not significant

(p = 0.113). The N$_2$O emissions were higher during the growing season compared to the non-growing season ($p < 0.05$) (Figure 3).

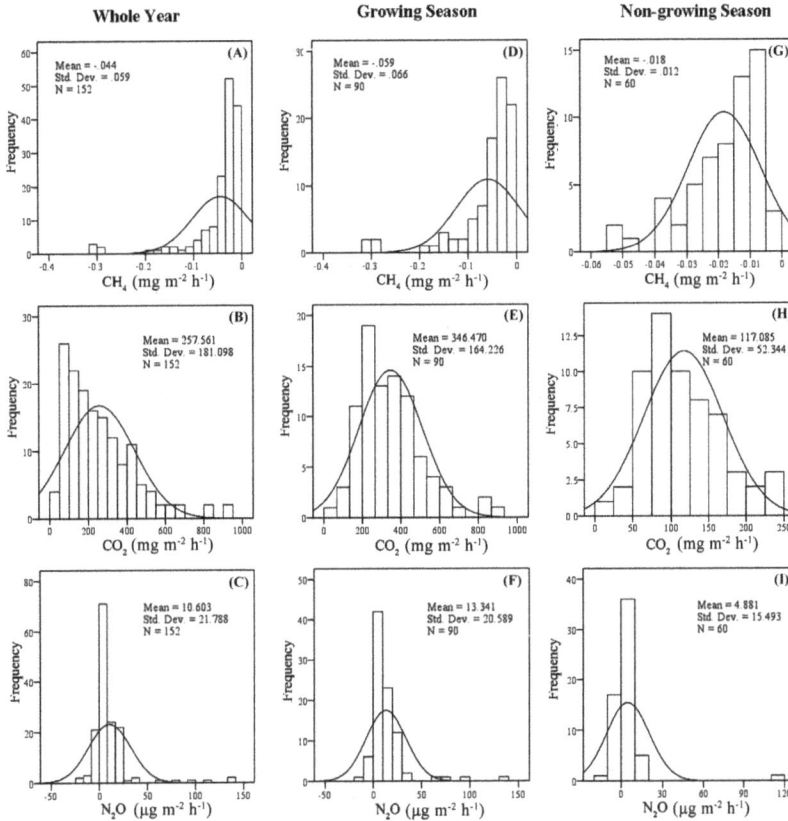

Figure 3. The frequency distribution histogram of CH$_4$, CO$_2$, and N$_2$O emissions during the whole year, growing season (from 1 May to 30 November), and non-growing season (from 1 December to 30 April), respectively. (**A**–**C**) in the upper-right corner represent the greenhouse gas (GHG) emissions during the whole year; (**D**–**F**) represent the GHG emissions during the growing season; and (**G**–**I**) represent the GHG emissions during the non-growing season.

3.4. Annual GHG Emissions

The annual CO$_2$ emissions were significantly higher in the 32-year-old stand compared to the other two younger stands ($p < 0.05$) (Figure 4). The emissions were 23.3% and 20.0% higher in the 32-year-old stand than those in the 10 and 17-year-old stands, respectively. Moreover, the annual soil CH$_4$ uptake had significant differences among the three stands. The annual CH$_4$ uptake was highest in the 32-year-old stand and lowest in the 10-year-old stand.

The highest annual soil N$_2$O emission was observed in the 32-year-old stand and we noted that the 32-year-old stand had a 56.8% higher annual N$_2$O emission than the 10-year-old stand and a 17.7% higher annual emission than the 17-year-old stand. However, the N$_2$O emissions among the three stands were not significantly different.

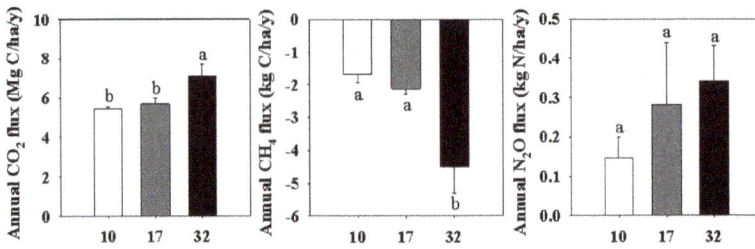

Figure 4. Annual cumulative CO_2, CH_4, and N_2O emissions measured during 2011 to 2013. Symbols on the x-axis (10, 17, and 32) mean the 10-year-old, 17-year-old, and 32-year-old stands. (Error bars in the figures means standard error, and different lower case letters indicate significant differences between the treatments, each with $p < 0.05$).

3.5. The Effect of Soil Temperature on GHG Emissions

In this research, soil CO_2 emissions increased exponentially with soil temperature both at 5 cm and at 10 cm soil depths ($R_S = 62.78e^{0.075T}$ at 5 cm soil depth, and $R_S = 61.89e^{0.077T}$ at 10 cm soil depth). The exponential model could explain 68% or 69% ($p < 0.001$) of the seasonal variation in soil CO_2 emissions (Table 3). The Q_{10} values were calculated to be 2.12 and 2.15 at 5 cm and at 10 cm soil depths, respectively (Table 3). Usually, Q_{10}-values were almost 3%–51% higher in the non-growing season than in the growing season.

Table 3. Parameters of the exponential model for soil CO_2 emissions as a function of soil temperature at 5 and 10 cm depths in the three stands.

Sites		10-Year-Old Stand		17-Year-Old Stand		32-Year-Old Stand		Three Stands	
Soil Depth (cm)		5	10	5	10	5	10	5	10
Whole Year	R^2	0.58	0.65	0.72	0.66	0.79	0.78	0.68	0.69
	α	67.61	59.71	62.01	65.99	58.23	60.60	62.78	61.89
	β	0.0659	0.0752	0.0732	0.0689	0.0872	0.0856	0.0752	0.0767
	Q_{10}	1.93	2.12	2.08	1.99	2.39	2.35	2.12	2.15
Growing Season	R^2	0.24	0.23	0.36	0.23	0.52	0.51	0.36	0.69
	α	132.36	129.40	89.30	113.75	77.02	84.09	96.82	61.89
	β	0.0383	0.0397	0.0569	0.0448	0.0746	0.0708	0.0568	0.0767
	Q_{10}	1.47	1.49	1.77	1.57	2.11	2.03	1.76	2.15
Non-growing Season	R^2	0.18	0.33	0.59	0.54	0.61	0.57	0.35	0.69
	α	71.56	51.86	55.85	55.52	53.15	52.70	65.24	61.89
	β	0.036	0.0811	0.0808	0.0824	0.0947	0.0986	0.0599	0.0767
	Q_{10}	1.43	2.25	2.24	2.28	2.58	2.68	1.82	2.15

CH_4 uptakes and N_2O emissions were significantly correlated with soil temperature at both 5 cm and 10 cm depths. There was a positive correlation between the CH_4 uptake and soil temperature (Pearson correlation, −0.3). In addition, N_2O and soil temperature had a positive correlation (Pearson Correlation, 0.3) (shown in Table 4).

Table 4. Pearson correlation coefficients between greenhouse gas and soil temperature and water content.

	CH$_4$	CO$_2$	N$_2$O	T 5 cm	T 10 cm	SWC 0–10 cm	SWC 10–20 cm
CH$_4$	1.000	−0.377 **	−0.041	−0.301 **	−0.317 **	−0.012	0.169
CO$_2$		1.000	0.380 **	0.765 **	0.776 **	−0.211 *	−0.276 **
N$_2$O			1.000	0.274 **	0.274 **	0.141	−0.047
T 5 cm				1.000	0.972 **	−0.319 **	−0.364 **
T 10 cm					1.000	−0.324 **	−0.385 **
SWC 0–10 cm						1.000	0.671 **
SWC 10–20 cm							1.000

Note: ** Correlation is significant at the 0.01 level (2-tailed). * Correlation is significant at the 0.05 level (2-tailed). T 5 cm and T 10 cm mean soil temperature at 5 cm soil depth and at 10 cm soil depth, respectively. SWC 0–10 cm and SWC 10–20 cm mean soil water content at 0–10 cm soil depth and at 10–20 cm soil depth, respectively.

3.6. Effects of Soil Water Content on GHG Emissions

Soil water content contributed substantially to the GHG emissions. The relationship between soil CO$_2$ emissions and soil water content at both 0–10 cm and 10–20 cm depths was negative. However, no significant relationship was found between CH$_4$ emission and soil water content, or N$_2$O emission and soil water content. (Table 4).

3.7. The Main Influencing Factors of Soil Greenhouse Gas Emissions

The variations in vegetation carbon, nitrogen, and soil properties were described by two significant canonical components (explaining 100% of the variance) (Figure 5). The first, Can 1, accounted for 98.65% of the total variance and was highly related to the trees' biomass, and C and N content in soil and foliage. Can 2 accounted for 1.21% of the total variance with close correlation among soil water content and soil temperature. The CO$_2$ and N$_2$O emissions, and CH$_4$ uptake all have positive correlations with Can 1 and negative correlations with Can 2.

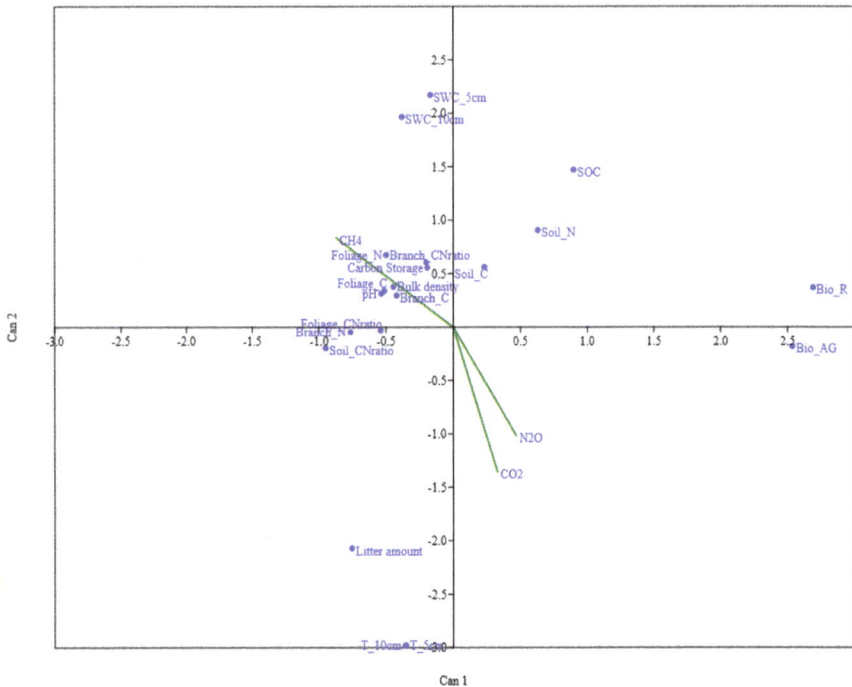

Figure 5. GHG emissions defined by the first two canonical variables (Can 1 and Can 2) extracted from the canonical correspondence analysis (CCA). In this plot, the position of points relative to the direction of vectors approximates correlations between soil GHG emissions and environmental factors. Vector length indicates the overall contribution of the variables to the analysis, and vector direction indicates the correlation of the variables with each axis.

4. Discussion

4.1. Soil Carbon Dynamic in Different-Age Stands

The soil was a source of CO_2 and sink of CH_4 in the three stands in both growing and non-growing seasons. The annual soil CO_2 emissions (5.5–7.1 Mg C ha^{-1} year^{-1}) were within the same range observed in other subtropical forests. For instance, annual soil CO_2 emission was 3.1–7.3 Mg C ha^{-1} year^{-1} in the seasonal tropical primary forests in Xishuangbanna region, southwest China, and from 3.1–7.3 to 11.1–12.0 Mg C ha^{-1} year^{-1} in the subtropical forests [9,38]. In subtropical and tropical forests, annual soil CH_4 uptake rates ranged from 0.8 kg C ha^{-1} year^{-1} to 4.3 kg C ha^{-1} year^{-1} [12,16,39]. Our study showed a similar uptake (1.7 kg C ha^{-1} year^{-1} to 4.5 kg C ha^{-1} year^{-1}) in plantations located in northern subtropical areas, thereby suggesting that annual CH_4 uptake does not significantly vary with subtropical or tropical biomes.

Soil CO_2 significantly varied with soil temperature and water content in the three stands in both growing and non-growing seasons. A positive relationship existed between soil temperature and CO_2 emission in these three stands, and a negative relationship was found between soil water content and CO_2 emission. The effects of soil temperature and soil water content on CO_2 emissions were statistically confounded. As such, we excluded the soil temperature effect through normalizing the soil respiration values with $R_S = 62.78e^{0.075T}$ at 5 cm soil depth and $R_S = 61.89e^{0.077T}$ at 10 cm soil depth, and found that the effect of soil water content on CO_2 emissions was not significantly negative (with Pearson correlation from −0.18 to −0.19). Respiration rates generally decreased with decreasing water content. Soil temperature was probably the key factor regulating soil respiration. However, soil water content also restricted soil respiration [40]. Both soil CO_2 emission and CH_4 uptake peaked in the period of May–November because of the wet-hot climate. The laboratory and field studies have verified that soil temperature and soil water content could account for most of the seasonal variation in soil CO_2 emission and CH_4 uptake [40–42].

Soil temperature and water content explained 76%–87% of soil CO_2 emission and 67%–75% of total annual emission in the wet season (April to September) of lower subtropical forests [6]. Q_{10}, an exponential relationship, has been commonly used to estimate soil respiration rates from soil temperature [36]. In previous literature, the mean Q_{10} values were 2.14 for tropical regions and 2.26 for temperate regions [43]. In our study, Q_{10} ranged from 1.9 to 2.4 during the whole year, and soil respiration in the non-growing season was more sensitive to soil temperature. The higher Q_{10} in the non-growing season could be associated with the phonological cycle of photosynthesis as compared to the growing season, which has consequences on the belowground carbon allocation. In the summer, about 50% or more of the soil CO_2 emissions could be originated from recently assimilated C, which trees allocate to the belowground system (root and rhizosphere) [44]. The values of Q_{10} increased with soil depth, and this result was the same as that obtained by Pavelka [45]. The seasonal variation in soil temperature was lower in the deeper layers and soil respiration rate was relatively more sensitive to temperature fluctuations [46]. During the growing and non-growing seasons the different values of Q_{10} were noted with different R^2 values, and the lower R^2 values were calculated in the growing season. During the growing season, soil temperature causes little changes in soil CO_2 emissions. The primary reason might be the low temperature amplitude during the growing season. Second, the other factors (except soil temperature) could explain the soil CO_2 emission such as the changes in photosynthesis and precipitation.

The soil temperature positively affected CH_4 uptake, and no significant relationship existed between CH_4 uptake and soil water content. Kiese and Werner observed that CH_4 uptake was negatively correlated with soil temperature and soil water content [38,39]. In mid-subtropical China, the highest CH_4 uptake (17.12 g C ha^{-1} day^{-1}) occurred in the summer-autumn season with increasing soil temperature and water content, but the relationships between CH_4 uptake and soil temperature and CH_4 uptake and soil water content were not significant [47]. In earlier studies, CH_4 uptake had decreased with increasing soil water content during the summer season [48,49]. Maximum CH_4 uptake rate was clearly associated with the lowest soil moisture and the highest soil temperature both in temperate and tropical forests [50]. Before oxidization by methanotrophs, the soil CH_4 was emitted from anaerobic environments to the atmosphere. In the forest's soil, a certain amount of CH_4 from the atmosphere was consumed by methanotrophs [51]. The optimum conditions for growth of methanotrophic bacteria and induction of methane oxidation activity were 20%–35% water contents and 25 °C–35 °C temperatures [52]. In our study, the water content ranged from 11% to 33%, which was almost in the optimum range, and temperatures showed a larger range from 1.4 °C to 30 °C. Soil temperature could be more important than water content in regulating CH_4 consumption in this study, which is in agreement with the results of previous reports [53].

4.2. Soil Nitrogen Dynamic in Different-Aged Stands

We observed highly dynamic N_2O emissions with low values in our study (i.e., 0.81–1.87 g N ha^{-1} day^{-1}), which were lower than some previously reported emissions. For example, our results are similar to the N_2O emissions from undrained forests in southern Sweden (i.e., 1.62 g N ha^{-1} day^{-1}) [54], but they are substantially lower than the 8.77 g N ha^{-1} day^{-1} previously recorded in the subtropical forest in southern China [12].

A seasonal variation in N_2O emissions has been reported in tropical and subtropical forests. For instance, the highest N_2O emissions have been observed during the spring and summer months with mean values of 2–5 g N ha^{-1} day^{-1}. The lowest emissions were obtained during winter seasons, with less than 0.5 g N ha^{-1} day^{-1} [9]. The higher N_2O emissions were emitted from temperate and tropical forest ecosystems during the wet and hot season [50]. The magnitude of

N_2O emissions was very closely linked to rainfall events [55]. The soil N_2O was produced by microbes through nitrification in aerobic conditions and through denitrification under anaerobic conditions [56]. Factor, such as precipitation, was observed to exert some influence on the soil aeration, but soil aeration could affect N_2O production. In our study, the highest soil N_2O emissions were observed between May and November when higher rainfall occurred with a mean value of 2.04 g N ha^{-1} day^{-1}. The lowest soil N_2O emissions were recorded between December and April with a mean value of 0.75 g N ha^{-1} day^{-1}.

N_2O emissions showed a positive correlation with soil temperature; no significant correlation with soil water content was observed, which was similar to a previous study in Japan [57]. However, some previous reports have shown that N_2O emissions have a positive correlation with soil temperature and soil water content [42].

4.3. Factors Affecting Soil Greenhouse Gas Emissions

The present study showed that soil GHG emissions differed among the three stands. The 32-year-old stand had significantly higher CO_2 emissions, CH_4 uptake, and N_2O emissions than the 10 and 17-year-old stands. Basically, these three stands differed in biomass/litter carbon storage, nitrogen content, and soil properties. The soil CO_2, N_2O, and CH_4 were produced by microbial activity, and these processes were controlled by environmental factors [58,59].

Forest soil CO_2 emissions were the sum of heterotrophic (microbes) and autotrophic respiration (roots), and the contribution of root respiration rates which were higher during the growing season [60]. The soil CO_2 emissions were a good indicator of total below-ground allocation of carbon and of ecosystem productivity. Among these stands, the older stands maintained higher productivity than the younger stands; it was not surprising that the older stand had the highest rates of soil respiration. Older stands released higher CO_2, and the major difference was that the older stand had higher soil carbon, which could probably reflect higher root and litter carbon storage [61]. The research in Loess Plateau of China [62] indicated that 48% of the variations in annual soil CO_2 emissions were explained by the combined carbon stock in top soil and litter, 77% by the root carbon stock, and 63% by the combined carbon stock in roots, litter, and top soil. The aboveground litter mineralization and decomposition contributed to about 8% of the soil respiration in a subtropical Montane cloud forest in Taiwan [63]. In our study, the total carbon storage of litter, soil, and roots in the older stands was higher than the two younger stands, which indicated higher annual CO_2 emissions in the older stands. Based on the principal component analyses, the litter composition was an important stimulator for soil CO_2 emissions because of the simultaneous effects on production and consumption of the soil surface organic matter [64].

Methane emissions of soils were correlated with microbial activities, and the upper soil layer were generally CH_4 sinks [65]. The rate of CH_4 uptake was regulated by the soil C and N levels as well as soil water content, and there was a close link between labile C, N, and CH_4 uptake in forest soils [66,67]. This research has shown that carbon and nitrogen contents of litter, soil, and root in older stands were higher than in younger stands, which indicates higher annual CH_4 uptake in older stands.

In contrast to the pattern of soil CO_2 and CH_4 emissions, no distinctly different trend in N_2O emission was observed among differently aged stands. According to the reported study, soil N_2O production and consumption were mainly influenced by the amount of mineral N in soils, and low N availability was linked with N_2O emissions [2]. Highly dynamic emissions of N_2O were found among different forest soil types [68]. The primary controlling factors of N_2O production were found to be soil pH and C/N ratio, and these soil properties could explain most of the variability of N_2O emissions [9,69]. However, we used three stands in our study but the results indicating similar annual N_2O emissions despite the different soil properties.

5. Conclusions

Soil respiration in each of the stands was strongly and positively related to soil temperature, and negatively related to soil water content. The soil CH_4 uptake was positively related to soil temperature, and soil N_2O emission had a positive relation with soil temperature. Affected by the annual climatic conditions (e.g., temperature and precipitation), soil respiration showed a clear seasonal variation, with high emissions in the wet-hot season (from May to November) and low emissions in the dry-cool season (from December to April).

Different stages of forest stands strongly affected soil respiration and CH_4 emission rates through root respiration and/or microbial activities, but had no significant relationship with soil N_2O emission. Carbon storage, nitrogen, and C/N ratio (soil, litter, and root) were the main factors affecting CH_4 uptake and N_2O emission. Soil properties such as soil water content and soil pH were important indicators for soil respiration.

Acknowledgments: We thank Umair Muhammad for his language assistance. This research was co-funded by the National Natural Science Foundation of China (31400605), National Key Technology R&D Program of China (2013BAD11B01), the National Natural Science Foundation of China (71333010), SJTU Agri-X Program (2014007), Shanghai Landscaping Administration Bureau (G141207), CFERN & GENE Award Funds on Ecological Paper, and the Strategic Priority Research Program of the Chinese Academy of Sciences (XDA05050200).

Author Contributions: S.Y., G.S., and C.L. conceived and designed the experiments; S.Y., X.Z., and F.X. performed the experiments; X.Z. analyzed the data; S.Y., X.Z., and J.P. wrote this manuscript.

Conflicts of Interest: The authors declare no conflict of interest.

References

1. Lal, R. Forest soils and carbon sequestration. *For. Ecol. Manag.* **2005**, *220*, 242–258. [CrossRef]
2. Chapuis-Lardy, L.; Wrage, N.; Metay, A.; Chotte, J.; Bernoux, M. Soils, a sink for N_2O? A review. *Glob. Chang. Biol.* **2007**, *13*, 1–17. [CrossRef]
3. Carle, J.; Vuorinen, P.; Del Lungo, A. Status and trends in global forest plantation development. *For. Prod. J.* **2002**, *52*, 12–23.
4. Dai, L.; Wang, Y.; Su, D.; Zhou, L.; Yu, D.; Lewis, B.J.; Qi, L. Major forest types and the evolution of sustainable forestry in China. *Environ. Manag.* **2011**, *48*, 1066–1078. [CrossRef] [PubMed]
5. Central Committee of the Communist Party of China and China State Council. *The Programming Outline for the Protection and Utilization of National Forestland (2010–2020)*; Central Committee of the Communist Party of China and China State Council: Beijing, China, 2010.
6. Zhang, D.; Sun, X.; Zhou, G.; Yan, J.; Wang, Y.; Liu, S.; Zhou, C.; Liu, J.; Tang, X.; Li, J. Seasonal dynamics of soil CO_2 effluxes with responses to environmental factors in lower subtropical forests of China. *Sci. China Ser. D Earth Sci.* **2006**, *49*, 139–149. [CrossRef]
7. Yang, Y.; Chen, G.; Guo, J.; Xie, J.; Wang, X. Soil respiration and carbon balance in a subtropical native forest and two managed plantations. *Plant Ecol.* **2007**, *193*, 71–84. [CrossRef]
8. Iqbal, J.; Ronggui, H.; Lijun, D.; Lan, L.; Shan, L.; Tao, C.; Leilei, R. Differences in soil CO_2 flux between different land use types in mid-subtropical China. *Soil Biol. Biochem.* **2008**, *40*, 2324–2333. [CrossRef]
9. Rowlings, D.; Grace, P.; Kiese, R.; Weier, K. Environmental factors controlling temporal and spatial variability in the soil-atmosphere exchange of CO_2, CH_4 and N_2O from an Australian subtropical rainforest. *Glob. Chang. Biol.* **2012**, *18*, 726–738. [CrossRef]
10. Hashimoto, S.; Tanaka, N.; Suzuki, M.; Inoue, A.; Takizawa, H.; Kosaka, I.; Tanaka, K.; Tantasirin, C.; Tangtham, N. Soil respiration and soil CO_2 concentration in a tropical forest, Thailand. *J. For. Res.* **2004**, *9*, 75–79. [CrossRef]
11. Davidson, E.; Ishida, F.; Nepstad, D. Effects of an experimental drought on soil emissions of carbon dioxide, methane, nitrous oxide, and nitric oxide in a moist tropical forest. *Glob. Chang. Biol.* **2004**, *10*, 718–730. [CrossRef]
12. Tang, X.; Liu, S.; Zhou, G.; Zhang, D.; Zhou, C. Soil-atmospheric exchange of CO_2, CH_4, and N_2O in three subtropical forest ecosystems in southern China. *Glob. Chang. Biol.* **2006**, *12*, 546–560. [CrossRef]

13. Yan, J.; Zhang, W.; Wang, K.; Qin, F.; Wang, W.; Dai, H.; Li, P. Responses of CO_2, N_2O and CH_4 fluxes between atmosphere and forest soil to changes in multiple environmental conditions. *Glob. Chang. Biol.* **2014**, *20*, 300–312. [CrossRef] [PubMed]

14. Dutaur, L.; Verchot, L.V. A global inventory of the soil CH_4 sink. *Glob. Biogeochem. Cycles* **2007**, *21*, 7949–7950. [CrossRef]

15. Megonigal, J.P.; Guenther, A.B. Methane emissions from upland forest soils and vegetation. *Tree Physiol.* **2008**, *28*, 491–498. [CrossRef] [PubMed]

16. Wolf, K.; Flessa, H.; Veldkamp, E. Atmospheric methane uptake by tropical montane forest soils and the contribution of organic layers. *Biogeochemistry* **2012**, *111*, 469–483. [CrossRef]

17. Smith, K.; Ball, T.; Conen, F.; Dobbie, K.; Massheder, J.; Rey, A. Exchange of greenhouse gases between soil and atmosphere: Interactions of soil physical factors and biological processes. *Eur. J. Soil Sci.* **2003**, *54*, 779–791. [CrossRef]

18. Yan, Y.; Sha, L.; Cao, M.; Zheng, Z.; Tang, J.; Wang, Y.; Zhang, Y.; Wang, R.; Liu, G.; Wang, Y. Fluxes of ch_4 and n_2o from soil under a tropical seasonal rain forest in xishuangbanna, Southwest China. *J. Environ. Sci.* **2008**, *20*, 207–215. [CrossRef]

19. Moore, K.; Fitzjarrald, D.; Sakai, R.; Goulden, M.; Munger, J.; Wofsy, S. Seasonal variation in radiative and turbulent exchange at a deciduous forest in central massachusetts. *J. Appl. Meteorol.* **1996**, *35*, 122–134. [CrossRef]

20. Livesley, S.; Grover, S.; Hutley, L.; Jamali, H.; Butterbach-Bahl, K.; Fest, B.; Beringer, J.; Arndt, S. Seasonal variation and fire effects on CH_4, N_2O and CO_2 exchange in savanna soils of northern Australia. *Agric. For. Meteorol.* **2011**, *151*, 1440–1452. [CrossRef]

21. LePage, B.; Williams, C.; Yang, H. *The Geobiology and Ecology of Metasequoia*; Springer: Dordrecht, The Netherlands, 2005; Volume 22.

22. Chu, K.; Cooper, W. An ecological reconnaissance in the native home of *Metasequoia glyptostroboides*. *Ecology* **1950**, *31*, 260–278. [CrossRef]

23. Bajpai, V.; Rahman, A.; Kang, S. Chemical composition and anti-fungal properties of the essential oil and crude extracts of *Metasequoia glyptostroboides* miki ex hu. *Ind. Crops Prod.* **2007**, *26*, 28–35. [CrossRef]

24. Polman, J.; Michon, S.; Militz, H.; Helmink, A. The wood of *Metasequoia glyptostroboides* (hu et cheng) of dutch origin. *Holz als Roh-und Werkstoff* **1999**, *57*, 215–221. [CrossRef]

25. Bureau C.S.S. Chapter 1 Comprehensive. In *Chongming Statistical Yearbook*; Shanghai Statistics press: Chongming, Shanghai, China, 2016.

26. SMC. Shanghai Meteorological data. Available online: http://www.smb.gov.cn/ (accessed on 27 Feberury 2014).

27. Xiaver, B. *Allometric Estimation of the Aboveground Biomass and Carbon in Metasequoia Glyptostroboide Plantations in Shanghai*; Cranfield University: Bedfordshire, UK, 2009.

28. Pedrotti, A.; Pauletto, E.; Crestana, S.; Holanda, F.; Cruvinel, P.; Vaz, C. Evaluation of bulk density of albaqualf soil under different tillage systems using the volumetric ring and computerized tomography methods. *Soil Tillage Res.* **2005**, *80*, 115–123. [CrossRef]

29. Schofield, R.; Taylor, A. The measurement of soil pH. *Soil Sci. Soc. Am. J.* **1955**, *19*, 164–167. [CrossRef]

30. Xiao, C. Characteristics and Carbon-Storage Estimation of *Metasequoia glyptostroboides* Plantation Ecosystems at Different Age Stages in Chongming Island. Master' Dissertation, Shanghai Jiao Tong University, Shanghai, China, 2010.

31. Wang, Y.; Wang, Y.; Ling, H. A new carrier gas type for accurate measurement of N_2O by GC-ECD. *Adv. Atmos. Sci.* **2010**, *27*, 1322–1330. [CrossRef]

32. Zheng, X.; Mei, B.; Wang, Y.; Xie, B.; Wang, Y.; Dong, H.; Xu, H.; Chen, G.; Cai, Z.; Yue, J. Quantification of n_2o fluxes from soil–plant systems may be biased by the applied gas chromatograph methodology. *Plant Soil* **2008**, *311*, 211–234. [CrossRef]

33. Yao, Z.; Wolf, B.; Chen, W.; Butterbach-Bahl, K.; Brüggemann, N.; Wiesmeier, M.; Dannenmann, M.; Blank, B.; Zheng, X. Spatial variability of N_2O, CH_4 and CO_2 fluxes within the xilin river catchment of inner mongolia, China: A soil core study. *Plant Soil* **2010**, *331*, 341–359. [CrossRef]

34. Hoff, J. *Lectures on Theoretical and Physical Chemistry*; Edward Arnold: London, UK, 1899.

35. Davidson, E.; Janssens, I. Temperature sensitivity of soil carbon decomposition and feedbacks to climate change. *Nature* **2006**, *440*, 165–173. [CrossRef] [PubMed]

36. Janssens, I.; Carrara, A.; Ceulemans, R. Annual Q_{10} of soil respiration reflects plant phenological patterns as well as temperature sensitivity. *Glob. Chang. Biol.* **2004**, *10*, 161–169.

37. Zimmermann, M.; Bird, M.; Zechmeister-Boltenstern, S. Q_{10} values of tropical forest soil respiration increases with ongoing decomposition. In Proceedings of the EGU General Assembly, Vienna, Austria, 22–27 April 2012; p. 4347.

38. Werner, C.; Zheng, X.; Tang, J.; Xie, B.; Liu, C.; Kiese, R.; Butterbach-Bahl, K. N_2O, CH_4 and CO_2 emissions from seasonal tropical rainforests and a rubber plantation in Southwest China. *Plant Soil* **2006**, *289*, 335–353. [CrossRef]

39. Kiese, R.; Hewett, B.; Graham, A.; Butterbach-Bahl, K. Seasonal variability of N_2O emissions and CH_4 uptake by tropical rainforest soils of Queensland, Australia. *Glob. Biogeochem. Cycles* **2003**, *17*, 469–474. [CrossRef]

40. Davidson, E.; Belk, E.; Boone, R. Soil water content and temperature as independent or confounded factors controlling soil respiration in a temperate mixed hardwood forest. *Glob. Chang. Biol.* **1998**, *4*, 217–227. [CrossRef]

41. Bowden, R.; Newkirk, K.; Rullo, G. Carbon dioxide and methane fluxes by a forest soil under laboratory-controlled moisture and temperature conditions. *Soil Biol. Biochem.* **1998**, *30*, 1591–1597. [CrossRef]

42. Schaufler, G.; Kitzler, B.; Schindlbacher, A.; Skiba, U.; Sutton, M.; Zechmeister-Boltenstern, S. Greenhouse gas emissions from european soils under different land use: Effects of soil moisture and temperature. *Eur. J. Soil Sci.* **2010**, *61*, 683–696. [CrossRef]

43. Tjoelker, M.; Oleksyn, J.; Reich, P. Modelling respiration of vegetation: Evidence for a general temperature-dependent Q_{10}. *Glob. Chang. Biol.* **2001**, *7*, 223–230. [CrossRef]

44. Högberg, P.; Read, D. Towards a more plant physiological perspective on soil ecology. *Trends Ecol. Evol.* **2006**, *21*, 548–554. [CrossRef] [PubMed]

45. Pavelka, M.; Acosta, M.; Marek, M.; Kutsch, W.; Janous, D. Dependence of the Q_{10} values on the depth of the soil temperature measuring point. *Plant Soil* **2007**, *292*, 171–179. [CrossRef]

46. Lloyd, J.; Taylor, J. On the temperature dependence of soil respiration. *Funct. Ecol.* **1994**, *8*, 315–323. [CrossRef]

47. Chen, C.; Yang, Z.; Xie, J.; Liu, X.; Zhong, X. Seasonal variations of soil CH_4 uptake rate in castanopsis carlesii forest in mid-subtropical China. *J. Appl. Ecol.* **2012**, *23*, 17–22.

48. Adamsen, A.; King, G. Methane consumption in temperate and subarctic forest soils: Rates, vertical zonation, and responses to water and nitrogen. *Appl. Environ. Microbiol.* **1993**, *59*, 485–490. [PubMed]

49. Singh, J.; Singh, S.; Raghubanshi, A.; Singh, S.; Kashyap, A.; Reddy, V. Effect of soil nitrogen, carbon and moisture on methane uptake by dry tropical forest soils. *Plant Soil* **1997**, *196*, 115–121. [CrossRef]

50. Luo, G.; Kiese, R.; Wolf, B.; Butterbach-Bahl, K. Effects of soil temperature and moisture on methane uptake and nitrous oxide emissions across three different ecosystem types. *Biogeosciences* **2013**, *10*, 3205–3219. [CrossRef]

51. Yavitt, J.; Downey, D.; Lang, G.; Sexston, A. Methane consumption in two temperate forest soils. *Biogeochemistry* **1990**, *9*, 39–52. [CrossRef]

52. Hanson, R.; Hanson, T.E. Methanotrophic bacteria. *Microbiol. Rev.* **1996**, *60*, 439–471. [PubMed]

53. Whalen, S.; Reeburgh, W. Moisture and temperature sensitivity of CH_4 oxidation in boreal soils. *Soil Biol. Biochem.* **1996**, *28*, 1271–1281. [CrossRef]

54. Von Arnold, K.; Nilsson, M.; Hånell, B.; Weslien, P.; Klemedtsson, L. Fluxes of CO_2, CH_4 and N_2O from drained organic soils in deciduous forests. *Soil Biol. Biochem.* **2005**, *37*, 1059–1071. [CrossRef]

55. Breuer, L.; Papen, H.; Butterbach-Bahl, K. N_2O emission from tropical forest soils of Australia. *J. Geophys. Res. Atmos.* **2000**, *105*, 26353–26367. [CrossRef]

56. Pihlatie, M.; Syväsalo, E.; Simojoki, A.; Esala, M.; Regina, K. Contribution of nitrification and denitrification to N_2O production in peat, clay and loamy sand soils under different soil moisture conditions. *Nutr. Cycl. Agroecosyst.* **2004**, *70*, 135–141. [CrossRef]

57. Morishita, T.; Aizawa, S.; Yoshinaga, S.; Kaneko, S. Seasonal change in N_2O flux from forest soils in a forest catchment in japan. *J. For. Res.* **2011**, *16*, 386–393. [CrossRef]

58. Gundersen, P.; Christiansen, J.; Alberti, G.; Brüggemann, N.; Castaldi, S.; Gasche, R.; Kitzler, B.; Klemedtsson, L.; Lobo-do-Vale, R.; Moldan, F. The response of methane and nitrous oxide fluxes to forest change in Europe. *Biogeosciences* **2012**, *9*, 3999–4012. [CrossRef]

59. Wang, H.; Liu, S.; Wang, J.; Shi, Z.; Lu, L.; Zeng, J.; Ming, A.; Tang, J.; Yu, H. Effects of tree species mixture on soil organic carbon stocks and greenhouse gas fluxes in subtropical plantations in China. *For. Ecol. Manag.* **2013**, *300*, 4–13. [CrossRef]

60. Hanson, P.; Edwards, N.; Garten, C.; Andrews, J. Separating root and soil microbial contributions to soil respiration: A review of methods and observations. *Biogeochemistry* **2000**, *48*, 115–146. [CrossRef]

61. Schlesinger, W.H.; Lichter, J. Limited carbon storage in soil and litter of experimental forest plots under increased atmospheric CO_2. *Nature* **2001**, *411*, 466–469. [CrossRef] [PubMed]

62. Zhou, Z.; Zhang, Z.; Zha, T.; Luo, Z.; Zheng, J.; Sun, O. Predicting soil respiration using carbon stock in roots, litter and soil organic matter in forests of Loess Plateau in China. *Soil Biol. Biochem.* **2013**, *57*, 135–143. [CrossRef]

63. Chang, S.; Tseng, K.; Hsia, Y.; Wang, C.; Wu, J. Soil respiration in a subtropical montane cloud forest in taiwan. *Agric. For. Meteorol.* **2008**, *148*, 788–798. [CrossRef]

64. Raich, J.; Tufekciogul, A. Vegetation and soil respiration: Correlations and controls. *Biogeochemistry* **2000**, *48*, 71–90. [CrossRef]

65. Le Mer, J.; Roger, P. Production, oxidation, emission and consumption of methane by soils: A review. *Eur. J. Soil Biol.* **2001**, *37*, 25–50. [CrossRef]

66. Koehler, B.; Corre, M.; Steger, K.; Well, R.; Zehe, E.; Sueta, J.; Veldkamp, E. An in-depth look into a tropical lowland forest soil: Nitrogen-addition effects on the contents of N_2O, CO_2 and CH_4 and N_2O isotopic signatures down to 2-m depth. *Biogeochemistry* **2012**, *111*, 695–713. [CrossRef]

67. Fender, A.; Pfeiffer, B.; Gansert, D.; Leuschner, C.; Daniel, R.; Jungkunst, H. The inhibiting effect of nitrate fertilisation on methane uptake of a temperate forest soil is influenced by labile carbon. *Biol. Fertil. Soils* **2012**, *48*, 621–631. [CrossRef]

68. Vanitchung, S.; Conrad, R.; Harvey, N.; Chidthaisong, A. Fluxes and production pathways of nitrous oxide in different types of tropical forest soils in Thailand. *Soil Sci. Plant Nutr.* **2011**, *57*, 650–658. [CrossRef]

69. Heil, J.; Liu, S.; Vereecken, H.; Brüggemann, N. Mechanisms of inorganic nitrous oxide production in soils during nitrification and their dependence on soil properties. In Proceedings of the EGU General Assembly, Vienna, Austria, 27 April–2 May 2014; p. 4208.

![forests logo] *forests*

Article

Heterotrophic Soil Respiration Affected by Compound Fertilizer Types in Red Pine (*Pinus densiflora* S. et Z.) Stands of Korea

Jaeyeob Jeong [1,2], Nanthi Bolan [3] and Choonsig Kim [4,*]

1   Centre for Environmental Risk Assessment and Remediation, CRC-CARE, University of South Australia, Adelaide 5095, SA, Australia; jeojy003@korea.kr
2   Forest Practice Research Center, National Institute of Forest Science, Pocheon 11186, Korea
3   Global Centre for Environmental Remediation, University of Newcastle, Callaghan 2308, NSW, Australia; nanthi.bolan@newcastle.edu.au
4   Department of Forest Resources, Gyeongnam National University of Science and Technology, Jinju 52725, Korea
*   Correspondence: ckim@gntech.ac.kr; Tel.: +82-55-751-3247

Academic Editors: Robert Jandl and Mirco Rodeghiero
Received: 1 October 2016; Accepted: 2 December 2016; Published: 7 December 2016

Abstract: This study was conducted to evaluate the effects of fertilizer application on heterotrophic soil respiration (Rh) in soil respiration (Rs) components in red pine stands. Two types of fertilizer ($N_3P_4K_1 = 113{:}150{:}37$ kg·ha^{-1}·year^{-1}; $P_4K_1 = 150{:}37$ kg·ha^{-1}·year^{-1}) were applied manually on the forest floor for two years. Rs and Rh rates were monitored from April 2011 to March 2013. Mean Rs and Rh rates were not significantly affected by fertilizer applications. However, Rh in the second year following fertilizer application fell to 27% for $N_3P_4K_1$ and 17% in P_4K_1 treatments, while there was an increase of 5% in the control treatments compared with the first fertilization year. The exponential relationships between Rs or Rh rates and the corresponding soil temperature were significant (Rh: $R^2 = 0.86$–0.90; $p < 0.05$; Rs: $R^2 = 0.86$–0.91; $p < 0.05$) in the fertilizer and control treatments. Q_{10} values (Rs increase per 10 °C increase in temperature) in Rs rates were lowest for the $N_3P_4K_1$ treatment (3.47), followed by 3.62 for the P_4K_1 treatment and 3.60 in the control treatments, while Rh rates were similar among the treatments (3.59–3.64). The results demonstrate the importance of separating Rh rates from Rs rates following a compound fertilizer application.

Keywords: autotrophic respiration; carbon cycle; heterotrophic respiration; pine forest; soil CO_2 efflux

1. Introduction

The quantitative evaluation of soil respiration (Rs) rates following a fertilizer application is a key process for understanding soil carbon (C) dynamics in forest ecosystem management [1–3]. However, contrasting effects of fertilizer application on Rs rates have been reported. Rs rates increased when nitrogen (N) was added to forest soils in Scot pine (*Pinus sylvestris* L.) in Sweden [4], while Rs rates were significantly lower for fertilized than for unfertilized plots due to reduced fine root production [5,6] and microbial respiration rates [7] in red pine plantations and boreal forest. Since Rs rates result from two main sources, autotrophic respiration (Ra: root respiration rates) and heterotrophic soil respiration (Rh: the microbial decomposition of soil organic matter), these conflicting reports could be due to fertilizer-induced differences in C fixation and allocation patterns among tree species, soil-specific differences in the microbial decomposition of soil organic matter [8–10], and mycorrhizal colonization of host tree species [4,7]. For example, N fertilization had a significant negative effect on Rs rates in a young *Cunninghamia lanceolata* forest [3], but no effect was observed in a coniferous plantation [2,8,11]. Rs and Rh responded differently to environmental resource variables such as nutrient availability.

Fertilizer applications result in a decrease or increase in Rh rates. For example, Rh rates were reduced after N applications in pine forests [12], while Ra rates would be expected to increase along with an increase in forest production following fertilizer application in N-limited forest stands [3,10,12]. However, reductions in Rh following N fertilizer application could be offset by increases in fine root production [8]. In contrast to this result, N fertilization increased Rh rates and microbial biomass C and microbial activity in a loblolly pine (*Pinus taeda* L.) plantation [11].

Fertilizer application effects on Rs or Rh rates in forest stands have mainly focused on the role of N addition [4,8,10]. However, there are a myriad of nutritional problems, such as multi-nutrient deficiency, in the forest stands [3,13,14]. The responses of Rs or Rh rates could be associated with the difference in nutrient availability induced by compound fertilizer types, which can influence favorable environmental conditions for microbial growth activity, soil organic matter decomposition, and root growth activity [2,11]. Although the influence of nutrient availability on Rs and Rh rates may depend on the variety of mechanisms, including changes in microbial biomass, microbial diversity, and root biomass, experimental data about compound types of fertilizer are limited in forest stands.

Red pine (*Pinus densiflora* S. et Z.) forests are the most important type of coniferous tree species and occupy more than 23.5% (1.5 million ha) of the Korean forest. Forest management practices, such as nutrient additions, are required to supply sufficient nutrients to optimize the growth of tree species because many studies have demonstrated the values of compound fertilizer applied to forest ecosystems for improving soil quality and tree growth in Korean forests [2,14]. Furthermore, despite the progress made in quantifying the C balance of many coniferous forests in Korea [15–17], there is a paucity of information about the underlying relationships of Rs and Rh rates, which may change in response to compound fertilizer types. More information that proves useful in evaluating the effects of compound types of fertilizer on Rs or Rh rates is needed. The overall objectives of this study were to 1) evaluate the effects of compound fertilizer application on Rs and Rh rates and 2) to determine the relationship between Rs and Rh rates and soil temperature using compound fertilizer types in red pine stands.

2. Materials and Methods

2.1. Experimental Design

This study was conducted in approximately 40-year-old natural red pine stands in the Wola National Experimental Forest, which is administered by the Southern Forest Resource Research Center, the National Institute of Forest Science, in Korea. The annual average precipitation and temperature in this area are 1490 mm·year^{-1} and 13.1 °C, respectively. The soil is a slightly dry, dark-brown forest soil (mostly Inceptisols, United States Soil Classification System) originating from sandstone or shale with a silt loam texture. The site index based on the height of dominant pine trees indicates low forest productivity (site index, 8–10 at 20-year-old base age), thus suggesting poor soil fertility (Figure 1). The experimental design consisted of a complete randomized block design with two blocks (35°12′32″ N, 128°10′23″ E; 180 m; 35°12′26″ N, 128°10′25″ E, 195 m) in the red pine stands, which were based on the homogeneity between the sites. The experiment involved 18 plots (3 treatments ($N_3P_4K_1$, P_4K_1, Control) × 3 replications × 2 blocks, plot size (a 10 m × 10 m square)). The treatment plots were established on the same facing slopes and aspects under similar environmental conditions to minimize spatial variation in site environmental properties.

Fertilizer applications were based on the guidelines ($N_3P_4K_1$ = 113:150:37 kg·ha^{-1}·year^{-1}) of fertilization in Korean forests [18] and without N fertilizer (P_4K_1 = 150:37 kg·ha^{-1}·year^{-1}). The compound types of fertilizer ($N_3P_4K_1$) are generally recommended for the improvement of growth in mature forests in the country. In addition, the compound types of fertilizer (P_4K_1) were selected based upon considering a myriad of nutritional problems such as phosphorus (P) deficiency in forest stands [13,14]. Urea, fused superphosphate, and potassium chloride fertilizers (Figure 1) were employed as sources of N, P, and K, respectively, and they were applied manually on the forest floor for two years, between 21 April 2011 and 9 April 2012 (total fertilizer amount: $N_3P_4K_1$ = 226:300:74 kg·ha^{-1}; P_4K_1 = 300:74 kg·ha^{-1}), respectively. The understory tree species in the study sites were lespedeza (*Lespedeza* spp.), cork oak (*Quercus variabilis* Bl.), konara oak (*Q. serrate* Thunb.), wild smilax (*Smilax china* L.), and grey blue spicebush (*Lindera glauca* (Siebold & Zucc.) Blume), etc.

Figure 1. Study site (**a**), fertilizer application ((**b**): white grains are urea) and trenching treatments (**c**) with polyvinyl chloride collars (**d**) to separate Rh rates from Rs rates.

2.2. Stand and Soil Characteristics

All trees with >6 cm diameter at breast height (DBH) in each plot were measured to determine stand density, basal area, and DBH among the treatments. Soil samples for the physical and chemical analysis before fertilizer treatment were collected through the top 20 cm at five randomly selected points in each treatment plot using an Oakfield soil sampler. These samples were air dried, passed through a 2 mm sieve, and used for particle size and soil chemical analyses. The distribution of particle size was determined by the hydrometer method. Soil pH (1:5 soil:water suspension) was measured with a glass electrode (Model-735, ISTEC, Seoul, Korea). The C and N content in the soil were determined using an elemental analyzer (Thermo Scientific, Flash 2000, Milan, Italy). Soil phosphorus (P) concentration extracted by NH_4F and HCl solutions was determined by a UV spectrophotometer (Jenway 6505, Staffordshire, UK). Exchangeable potassium (K^+), calcium (Ca^{2+}), and magnesium (Mg^{2+}) concentrations were determined through ICP-OES (Perkin Elmer Optima 5300DV, Shelton, CT, USA). To measure the change of inorganic soil N concentrations following fertilizer applications, a 5-gram subsample of fresh mineral soil was extracted with 50 mL of 2 *M* KCl solution immediately after sampling. The soil extract solutions were stored at 4 °C in a cooler. Ammonium (NH_4^+) and nitrate (NO_3^-) concentrations in the soil extract samples were determined using an Ion Chromatography (AQ2 Discrete Analyzer, Southampton, UK).

2.3. Soil Respiration Rates

A root exclusion collar used for trenching was used to separate Rh rates [16,19,20] from Rs rates (Figure 1). Trenching in the central part of each plot was completed by excavating the outside edges of a columnar soil that was 50 cm diameter and 30 cm deep about one month (24 March 2011) before fertilizer was applied. The soil depth to 30 cm involved the bottom of the B horizon and top of the C horizon in a shallow soil at the study site. In addition, the trenching depth was found to cut down most live roots. Polyvinyl chloride (PVC) collars (50 cm inner diameter and 30 cm height with 4 mm thickness) were inserted into the columnar soil ($N_3P_4K_1$: six plots; P_4K_1: six plots; control: six plots) and backfilled with the excavated soil. Seedlings and herbaceous vegetation inside the collars were manually removed, while litter fall was retained within the collars during the study period.

In this study, Rs rates were regarded as soil CO_2 efflux emitted from the outside of the trenched location, while Rh, in the absence of root respiration, was regarded as soil CO_2 efflux emitted inside of the PVC collars in each plot [16,20]. Four measurements with two repetitions (two inside the PVC collars and two outside the trenched locations) of each plot were taken monthly between 10:00 and 12:30 h during the study period (April 2011–March 2013) with an infrared gas analyzer system (Model EGM-4 environmental gas monitor systems, PP systems, Hitchin, UK). It was equipped with a flow-through closed soil respiration chamber (Model SRC-2, same manufacturer). Although the two-year study period may not be long enough to detect fertilization effects on Rs and Rh, other studies found the Rs and Rh changes in response to fertilization treatments over the duration of two years of study in forest stands [6,8,11]. Soil temperature was measured at 8 cm depth adjacent to the soil respiration chamber using a digital soil temperature probe (K-type, Summit SDT 200, Seoul, Korea).

2.4. Data Analysis

Data after testing for normality and homogeneity of variances were examined via two-way analysis of variance (ANOVA) to determine the significance of the main effects (year (Y), compound types of fertilizer (F)) and their interactions (Y × F). The model describing the data analysis is as follows (Equation (1)):

$$Y_{ij} = u + Y_i + F_j + (Y \times F)_{ij} + e_{ij} \tag{1}$$

where u is the overall mean effect, Y is year ($i = 1, 2$), and F is fertilizer treatment ($j = 1, 2, 3$). All ANOVA were executed using the General Linear Models procedure in SAS [21]. Treatment means were compared using Tukey's test. Rs and Rh data collected for the two-year period served to test exponential functions [22] between soil CO_2 efflux rates and soil temperature (Equation (2)):

$$\text{Soil } CO_2 \text{ efflux rates} = B_0 e^{B1ST} \tag{2}$$

where B_0 and $B1$ are coefficients estimated through regression analysis and ST is the soil temperature. The Q_{10} values (Equation (3)) were calculated using the $B1$ coefficient which is used in the multiplier for soil CO_2 efflux rates given an increase of 10 °C in soil temperature:

$$Q_{10} = e^{10 \times B1} \tag{3}$$

3. Results

3.1. Stand and Soil Characteristics

Mean stand densities, DBH, and basal area were not significantly different between the control and fertilizer treatments (Table 1). The distribution of soil particles, such as sand, silt, and clay, was not significantly different among the treatments. While soil nutrient concentrations, such as C, N, P, and K^+ were not significantly different between the fertilizer and control treatments, exchangeable Ca^{2+} and Mg^{2+} were significantly higher in the P_4K_1 than in the control treatments (Table 1).

Table 1. General stand and soil characteristics of the study site before fertilizer application.

Treatment	Stand Density (trees·ha⁻¹)	DBH (cm)	Basal Area (m²·ha⁻¹)	Sand (%)	Silt (%)	Clay (%)	C (%)	N (%)	P (mg·kg⁻¹)	K⁺ (cmolc·kg⁻¹)	Ca²⁺ (cmolc·kg⁻¹)	Mg²⁺ (cmolc·kg⁻¹)
Control	1217 (133) a	15.74 (0.84) a	22.37 (1.95) a	45 (3.5) a	43 (3.0) a	12 (1.0) a	2.40 (0.28) a	0.07 (0.01) a	3.9 (0.40) a	0.09 (0.01) a	1.35 (0.19) b	0.43 (0.05) b
$N_3P_4K_1$	1150 (193) a	15.89 (1.10) a	20.56 (2.42) a	42 (2.9) a	44 (1.8) a	14 (1.0) a	2.82 (0.21) a	0.09 (0.01) a	6.5 (0.73) a	0.09 (0.01) a	1.77 (0.17) ab	0.54 (0.04) ab
P_4K_1	1150 (152) a	16.46 (1.46) a	22.62 (2.00) a	42 (2.0) a	45 (1.9) a	13 (1.8) a	2.66 (0.27) a	0.08 (0.01) a	5.8 (1.61) a	0.09 (0.01) a	2.10 (0.26) a	0.65 (0.05) a

Values in parenthesis represent standard errors ($n = 6$). DBH: diameter at breast height at 1.2 m. The same letters among the treatments are not significantly different at $p < 0.05$.

3.2. Monthly Variation of Rh and Rs Rates

Monthly variations in Rh rates were not significantly affected ($p > 0.05$) by the compound fertilizer types over the two-year study period except for July 2012 (Figure 2). However, Rs rates during early growing season (March–May 2012) were significantly lower in the control treatment than in the $N_3P_4K_1$ treatment (Figure 2). Rh and Rs rates in all treatments showed clear seasonal variation in which the rates increased during spring and summer, and reached their maximum values in July and September (Figure 2). In addition, temporal variation in Rs and Rh rates had a similar seasonal pattern to soil temperature, whereas the variations were not related to extractable soil NH_4^+ and NO_3^- concentrations regardless of the compound of fertilizer types (Figure 3).

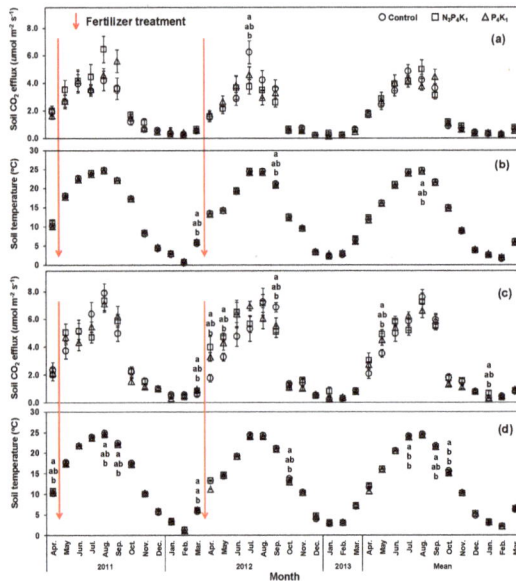

Figure 2. Monthly variation of Rh rates (**a**) and soil temperature (**b**) or Rs rates (**c**) and soil temperature (**d**) for fertilizer and control treatments in red pine stands. Vertical bars represent standard errors ($n = 12$). Different letters at each month indicate a significant difference among treatments at $p < 0.05$.

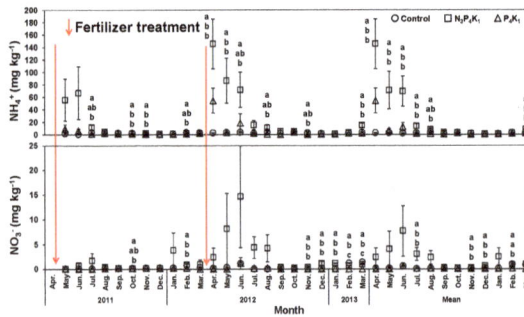

Figure 3. Monthly variation of extractable soil NH_4^+ and NO_3^- concentrations for fertilizer and control treatments in red pine stands. Vertical bars represent standard errors ($n = 6$). Different letters at each month indicate a significant difference among treatments at $p < 0.05$.

3.3. Annual Rh and Rs Rates

Annual Rh rates had a significant main effect on year with no significant fertilizer treatment and interaction effects, but annual Rs rates and fertilizer treatment did not generate any significant main and interaction effects (Table 2). There was a significant effect on mean annual soil temperature during the study period, but soil temperature was not affected by fertilizer application. Annual Rs rates were not significantly affected by the compound fertilizer types for two years, although the rates were slightly higher in the $N_3P_4K_1$ than in the P_4K_1 or the control treatments (Table 2). Additionally, mean annual Rh rates were not significantly different among the compound fertilizer types and the control. Mean annual Rs rates ($\mu mol \cdot m^{-2} \cdot s^{-1}$) were same between 2011 (3.02) and 2012 (3.02), but Rh rates were significantly lower in 2012 (1.77) than in 2011 (2.17).

Table 2. Mean annual Rh or Rs rates and soil temperature for fertilizer and control treatments in red pine stands (2011: April 2011–March 2012; 2012: April 2012–March 2013) with *p*-value by a two-way analysis of variance (ANOVA) on soil respiration rates and soil temperature.

Year	Treatment	df	Soil Respiration Rates ($\mu mol \cdot m^{-2} \cdot s^{-1}$)		Soil Temperature (°C)	
			Rh	Rs	Rh	Rs
2011	Control	-	1.93 (0.20)	3.08 (0.21)	13.4 (0.10)	13.7 (0.06)
	$N_3P_4K_1$	-	2.37 (0.26)	3.05 (0.11)	13.5 (0.04)	13.7 (0.05)
	P_4K_1	-	2.20 (0.13)	2.95 (0.12)	13.4 (0.05)	13.7 (0.03)
	Mean	-	2.17 (0.12)	3.02 (0.88)	13.4 (0.04)	13.7 (0.03)
2012	Control	-	1.91 (0.19)	2.82 (0.20)	12.8 (0.10)	13.1 (0.07)
	$N_3P_4K_1$	-	1.64 (3.19)	3.19 (0.23)	12.8 (0.10)	13.2 (0.08)
	P_4K_1	-	1.77 (0.12)	3.50 (0.16)	12.8 (0.14)	12.9 (0.15)
	Mean	-	1.77 (0.09)	3.02 (0.11)	12.8 (0.06)	13.1 (0.06)
Mean	Control	-	1.92 (0.14)	2.95 (0.14)	13.1 (0.07)	13.4 (0.04)
	$N_3P_4K_1$	-	2.01 (0.16)	3.12 (0.12)	13.1 (0.03)	13.4 (0.04)
	P_4K_1	-	1.98 (0.10)	3.00 (0.10)	13.1 (0.07)	13.3 (0.07)
p-value	Year (Y)	1	0.009	0.973	<0.001	<0.001
	Treatment (F)	2	0.882	0.598	0.733	0.504
	Y × F	2	0.152	0.447	0.806	0.253

Values in parenthesis represent standard errors (*n* = 12).

Rh rates of the fertilizer treatments in 2011 represented 78% for the $N_3P_4K_1$ treatment, 75% for the P_4K_1 treatments, and 63% of Rs rates in the control treatment. In comparison, the Rh rates of the fertilizer treatments in 2012 were 51% for the $N_3P_4K_1$, 58% for the P_4K_1, and 68% of Rs rates in the control treatment (Table 3). Rh rates after the second year following fertilizer application fell to 27% for the $N_3P_4K_1$ and 17% in P_4K_1 treatments, respectively, while an increase of 5% in the control treatment was comparable to the first year's fertilizer application. Ra rates (Rs–Rh) in 2011 were 22%–25% in the fertilizer treatments and 37% of Rs rates in the control treatment, respectively, while the rates in 2012 were 42%–49% in the fertilizer treatments and 32% of Rs rates in the control treatment, respectively (Table 3).

Table 3. Proportion of Rh rates from Rs rates (2011: April 2011–March 2012; 2012: April 2012–March 2013).

Year	Treatment	Rh	Ra	Rs
			(%)	
2011	Control	63	37	100
	$N_3P_4K_1$	78	22	100
	P_4K_1	75	25	100
	Mean	72	28	100

Table 3. *Cont.*

Year	Treatment	Rh	Ra	Rs
			(%)	
	Control	68	32	100
2012	$N_3P_4K_1$	51	49	100
	P_4K_1	58	42	100
	Mean	59	41	100

3.4. Temperature Dependency of Rh and Rs

The exponential relationships between Rs and Rh rates and the corresponding soil temperature (Figure 4) were significant (Rh: $R^2 = 0.86$–0.90, $p < 0.05$; Rs: $R^2 = 0.86$–0.91, $p < 0.05$) in the fertilizer and control treatments. Soil temperature explained 86% to 91% of the variation in Rs and Rh rates in the fertilizer and control treatments. Q_{10} values in Rs rates were 3.47 for the $N_3P_4K_1$ treatment, 3.62 for the P_4K_1 treatment, and 3.60 in the control treatments, while Q_{10} values in Rh rates were 3.60 for the $N_3P_4K_1$ treatment, 3.64 for the P_4K_1 treatment, and 3.59 in the control treatment.

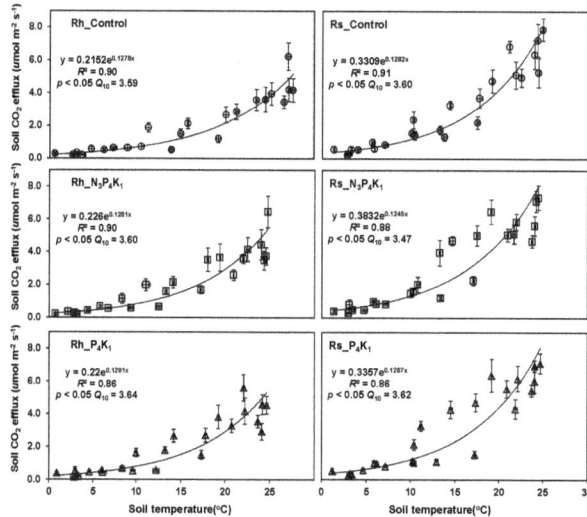

Figure 4. Exponential regressions showing the relationship between soil temperatures and Rh or Rs rates for fertilizer ($N_3P_4K_1$: squares; P_4K_1: triangles) and control (circles) treatments in red pine stands. Vertical bars represent standard error ($n = 12$).

3.5. Relationships between Rs and Rh Rates

Rh rates were positively correlated ($r = 0.91$–0.95, $p < 0.05$) to Rs for all treatments (Figure 5). The correlation coefficient between Rs and Rh rates in the P_4K_1 treatment ($r = 0.95$, $p < 0.05$) was slightly higher than the other treatments ($r = 0.91$–0.92, $p < 0.05$). In addition, the regressions in the fertilizer and control treatments represented a linear relationship with similar slopes among the treatments.

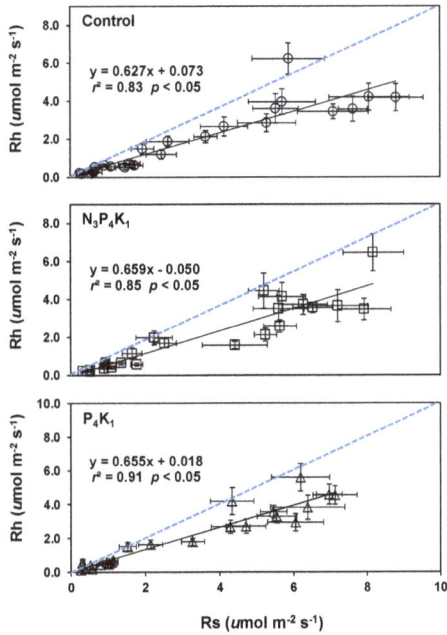

Figure 5. Relationships between Rs and Rh rates for fertilizer ($N_3P_4K_1$: squares; P_4K_1: triangles) and control (circles) treatments in red pine stands. Vertical bars represent standard error (n = 12). Dashed line represents a 1:1 relationship (Intercept = 0; slope = 1) between Rs and Rh rates in red pine stands.

4. Discussion

Fertilizer application revealed a significant effect on the monthly Rs rates, but the monthly Rh rates were generally less influenced by the compound fertilizer types. The monthly Rs rates in the early growing season were significantly higher in the $N_3P_4K_1$ fertilizer application compared with the control treatment. This finding may be related to Ra rates induced by root growth activities in spring [13] following the $N_3P_4K_1$ treatment. However, less monthly variations in Rh rates indicate that microbial activity in fertilizer treatments could be limited by other environmental factors, rather than by the changes in N availability after fertilizer application because extractable soil NH_4^+ concentration was consistently greater from the $N_3P_4K_1$ treatments than from the P_4K_1 or control treatments (Figure 3). For example, soil temperature explained the majority of temporal variations in Rs and Rh rates in the fertilizer and control treatments because microbial decay and root growth activities were temperature-dependent [22–24]. Studies in Korean forest stands have reported that seasonal Rs and Rh rates correlated strongly to seasonal fluctuations in soil temperature [2,15] because of no monthly fluctuation in soil water content [2,25]. In addition, the variation in Rs and Rh rates at the seasonal scale was affected by limiting soil water content, such as decreasing soil matric potential or high soil water content [2,23,24,26].

Fertilizer application induced a decrease in the proportion of Rh, while annual rates of Rs were not affected by the compound fertilizer types. Rh rates of fertilizer treatments in 2012 had rapidly declined by 17%–27% compared to 2011. The rapid decline in the proportion of Rh rates following the fertilizer treatments could be associated with the decreased decomposition of dead roots because of a similar soil temperature between the control and fertilizer treatments. It has also attributed to reduced diffusion for Rh rates by decreased air-filled pore space because many studies have reported increases in soil water content following trenching due to the elimination of root uptake of soil water [16,20]. However, changes in soil water content due to trenching may have no significant effects because of little change was observed in Rh rates in the control treatment during the two-year study period (Table 2). In addition, the results need to be interpreted cautiously because the effect of soil water content and root decomposition on the Rs and Rh rates was not measured in this study. This reduction in the proportion of Rh rates after fertilizer application concurs with many previous studies describing the negative effects of high soil N concentrations on soil organic decomposition rates [3,6,12]. For example, the decrease in Rh rates with fertilizer application may be due to the rapid change in quality and quantity of substrates, which could have attributed to changes in nutrient availability for microbial decay by decreased soil microbial biomass following fertilizer application [10]. In contrast to this result, Samuelson et al. [11] observed that applying fertilizer increased Rh rates, microbial biomass, and microbial activity, with reduced fine root biomass in a loblolly pine plantation in the USA. The proportion of Rh rates (63%–68%) in the control treatment of this study was comparable to approximately 66% of Rs rates observed in temperate coniferous forests in Korea [16]. In contrast to Rh rates, Ra rates in the fertilizer treatments in 2011 were 22%–25% of Rs in 2011 and 42%–49% of Rs in 2012, respectively. The increase in the proportion of Ra rates in fertilizer treatments in 2012 could be due to the changing of C allocation to the roots in response to increased nutrient availability [5,7]. Additionally, soil environmental changes in response to fertilizer application are closely linked to root growth activities and nutrient availability [11]. The Ra values of the control treatment in this stand were comparable to those of 33% and 62% for pine forests [9].

An exponential regression has been widely used to describe the relationship between Rs rates and temperature following fertilizer application in forest stands [2,27]. In this study, significant exponential relationships were obtained between Rs or Rh rates and the corresponding soil temperature in the fertilizer (R^2 = 0.86–0.90, $p < 0.05$) and control (R^2 = 0.90–0.91, $p < 0.05$) treatments. The effect of soil temperature on Rs rates was commonly expressed by the coefficient Q_{10} which could indicate sensitivity to soil temperature [19]. Q_{10} values in Rs rates were lower in the $N_3P_4K_1$ treatment (3.47) than in the P_4K_1 (3.62) and the control (3.60) treatments. A decreased temperature sensitivity of Rs rates under $N_3P_4K_1$ treatment may be attributed to increased Ra rates resulting from the change in nutrient availability by the N supply compared with the P_4K_1 or control treatments. For example, mean annual extractable soil NH_4^+ and NO_3^- concentrations during the study period were significantly higher for the $N_3P_4K_1$ (7.65 mg·kg^{-1}) than for the control (0.27 mg·kg^{-1}) treatments (Figure 3). Q_{10} values of Rs rates in this study were comparable to those of other red pine forests (3.45–3.77 at 12 cm soil depth) in Korea [15]. In contrast to Rs rates, the effects of soil temperature on Rh rates were not influenced by compound fertilizer types. This result indicates that Rh rates in red pine stands might be independent of compound fertilizer types because soil temperature is more likely to control Rh activity [28] compared with nutrient availability at a given site. However, the high Q_{10} value of the Rh rates (3.60) compared with that of Rs rates (3.47) of the $N_3P_4K_1$ treatment may have resulted from the high availability of N for microbial decay due to the increased dead root biomass following trenching.

5. Conclusions

Mean annual Rs rates were minimally affected by the change in nutrient availability with compound fertilizer types in red pine stands. It is also evident that Rh rates were independent of compound fertilizer types because soil temperature is likely to control Rh activity. The proportion of Rh rates fell to 27% for $N_3P_4K_1$ and 17% in P_4K_1 treatments in the second year compared with the first fertilization year. The results demonstrate the importance of separating Rh rates from Rs rates following the application of compound fertilizer.

Acknowledgments: This study was partially supported by the Centre for Environmental Risk Assessment and Remediation (CERAR), the University of South Australia, by Basic Science Research Program through the National Research Foundation of Korea (NRF) funded by the Ministry of Education (2010-0022193; No. 2014R1A1A2054994) and by Forest Science and Technology Project (S211212L030320) provided by Korea Forest Service.

Author Contributions: Jaeyeob Jeong performed the experiment and wrote the paper; Nanthi Bolan provided constructive suggestions on the study; Choonsig Kim conceived and designed the experiments.

Conflicts of Interest: The authors declare no conflict of interest.

References

1. Jandl, R.; Lindner, M.; Vesterdal, L.; Bauwens, B.; Baritz, R.; Hagedorn, F.; Johnson, D.W.; Minkkinen, K.; Byrne, K.A. How strongly can forest management influence soil carbon sequestration? *Geoderma* **2007**, *137*, 253–268. [CrossRef]
2. Kim, C. Soil carbon storage, litterfall and CO_2 efflux in fertilized and unfertilized larch (*Larix leptolepis*) plantations. *Ecol. Res.* **2008**, *23*, 757–763. [CrossRef]
3. Wang, Q.; Zhang, W.; Sun, T.; Chen, L.; Pang, X.; Wang, Y.; Xiao, F. N and P fertilization reduced soil autotrophic and heterotrophic respiration in young *Cunninghamia lanceolate* forest. *Agric. For. Meteorol.* **2017**, *232*, 66–73. [CrossRef]
4. Hasselquist, N.J.; Metcalfe, D.B.; Högberg, P. Contrasting effects of low and high nitrogen additions on soil CO_2 flux components and ectomycorrhizal fungal sporocarp production in a boreal forest. *Glob. Chang. Biol.* **2012**, *18*, 3596–3605. [CrossRef]
5. Haynes, B.E.; Gower, S.T. Belowground carbon allocation in unfertilized and fertilized red pine plantations in northern Wisconsin. *Tree Physiol.* **1995**, *15*, 317–325. [CrossRef] [PubMed]
6. Olsson, P.; Linder, S.; Giesler, R.; Högberg, P. Fertilization of boreal forest reduces both autotrophic and heterotrophic soil respiration. *Glob. Chang. Biol.* **2005**, *11*, 1745–1753. [CrossRef]
7. Phillips, R.P.; Fahey, T.J. Fertilization effects on fineroot biomass, rhizosphere microbes and respiratory fluxes in hardwood forest soils. *New Phytol.* **2007**, *176*, 655–664. [CrossRef] [PubMed]
8. Lee, K.H.; Jose, S. Soil respiration, fine root production, and microbial biomass in cottonwood and loblolly pine plantations along a nitrogen fertilization gradient. *For. Ecol. Manag.* **2003**, *185*, 263–273. [CrossRef]
9. Raich, J.W.; Tufekciogul, A. Vegetation and soil respiration: Correlations and controls. *Biogeochemistry* **2000**, *48*, 71–90. [CrossRef]
10. He, T.; Wang, Q.; Wang, S.; Zhang, F. Nitrogen addition altered the effect of belowground C allocation on soil respiration in a subtropical forest. *PLoS ONE* **2016**, *11*, e0155881. [CrossRef] [PubMed]
11. Samuelson, L.; Mathew, R.; Stokes, T.; Feng, Y.; Aubrey, D.; Coleman, M. Soil and microbial respiration in a loblolly pine plantation in response to seven years of irrigation and fertilization. *For. Ecol. Manag.* **2009**, *258*, 2431–2438. [CrossRef]
12. Franklin, O.; Högberg, P.; Ekblad, A.; Ågren, G.I. Pine forest floor carbon accumulation in response to N and PK additions: Bomb ^{14}C modelling and respiration studies. *Ecosystems* **2003**, *6*, 644–658. [CrossRef]
13. Hwang, J.; Son, Y.; Kim, C.; Yi, M.J.; Kim, Z.S.; Lee, W.K.; Hong, S.K. Fine root dynamics in thinned and limed pitch pine and Japanese larch plantations. *J. Plant Nutr.* **2007**, *30*, 1821–1839. [CrossRef]
14. Kim, C.; Jeong, J.; Park, J.H.; Ma, H.S. Growth and nutrient status of foliage as affected by tree species and fertilization in a fire-disturbed urban forest. *Forests* **2015**, *6*, 2199–2213. [CrossRef]
15. Noh, N.J.; Son, Y.; Lee, S.K.; Yoon, T.K.; Seo, K.W.; Kim, C.; Lee, W.K.; Bae, S.W.; Hwang, J. Influence of stand density on soil CO_2 efflux for a *Pinus densiflora* forest in Korea. *J. Plant Res.* **2010**, *123*, 411–419. [CrossRef] [PubMed]
16. Lee, N.; Koo, J.W.; Noh, N.J.; Kim, J.; Son, Y. Autotrophic and heterotrophic respiration in needle fir and *Quercus*-dominated stands in a cool-temperate forest, central Korea. *J. Plant Res.* **2010**, *123*, 485–495. [CrossRef] [PubMed]
17. Kim, C.; Jeong, J.; Bolan, N.S.; Naidu, R. Short-term effects of fertilizer application on soil respiration in red pine stands. *J. Ecol. Field Biol.* **2012**, *35*, 307–311. [CrossRef]
18. Joo, J.H.; Lee, W.K.; Kim, T.H.; Lee, C.Y.; Jin, I.S.; Park, S.K.; Oh, M.Y. Studies on fertilization in pruning and thinning stands. *The Res. Rep. For. Res. Insti., Korea* **1983**, *30*, 155–189.
19. Bond-Lamberty, B.; Bronson, D.; Bladyka, E.; Gower, S.T. A comparison of trenched plot techniques for partitioning soil respiration. *Soil Biol. Biochem.* **2011**, *43*, 2108–2114. [CrossRef]
20. ArchMiller, A.A.; Samuelson, L.J. Partitioning longleaf pine soil respiration into its heterotrophic and autotrophic components through root exclusion. *Forests* **2016**, *7*, 39. [CrossRef]
21. Statistical Analysis System (SAS) Institute Inc. *SAS/STAT Statistical Software*, version 9.1; SAS Publishing: Cary, NC, USA, 2003.

22. Davidson, E.A.; Janssens, I.A. Temperature sensitivity of soil carbon decomposition and feedbacks to climate change. *Nature* **2006**, *440*, 165–173. [CrossRef] [PubMed]

23. Suseela, V.; Conant, R.T.; Wallenstein, M.D.; Dukes, J.S. Effects of soil moisture on the temperature sensitivity of heterotrophic respiration vary seasonally in an old-field climate change experiment. *Glob. Chang. Biol.* **2012**, *18*, 336–348. [CrossRef]

24. Yuste, J.C.; Baldocchi, D.D.; Gershenson, A.; Goldstein, A.; Misson, L.; Wong, S. Microbial soil respiration and its dependency on carbon inputs, soil temperature and moisture. *Glob. Chang. Biol.* **2007**, *13*, 2018–2035.

25. Kim, C.; Jeong, J. Comparison of soil CO_2 efflux rates in *Larix leptolepis*, *Pinus densiflora* and *P. rigitaeda* plantations in Southern Korea. *Dendrobiology* **2016**, *76*, 51–60. [CrossRef]

26. Davidson, E.A.; Belk, E.; Boone, R.D. Soil water content and temperature as independent or confounded factors controlling soil respiration in a temperate mixed hardwood forest. *Glob. Chang. Biol.* **1998**, *4*, 217–227. [CrossRef]

27. Bowden, R.D.; Davidson, E.; Savage, K.; Arabia, C.; Steudler, P. Chronic nitrogen additions reduce total soil respiration and microbial respiration in temperate forest soils at the Harvard Forest. *For. Ecol. Manag.* **2004**, *196*, 43–56. [CrossRef]

28. Wei, W.; Weile, C.; Shaopeng, W. Forest soil respiration and its heterophic and autotrophic components: Global patterns and responses to temperate and precipitation. *Soil Biol. Biochem.* **2010**, *42*, 1236–1244. [CrossRef]

forests

MDPI

Article

Spatial Upscaling of Soil Respiration under a Complex Canopy Structure in an Old-Growth Deciduous Forest, Central Japan

Vilanee Suchewaboripont [1], Masaki Ando [2], Shinpei Yoshitake [3], Yasuo Iimura [4], Mitsuru Hirota [5] and Toshiyuki Ohtsuka [1,3,*]

[1] United Graduate School of Agricultural Science, Gifu University, 1-1 Yanagido, Gifu 501-1193, Japan; vilanee.s@hotmail.com

[2] Laboratory of Forest Wildlife Management, Faculty of Applied Biology Sciences, Gifu University, 1-1 Yanagido, Gifu 501-1193, Japan; m_ando@gifu-u.ac.jp

[3] River Basin Research Center, Gifu University, 1-1 Yanagido, Gifu 501-1193, Japan; syoshi@green.gifu-u.ac.jp

[4] School of Environmental Science, The University of Shiga Prefecture, Hikone, Shiga 522-8533, Japan; iimura.y@ses.usp.ac.jp

[5] Faculty of Life and Environmental Sciences, University of Tsukuba, Tsukuba, 305-8577, Japan; hirota@biol.tsukuba.ac.jp

* Correspondence: toshi@green.gifu-u.ac.jp; Tel.: +81-58-293-2065

Academic Editors: Robert Jandl and Mirco Rodeghiero
Received: 6 December 2016; Accepted: 24 January 2017; Published: 30 January 2017

Abstract: The structural complexity, especially canopy and gap structure, of old-growth forests affects the spatial variation of soil respiration (R_s). Without considering this variation, the upscaling of R_s from field measurements to the forest site will be biased. The present study examined responses of R_s to soil temperature (T_s) and water content (W) in canopy and gap areas, developed the best fit model of R_s and used the unique spatial patterns of R_s and crown closure to upscale chamber measurements to the site scale in an old-growth beech-oak forest. R_s increased with an increase in T_s in both gap and canopy areas, but the effect of W on R_s was different between the two areas. The generalized linear model (GLM) analysis identified that an empirical model of R_s with the coupling of T_s and W was better than an exponential model of R_s with only T_s. Moreover, because of different responses of R_s to W between canopy and gap areas, it was necessary to estimate R_s in these areas separately. Consequently, combining the spatial patterns of R_s and the crown closure could allow upscaling of R_s from chamber-based measurements to the whole site in the present study.

Keywords: soil respiration; spatial variation; gap/canopy structure; upscaling; old-growth forest

1. Introduction

Studies of soil respiration (R_s) have been conducted in many temperate forests and show the temporal variation at both diurnal [1] and seasonal [2–4] time scales. The diurnal variation in R_s is explained by plant physiology, especially photosynthesis [1], and soil temperature (T_s) [5]. The seasonal variation in R_s is primarily controlled by T_s when the soil water content (W) is not limited, and the response of R_s to T_s is usually explained by an exponential (Q_{10}) relationship [6–8]. Under drought conditions, W and T_s are considered to be coupled factors in relation to their effect on R_s [9,10]. Continuous measurement of the effects of T_s and W on R_s is required for the scaling up of R_s from chamber-based measurements to a forest ecosystem. Some studies on the effects of T_s and W on R_s have been conducted using an automatic opening and closing chamber (AOCC) system. Although this approach is based on the closed dynamic method, the AOCC system allows continuous measurement of R_s on both short- and long-term scales and provides the detail needed to develop our understanding of the relationships between R_s and environmental variables [3,11–14]. Moreover, this system minimizes disturbance of the soil surface during R_s measurements.

Many old-growth forests show canopy structural complexity, particularly canopies and gaps [15–20]. The complex structure of canopies related to forest age facilitates a greater harvesting of light than a simple structure, and thus it increases net primary production [21,22]. This structural complexity also reflects the spatial variation in R_s, which is greater in canopy areas than in gap areas [23,24]. Despite the fact that these R_s patterns impact the upscaling of R_s to the forest ecosystem level, only a few publications focusing on the upscaling method based on spatial variation in R_s are available. For example, Tang and Baldocchi [25] used crown closure and different rates of R_s between areas under trees and open areas to spatially upscale R_s to a whole site in an oak-grass savanna ecosystem in California.

In a recent study of R_s in an old-growth beech-oak forest in central Japan, the effect of the complexity of the vertical structure, especially the canopy/gap structure, on the spatial variation of R_s was investigated [24]. The R_s was greater in canopy areas than in gap areas during the growing season, and there was no significant difference in T_s or W between canopy and gap areas. However, diurnal and seasonal changes in R_s and these environmental factors have not been studied, and the responses of R_s to changes in T_s and W in canopy and gap areas are unclear. Consequently, the present study aims to (1) quantify the temporal variation of R_s in canopy and gap areas; (2) characterize the response of R_s to T_s and W in canopy and gap areas; (3) develop models for R_s determined as a function of T_s and W; and (4) upscale chamber measurements of R_s to the site scale based on the spatial patterns of R_s between canopy and gap areas and crown closure. We used the AOCC system to measure R_s continuously in order to understand the relationships between R_s and environmental variables and to develop suitable models of R_s for estimation of annual R_s.

2. Materials and Methods

2.1. Study Site

The study site (36°9′ N, 136°49′ E, 1330 m above sea level) is located in primary deciduous broad-leaved forests around the Ohshirakawa river basin (840–1600 m above sea level) on the mid slope of Mt. Hakusan. The forests became established since the last eruption of Mt. Hakusan in 1659 [26] and are protected by the Hakusan National Park under the management of the Forest Agency of Japan and the Ministry of Environment, Japan. No evidence of human disturbance in this area was found before the 1960s. After the construction of the Ohshirakawa dam (approximately 800 m from the study site) in the 1960s, this area has been accessed by local people for the collection of mushrooms, bamboo shoots and chestnuts [20].

A permanent 1-ha (100 m × 100 m) plot of primary forest was reconstructed to examine carbon cycling of the forest ecosystem during July 2011 after a first construction of the plot in 1993 [27]. This plot is on an east-facing gentle slope with an average slope of 3 degrees. In the study plot, the canopy layer is dominated by *Fagus crenata* (beech, 47.6% of basal area) and *Quercus mongolica* var. *crispula* (oak, 37.4% of basal area) trees with diameters at breast height (DBH) of ≥25 cm and heights of approximately 25–30 m. The sub-tree layer comprises beech, *Acer tenuifolium*, and *Vibrunum furcatum*, with high stem densities and heights of 10–15 m. Evergreen dwarf bamboo (*Sasa kurilensis*) of 1.5–2.0 m in height sparsely covers the forest floor. The DBHs of all trees were measured in 2014, and tree biomasses were estimated following Suchewaboripont et al. [20]. The aboveground biomass (except for leaves) of canopy trees was large, i.e., 475.9 Mg ha^{-1}. Most of this biomass consisted of canopy trees of beech (45.9%) and oak (46.7%), and the DBH of these trees ranged from 25.75 to 101.28 cm and 44.31 to 195.04 cm, respectively. The forest age is >250 years, and the age of a dead oak tree (DBH = 74.5 cm) near the study plot was determined as being over 258 years. The soil in the study plot is volcagogenous regosol with thin A (0–18 cm) and B (19–24 cm) horizons [20].

Air temperature during 2013–2014 in the study plot was monitored using a data logger (HOBO weather station, Onset Computer, Bourne, MA, USA). Air temperature was measured every 30 min, and the data were processed to provide an average every 24 h. The annual mean air-temperature was 5.8 °C, with a maximum daily temperature of 22.7 °C during August and a minimum daily temperature of −13.2 °C during February. The average annual precipitation during 2013–2014 was 3289 mm, measured at Miboro weather station (36°9′ N, 136°54′ E, 640 m above sea level). Heavy snowfall from November to April resulted in accumulated snow depths of >4 m in 2013.

2.2. Measurement of the Soil CO_2 Efflux

To define the gap and canopy areas for measurement of R_s in the study plot, we followed the definitions used in our previous study [24]. Ground areas under canopy openings (≥5 m^2 in area) caused by canopy tree deaths were defined as gap areas [28], and the areas under canopy trees were defined as canopy areas. In the study plot, it was difficult to choose the most suitable areas for placing an AOCC system, in particular, and measuring R_s and environmental factors due to varying density of sub-trees (shown as their basal areas in Figure 1b) and *Sasa* (Figure 1c). However, we selected three canopy (C1, C2, and C3) and two gap (G1 and G2) areas for measurement of their R_s and environmental factors.

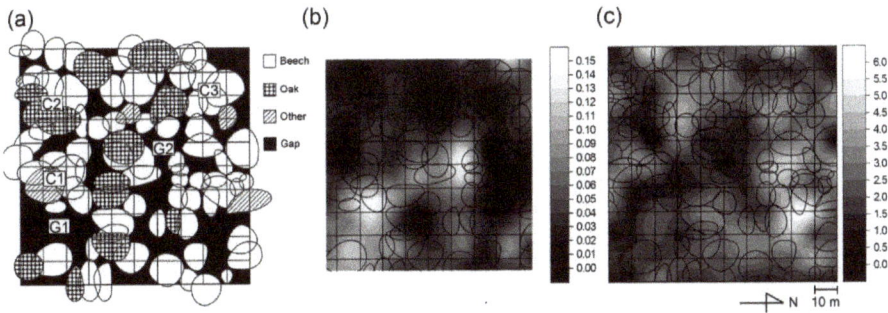

Figure 1. Five locations for measurement of soil respiration (R_s) and environmental factors based on the crown projection diagram (**a**); the basal area in the sub-tree layer (**b**); and the density of dwarf bamboo (**c**) (from Suchewaboripont et al. [24]). Basal areas of sub-trees (DBH < 25 cm) and stem densities of *Sasa* are presented as m^2 and stem numbers m^{-2}, respectively.

Changes in CO_2 efflux from soil (R_s) were measured continuously for 24–48 h once a month by the closed-chamber method with an infrared gas analyzer (IRGA) (GMP343, Vaisala Ltd., Vantaa, Finland). Three chambers (25 cm internal diameter and 25 cm height) were installed in each area. R_s in all chambers in each area was measured using an AOCC system. The AOCC system had a similar concept to that used by Hirota et al. [29]. Without electricity at the study site, this system was connected to two 12-volt DC car batteries as its electric power for continuous measurement around 2–3 days. The system comprised an automated lid arm subsystem (Figure 2) and a control system for timing the opening and closing of the lid arm. The lid was attached to an IRGA for the measurement of R_s in the closed chambers. Rotation of the automated lid arm progressed from chamber 1 to chamber 2 to chamber 3 (Figure 2), with one such cycle taking approximately 16 min. After the closure of chamber 3, the lid arm returned to chamber 1 and started a new rotation cycle. During chamber closure for 5 min, R_s was measured, and all data were recorded by a data logger (Thermic 2300, Etodenki, Tokyo, Japan) at intervals of 10 s.

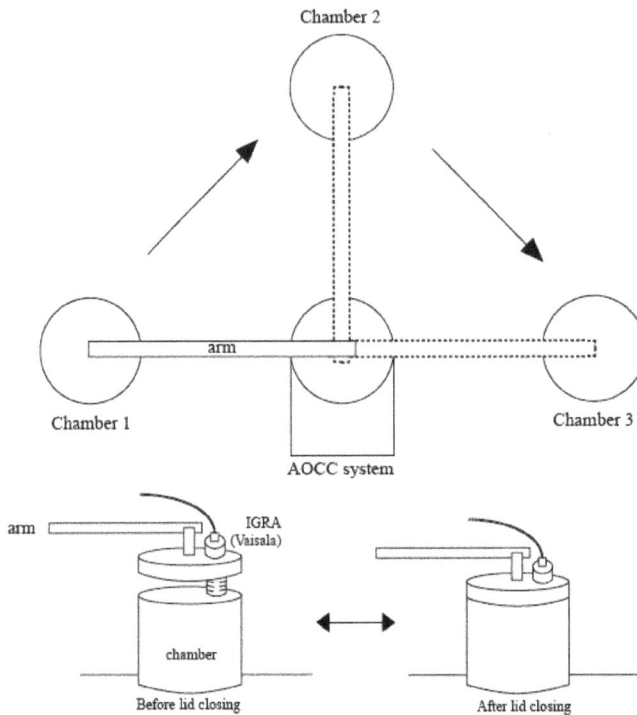

Figure 2. Automatic opening and closing chamber (AOCC) system for measurement of R_s in three chambers. The lid connected to the arm of the AOCC system closed the chamber for a 5-min interval. The arm rotated from chamber 1 to chamber 2 to chamber 3. After measurement completion in chamber 3, the arm rotated back to chamber 1 for the next cycle. R_s in a chamber was continuously measured using an infrared gas analyzer (IRGA) (GMP343, Vaisala Ltd., Vantaa, Finland).

In 2013, we measured R_s in C1 and G1 areas during 25–28 June and in all areas from July to October (16–19 July, 19–23 August, 24–28 September, and 22–26 October). In 2014, we measured R_s in C1 and G1 areas from June to October (16–18 June, 15–17 July, 2–4 September, 8–10 October, and 29–31 October). In August 2014, R_s was not measured because of an extended period of rainy days.

2.3. Measurement of Soil Temperature and Water Content

Soil temperature (T_s) in all canopy and gap areas was measured at a depth of 5 cm in close proximity to the AOCC system using a copper–constantan thermocouple during the same experimental period as the measurement of R_s. From 23 June 2013 to 30 October 2014, T_s was monitored every 10 min by temperature sensors and data loggers (TidbiT v1 Temp logger, Onset Computer Co., Ltd., Massachusetts, USA). The T_s data were processed to provide an average every 1 h.

Soil water content (W) was measured continuously at a depth of 5 cm near the AOCC system in C1, C3, G1, and G2 from 27 June to 27 October 2013 using a thetaprobe ML2x-L5 (Delta-T Devices, Camblidge, UK) connected to a data logger (THLOG-2, Dynamax Inc., Texas, USA). In C2, W near the AOCC system was measured from 25 September to 27 October 2013 using a SM200 soil moisture sensor (Delta-T Devices, England) connected to a data logger (3635-25 voltage logger, Hioki E.E. Corporation, Nagano, Japan). During the growing season (3 June–29 October) of 2014, W in G1 and C1 was continuously measured using an SM200 soil moisture sensor. Because of the use of different sensors for measuring W, each sensor was calibrated using the field calibration equation reported by

Abbas et al. [30]. The measurements of W in all areas were taken every 15 min, and the data were processed to provide an average every 1 h.

2.4. Data Analysis

R_s was calculated on the basis of the linear increase in CO_2 concentration in the closed chamber using equation [14]:

$$R_{s\_hourly} \left(g\,C\,m^{-2}\,h^{-1} \right) = \Delta CO_2 \times 10^{-6} \times \rho \times V \times A^{-1} \times 60 \times 60 \times 12/44, \tag{1}$$

where ΔCO_2 is the increase in CO_2 concentration in the chamber (ppm·s^{-1}; Figure 3), ρ is the density of CO_2 in the air, V is the volume of the chamber (m^3), and A is the soil surface area (m^2). The average R_s in each area was calculated as the mean of all three chambers during 1 cycle of the AOCC system.

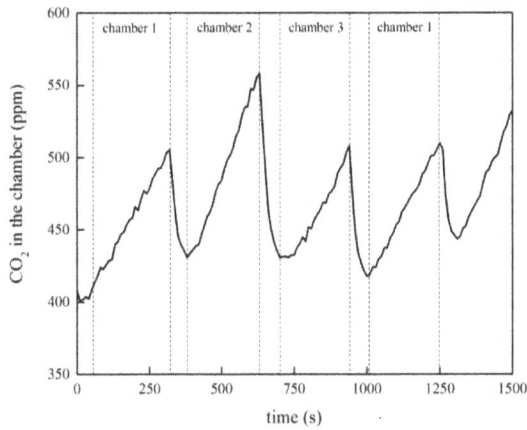

Figure 3. Changes in CO_2 concentration in chambers during the measurements using the AOCC system. ΔCO_2 was calculated from the intervals, which are evident from the high and low values that intersect the dotted lines.

To examine the response of R_s to T_s and W, these parameters were fitted using an exponential equation [31] as follows

$$R_s = ae^{bT_s}W^c, \tag{2}$$

where R_s is soil respiration (g·C·m^{-2}·h^{-1}), T_s is soil temperature at a 5 cm depth ($^\circ$C), W is soil water content at a 5 cm depth (%) and a, b and c are fitted coefficients. To estimate the coefficients a, b and c and fit the model using generalized linear models (GLM) analysis, Equation (2) was transformed as

$$\log R_s = \log a + bT_s + c\log W, \tag{3}$$

Thus, the full model for GLM analysis with a Gamma distribution and a log link function was expressed as

$$R_s \sim a' + bT_s + c\log W, \tag{4}$$

Moreover, the temperature sensitivity parameter (Q_{10}) was calculated as follows:

$$Q_{10} = e^{10b}, \tag{5}$$

Akaike's information criterion (AIC) was used to verify the accuracy of the R_s models. The best fitting model with the minimum AIC represented the accuracy of the best-fit R_s model. The analysis of models was performed using R version 3.0.3 [32] and library lme4 version 1.1-6 [33].

Annual R_s was estimated by summation of hourly R_s using W and T_s of areas C1 and G1 because both W and T_s in these areas were continuously recorded during the experimental period. However, the continuous measurement of W in the growing season could not be completed in both 2013 and 2014. Thus, annual R_s was estimated from 27 September 2013 to 26 September 2014. The data from 27 September to 27 October 2013 and from 4 June to 26 September 2014 were used to estimate R_s for the growing season, and R_s during the snow period was summed from 28 October 2013 to 3 June 2014. Because of no effect of W on R_s during the snow period, R_s during this period was estimated using the exponential relationship with T_s.

Upscaling R_s from the chamber measurements to the site scale was conducted on the basis of the spatial pattern of R_s between gap and canopy areas and the area of crown closure at the study site [25]. Model-estimated R_s in gap and canopy areas was used to represent R_s in canopy and gap areas in the study plot. Then, crown closure was used as a weighting factor to spatially transfer R_s over the whole study plot. This simple equation is defined as

$$F = F_c \times c + F_g \times (1 - c),$$

where F is R_s over the whole study site, F_g and F_c are the R_s values for the gap and canopy areas, respectively, and c is the crown closure measured by the vertically projected crown area divided by the whole study area. The crown closure of the study site was estimated on the basis of the crown projection diagram using Adobe Photoshop (Adobe Systems Incorporated, San Jose, California, USA) [24]. The areas of gap and canopy were estimated to be 25.6% and 74.4%, respectively, of the entire area.

3. Results

3.1. Temporal Changes in Soil Temperature and Soil Water Content

Clear temporal changes of T_s in gap and canopy areas were observed in the study plot (Figure 4). T_s rapidly increased after snow melt in May, and it trended to be higher in canopy areas than in gap areas. T_s in all areas peaked in August. After September, T_s in all areas dropped to nearly 0 °C during snowfall. There were no clear differences in T_s between canopy and gap areas along the time course, although T_s in canopy areas trended to be lower than in gap areas under snow cover.

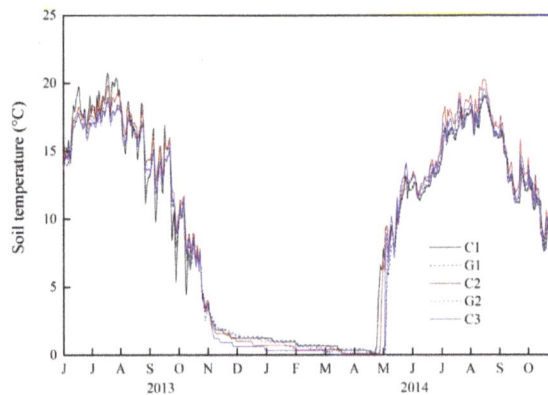

Figure 4. Temporal changes in soil temperature (T_s) at a depth of 5 cm in gap and canopy areas.

In 2013 (Figure 5a) and 2014 (Figure 5b), W did not show clear temporal changes in either the canopy or gap areas. In 2013, W was low in August, late September, and early October, although precipitation was recorded during these periods. In 2014, W was low in late June, late July, and September, but it was high in August because of a long period of rainy days.

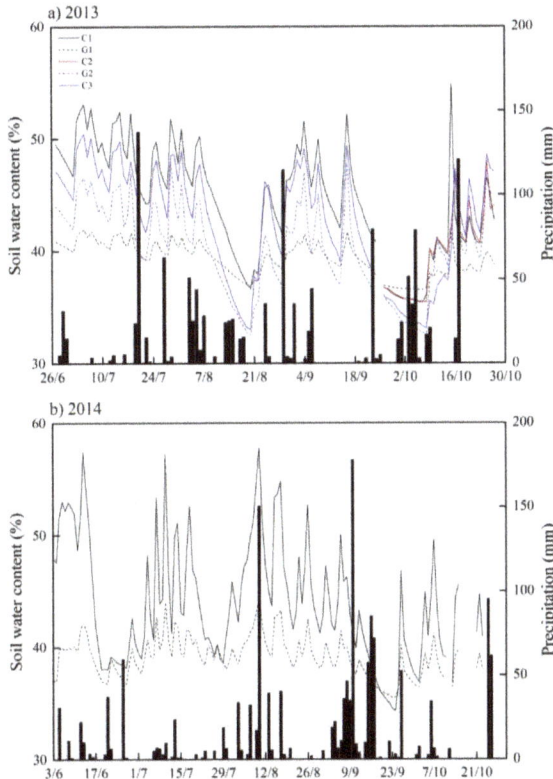

Figure 5. Temporal changes in soil water content (W) at a depth of 5 cm during the growing seasons of 2013 (**a**) and 2014 (**b**). Precipitation data were provided by Miboro weather station.

3.2. Diurnal Changes in Soil Temperature, Soil Water Content, and Soil Respiration

Diurnal R_s, T_s, and W at C1 and G1 were continuously monitored on measurement days during the 2014 growing season (Figure 6). T_s in both canopy and gap areas showed a diurnal pattern that was highest in the afternoon (around 1–3 p.m.) and lowest in the early morning (around 6 a.m.). R_s also showed a similar diurnal pattern to T_s, i.e., high in the afternoon and low in the early morning. However, except in late October, W did not show a clear diurnal pattern, although this was not investigated thoroughly in October. Diurnal W was likely similar between canopy and gap areas in the middle of June and July because of typical soil drying during these months. In October, diurnal W was higher in canopy areas than in gap areas because of the effect of rain a few days before the measurement day.

Canopy   Gap

a) 19 June   b) 19 June

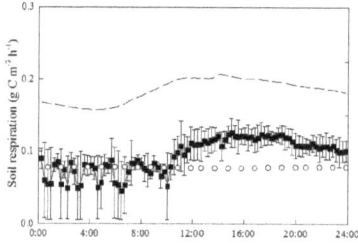 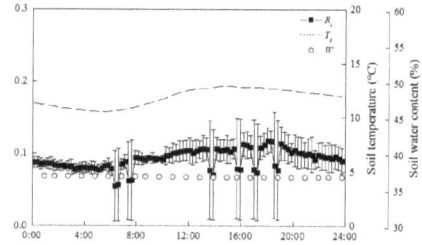

c) 15-16 July   d) 15-16 July

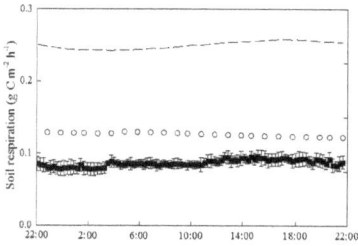 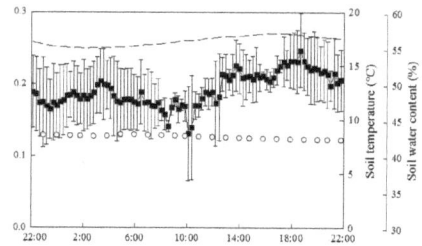

e) 8-9 October   f) 8-9 October

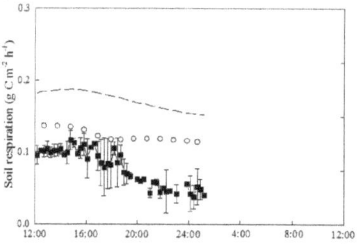 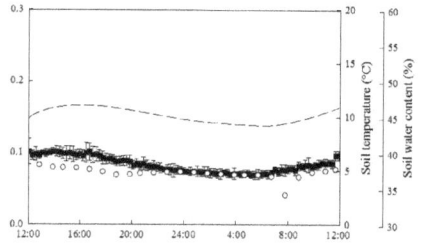

g) 29-30 October   h) 29-30 October

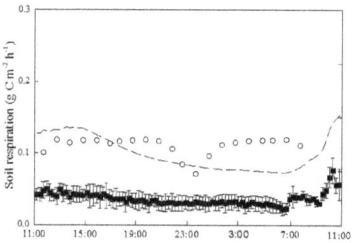 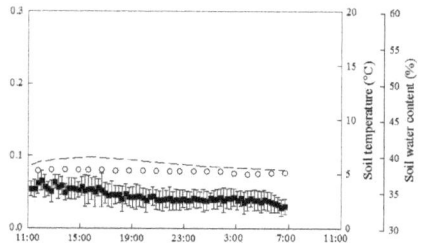

Figure 6. Diurnal patterns of soil respiration (R_s) measured using the AOCC system, soil temperature (T_s), and soil water content (W) for canopy and gap areas during 19 June (**a**, **b**), 15–16 July (**c**, **d**), 8–9 October (**e**, **f**) and 29–30 October (**g**, **h**) 2014. Diurnal R_s at C1 in September and early October could not be measured because of problems with automatic chamber closure. Error bar represents the standard error of the mean.

3.3. Effect of Soil Temperature and Soil Water Content on Soil Respiration

Canopy and gap areas showed temporal changes in R_s, which were related to the temporal pattern of T_s (Figure 7). R_s increased with an increase in T_s during June and July, and it peaked with the highest T_s in August. Then, R_s declined with decreasing T_s during August to October. The relationship between R_s and T_s was fitted as an exponential equation using GLM analysis: $R_s = 0.0207e^{0.1258Ts}$ for canopy areas, $R_s = 0.0177e^{0.1296Ts}$ for gap areas, and $R_s = 0.0198e^{0.1260Ts}$ for all areas. The Q_{10} values for canopy, gap, and all areas were calculated to be 3.52, 3.65, and 3.52, respectively.

Although the effect of T_s on R_s was similar in the two areas (Figure 7), the effect of W on R_s showed different patterns for canopy and gap areas. In canopy areas, R_s increased with W when W ranged from 32% to 56%. Meanwhile, R_s in gap areas decreased when W increased during the same range.

The analysis of the full model using GLM and AIC is summarized in Table 1. The best fit for both canopy and gap areas was found with the coupling of T_s and W and the minimum AIC. All parameters related to the factors of this model were estimated (Table 2) and transformed to the equations displayed in Table 3. Figure 7 shows the relationship among R_s, T_s and W based on the best equations in Table 3; R_s in canopy areas tended to increase with T_s and W, and R_s in gap areas tended to increase with T_s and decreasing W.

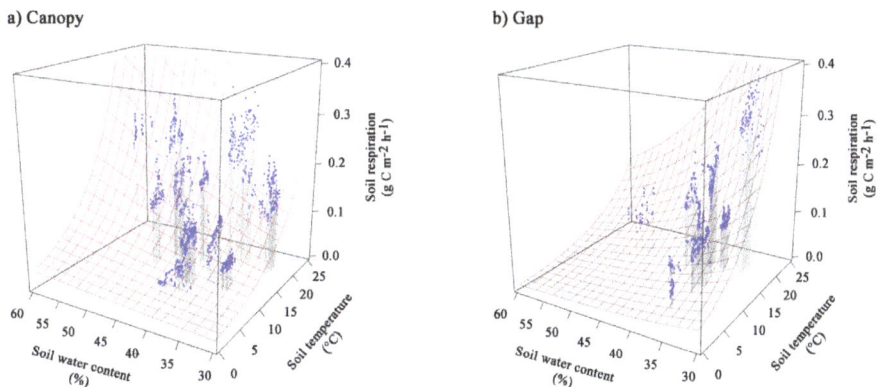

Figure 7. Relationships among soil respiration (R_s), soil temperature (T_s) and soil water content (W) in canopy (**a**) and gap (**b**) areas. The regression surface shows the best fitted model in each area.

Table 1. Summary of the results of the GLMs sorted in the analysis in gap and canopy areas.

Model	Formula	AIC	ΔAIC	Deviance	*df*	Dispersion parameter
		Canopy				
1	$R_s \sim a'$	−5167.4	3152.4	609.3	2025	0.301
2	$R_s \sim a' + T_s$	−7236.2	1083.5	226.2	2024	0.112
3	$R_s \sim a' + c \log W$	−5270.1	3049.6	580.0	2024	0.287
4	$R_s \sim a' + bT_s + c \log W$	−8319.7	0.0	133.4	2023	0.066
		Gap				
1	$R_s \sim a'$	−3714.9	1551.9	473.5	1457	0.325
2	$R_s \sim a' + T_s$	−4865.5	401.3	221.0	1456	0.125
3	$R_s \sim a' + c \log W$	−3964.1	1302.7	401.8	1456	0.276
4	$R_s \sim a' + bT_s + c \log W$	−5266.8	0.0	168.6	1455	0.116

Table 2. Summary of the estimated coefficients in model 4, which was selected as the best model based on Akaike's information criteria (AIC) in Table 1.

Factor	Estimate	Standard Error	*t* Value	*p* Value
		Canopy		
(Intercept)	−10.2589	0.1646	−62.33	<0.001
T_s	0.1349	0.0015	88.11	<0.001
log W	1.6838	0.0430	39.20	<0.001
		Gap		
(Intercept)	4.6302	0.3816	12.14	<0.001
T_s	0.1284	0.0023	55.09	<0.001
log W	−2.3719	0.1038	−22.86	<0.001

Table 3. Soil respiration (R_s) estimated from the model using soil temperature (T_s) and soil water content (W). R_s during the growing season was estimated from 27 September to 27 October 2013 and from 4 June to 26 September 2014, and R_s during the snow season was estimated from 28 October 2013 to 3 June 2014. Total R_s was calculated from the R_s values of the canopy and gap areas based on their relative sizes (canopy: 74.4%, gap: 25.6%).

Model		Estimated R_s ($g \cdot C \cdot m^{-2}$)	
		Growing Season	Snow Season
T_s			
All	$R_s = 0.0198e^{0.1260T_s}$	437.4	153.7
Canopy	$R_s = 0.0207e^{0.1258T_s}$	507.7	111.8
Gap	$R_s = 0.0177e^{0.1296T_s}$	451.8	97.0
Total		493.4	108.0
T_s & W			
Canopy	$R_s = 3.5044 \times 10^{-5}e^{0.1349T_s}W^{1.684}$	541.4	No data
Gap	$R_s = 102.5405e^{0.1284T_s}W^{-2.372}$	384.6	No data
Total		501.3	

3.4. Estimation and Upscaling of R_s to the Whole Site

The estimated R_s values from the T_s dependence model and the model incorporating both T_s and W are shown in Table 3. For the growing season, estimated R_s from the exponential relationship was lower than that from the coupled model in canopy areas, but this comparison was reversed in gap areas. During the snow season, R_s was not affected by W. Thus, R_s during this season was estimated from the specific exponential model with T_s.

On the basis of the crown closure of the study plot (canopy: 74.4% of the whole area; gap: 25.6% of the whole area), the R_s values estimated using the model with T_s dependence and the model with the coupling of T_s and W were 493.4 and 501.3 $g \cdot C \cdot m^{-2}$, respectively, for the growing season. The R_s value for the snow season estimated using the T_s dependence model was 108.0 $g \cdot C \cdot m^{-2}$. Therefore, the annual R_s, in which the estimated R_s for the snow season was included, ranged between 601.4 and 609.3 $g \cdot C \cdot m^{-2} \cdot year^{-1}$. This range based on the crown closure was greater than the annual R_s estimated using the T_s dependence model across all areas (591.1 $g \cdot C \cdot m^{-2} \cdot year^{-1}$).

4. Discussion

4.1. Effect of Soil Temperature and Soil Water Content on Soil Respiration

Temporal change in R_s is primarily controlled by T_s in many temperate deciduous forests [6–8]; R_s generally increases with an exponential increase in T_s within the range 0–25 °C. In concurrence with these previous investigations, the present study showed a pattern of increasing R_s with increasing T_s (Figure 7).

In some temperate forests, R_s is influenced not only by T_s but also by W (e.g., [3,32,34–38]). During the dry season, W decreases and the diffusion of soluble substrates slows down, resulting in low R_s [34]. This change in R_s caused by low W might support the result of the effect of low W on low R_s in canopy areas in the present study. When the soil pore space is filled with water and approaches saturation, the movement of oxygen is limited. As a result, the metabolic activity of aerobic organisms in soil decreases; thus, R_s also decreases [38]. This probably explained why high W affected low R_s in gap areas. Without the effect of T_s, the relationship between R_s and W is generally described by a curve, which has minima at the extreme low and high values of W and its maximum at the value of W where the balance of water and oxygen is optimal [9,25]. However, in the present study, there is a possibility that W values producing maximum R_s would be quite different between canopy and gap areas. In addition, the appearance of tree coarse roots partly contributed to the increasing R_s with W in canopy areas. There is evidence that coarse root (diameter > 5 mm) respiration of *Pinus taeda* declines at low soil water availability [39]. Therefore, because of high root biomass of large trees in canopy areas, high W might induce more metabolic activity and a higher root respiration rate in canopy areas than in gap areas.

The sensitivity of the respiratory process, described by Q_{10}, is known to be related to changes in temperature. Moreover, Q_{10} largely results from the confounding effect of temperature on multiple processes with covarying variables such as W [9,40]. The factors controlling R_s and Q_{10} across sites with varying drainage classes were investigated in mixed hardwood forests of the USA by Davidson et al. [9]. This study showed that R_s at well-drained sites was greater than that at wetter sites, whereas Q_{10} at the wetter sites was greater than that at the well-drained sites. R_s would be less responsive to T_s at the wetter sites than the well-drained sites because, as previously mentioned, high W limits air diffusion and decreases R_s. In the present study, however, the Q_{10} of R_s in canopy (3.52) and gap (3.65) areas was not different, whereas W in canopy areas tended to be greater than that in gap areas (Figure 5). In addition to W, several studies reported that the Q_{10} of R_s varied among different components in soil, for example, litter and roots [41,42]. It is possible that both Q_{10} and R_s between canopy and gap areas were contributed to by the different sensitivity of soil components. For example, the gap areas had the high density of dwarf bamboo (Figure 1) but this litter is not decomposed easily because of the silica content, partly contributing to the low R_s in gap areas. Thus, further study on the contribution of respiration from various soil components to R_s in canopy and gap areas is needed to explain the Q_{10} results in the present old-growth forest.

4.2. Estimation of Soil Respiration in the Old-Growth Forest

The continuous measurement of temporal R_s is necessarily required to develop models for estimation of annual R_s. Although the previous study of soil respiration in the present old-growth forest was conducted using the soda-lime method [24], this technique could not provide enough detail of diurnal and seasonal changes in R_s. The AOCC system based on closed dynamic methods allows continuous measurement of R_s, creating enough data to understand the relationship between diurnal and seasonal R_s and environmental factors. This method has been used in some forests [3,11–14] but is rarely used in old-growth forests. Using this system in the present study site sometimes resulted in some problems, in addition to no electricity, for example, the chambers could not fully close due to heavy rain. However, the AOCC system provided sufficient and detailed R_s measurements in the present study.

For evaluation of annual R_s using equations, seasonal R_s has often been estimated using only T_s dependence as an exponential relationship in forests where W is not limiting (e.g., [6–8,43–45]). However, in forests where there is a drying period, R_s has been estimated by the coupling of T_s and W, although an empirical model using these variables has been unclear [3,31,35–39]. In the present study, the GLM analysis identified the model with the coupling of T_s and W as the best model (Table 1). This model also had more accuracy than the model with only T_s because the model with the coupling of T_s and W included responses of R_s to W, which differed between gap and canopy areas (Figure 7). Therefore, the empirical model with the coupling of T_s and W was suitable for the estimation of R_s for the growing season at the present study site.

At the stand level, the high spatial variability in R_s resulted from large variations in, for example, W [46], soil physical properties [24,47–49], fine root biomass [23,24,36,37,50,51], and stand structure [25,46]. A few studies have addressed this spatial variation for upscaling R_s from different measurement points to the site scale [25,46]. In the present old-growth forest, there was a clear gap and canopy structure, and the high R_s in the canopy areas could represent the high root respiration because of high root biomass [24]. Additionally, different responses of R_s to W between canopy and gap areas were found in the present study. Thus, the estimation of R_s should be conducted separately for canopy and gap areas. The estimation of annual R_s with the crown closure could reflect a respective range of R_s values across all gaps to all canopies as 481.6 to 653.2 $g \cdot C \cdot m^{-2} \cdot year^{-1}$. Therefore, the combination of the spatial patterns of R_s, particularly canopy and gap areas, and the crown closure could be used for upscaling R_s from chamber measurements to the stand level in old-growth forests.

5. Conclusions

In the present old-growth forest, the structural complexity created by canopy and gap spaces induced different responses of R_s to T_s and W. R_s increased with increasing T_s in both canopy and gap areas, and Q_{10} values in these areas were not different. In terms of the effect of W on R_s, this relationship differed between canopy and gap areas; R_s increased with W in canopy areas, but R_s tended to decrease with W in gap areas. Consequently, to understand the influence of T_s and W on R_s estimation, the GLM analysis identified that an empirical model that couples these factors (T_s and W) was better than a simple exponential model with only T_s. Because of these results, it was necessary to estimate R_s in canopy and gap areas separately. Additionally, combining the unique spatial patterns of R_s and the area of the crown closure could allow for upscaling of R_s from field measurements to the whole site in old-growth forests.

Acknowledgments: We thank members of the Takayama Forest Research Station, the Institute for Basin Ecosystem Studies, Gifu University, for providing their facilities during the field survey. We also thank members of the laboratory of the Institute for Basin Ecosystem Studies, Gifu University, for their assistance with field measurements.

Author Contributions: V.S. and T.O. conceived and designed the experiments; V.S., S.Y., Y.I. and T.O. performed the experiments; V.S. and M.A. analyzed the data; M.H. contributed materials tools; V.S. wrote the paper.

Conflicts of Interest: The authors declare no conflict of interest.

References

1. Liu, Q.; Edwards, N.T.; Post, W.M.; Gu, L.; Ledford, J.; Lenhart, S. Temperature-independent diel variation in soil respiration observed from a temperate deciduous forest. *Glob. Chang. Biol.* **2006**, *12*, 2136–2145. [CrossRef]
2. Hashimoto, T.; Miura, S.; Ishizuka, S. Temperature controls temporal variation in soil CO_2 efflux in a secondary beech forest in Appi Highlands, Japan. *J. For. Res.* **2009**, *14*, 44–50. [CrossRef]
3. Joo, S.J.; Park, S.U.; Park, M.S.; Lee, C.S. Estimation of soil respiration using automated chamber systems in an oak (*Quercus mongolica*) forest at the Nam-San site in Seoul, Korea. *Sci. Total Environ.* **2012**, *416*, 400–409. [CrossRef] [PubMed]
4. Kishimoto-Mo, A.W.; Yonemura, S.; Uchida, M.; Kondo, M.; Murayama, S.; Koizumi, H. Contribution of soil moisture to seasonal and annual variations of soil CO_2 efflux in a humid cool-temperate oak-birch forest in central Japan. *Ecol. Res.* **2015**, *30*, 311–325. [CrossRef]
5. Xu, M.; Qi, Y. Soil-surface CO_2 efflux and its spatial and temporal variations in a young ponderosa pine plantation in Northern California. *Glob. Chang. Biol.* **2001**, *7*, 667–677. [CrossRef]

6. Mo, W.; Lee, M.; Uchida, M.; Inatomi, M.; Saigusa, N.; Mariko, S.; Koizumi, H. Seasonal and annual variations in soil respiration in a cool-temperate deciduous broad-leaved forest in Japan. *Agric. For. Meteorol.* **2005**, *134*, 81–94. [CrossRef]

7. Ohtsuka, T.; Shizu, Y.; Nishiwaki, A.; Yashiro, Y.; Koizumi, H. Carbon cycling and net ecosystem production at an early stage of secondary succession in an abandoned coppice forest. *J. Plant Res.* **2010**, *123*, 393–401. [CrossRef] [PubMed]

8. Inoue, T.; Nagai, S.; Inoue, S.; Ozaki, M.; Sakai, S.; Muraoka, H.; Koizumi, H. Seasonal variability of soil respiration in multiple ecosystems under the same physical-geographical environmental conditions in central Japan. *For. Sci. Technol.* **2012**, *8*, 52–60. [CrossRef]

9. Davidson, E.A.; Belk, E.; Boone, R. Soil water content and temperature as independent of confounded factors controlling soil respiration in a temperate mixed hardwood forest. *Glob. Chang. Biol.* **1998**, *4*, 217–227. [CrossRef]

10. Wang, Y.; Hao, Y.; Cui, X.Y.; Zhao, H.; Xu, C.; Zhou, X.; Xu, Z. Responses of soil respiration and its components to drought stress. *J. Soil Sediments* **2013**, *14*, 99–109. [CrossRef]

11. Irvine, J.; Law, B.E. Contrasting soil respiration in young and old-growth ponderosa pine forests. *Glob. Chang. Biol.* **2002**, *8*, 1183–1194. [CrossRef]

12. Edwards, N.T.; Riggs, J.S. Automated monitoring of soil respiration: A moving chamber design. *Soil Sci. Soc. Am. J.* **2003**, *67*, 1266–1271. [CrossRef]

13. Liang, N.; Inoue, G.; Fujinuma, Y. A multichannel automated chamber system for continuous measurement of forest soil CO_2 efflux. *Tree Physiol.* **2003**, *23*, 825–832. [CrossRef] [PubMed]

14. Suh, S.U.; Chun, Y.M.; Chae, N.; Kim, J.; Lim, J.H.; Yokozawa, M.; Lee, M.S.; Lee, J.S. A chamber system with automatic opening and closing for continuously measuring soil respiration based on an open-flow dynamic method. *Ecol. Res.* **2006**, *21*, 405–414. [CrossRef]

15. Nakashizuka, T.; Numata, M. Regeneration process of climax beech forests: I. Structure of a beech forest with the undergrowth of *Sasa*. *Jpn. J. Ecol.* **1982**, *32*, 57–67.

16. Tanouchi, H.; Yamamoto, S. Structure and regeneration of canopy species in an old-growth evergreen broad-leaved forest in Aya district, Southwestern Japan. *Vegetation* **1995**, *117*, 51–60. [CrossRef]

17. Manabe, T.; Nishimura, N.; Miura, M.; Yamamoto, S. Population structure and spatial patterns for trees in a temperate old-growth evergreen broad-leaved forest in Japan. *Plant Ecol.* **2000**, *151*, 181–197. [CrossRef]

18. Miura, M.; Manabe, T.; Nishimura, N.; Yamamoto, S. Forest canopy and community dynamics in a temperate old-growth evergreen broad-leaved forest, south-western Japan: A 7-year study of a 4-ha plot. *J. Ecol.* **2001**, *89*, 841–849. [CrossRef]

19. Matsushita, M.; Hoshino, D.; Yamamoto, S.; Nishimura, N. Twenty-three years and dynamics in an old-growth *Chamaecyparis* forest in central Japan. *J. For. Res.* **2014**, *19*, 134–142. [CrossRef]

20. Suchewaboripont, V.; Iimura, Y.; Yoshitake, S.; Kato, S.; Komiyama, A.; Ohtsuka, T. Change in biomass of an old-growth beech-oak forest of Mt. Hakusan over a 17-year period. *Jpn. J. For. Environ.* **2015**, *57*, 33–42.

21. Hardiman, B.S.; Bohrer, G.; Gough, C.M.; Vogel, C.S.; Curtis, P.S. The role of canopy structural complexity in wood net primary production of a maturing northern deciduous forest. *Ecology* **2011**, *92*, 1818–1827. [CrossRef] [PubMed]

22. Hardiman, B.S.; Gough, C.M.; Halperin, A.; Hofmeister, K.L.; Nave, L.E.; Bohrer, G.; Curtis, P.S. Maintaining high rates of carbon storage in old forests: A mechanism lining canopy structure to forest function. *For. Ecol. Manag.* **2013**, *298*, 111–119. [CrossRef]

23. Hojjati, S.M.; Lamersdorf, N.P. Effect of canopy composition on soil CO_2 emission in a mixed spruce-beech forest at Solling, Central Germany. *J. For. Res.* **2010**, *21*, 461–464. [CrossRef]

24. Suchewaboripont, V.; Ando, M.; Iimura, Y.; Yoshitake, S.; Ohtsuka, T. The effect of canopy structure on soil respiration in an old-growth beech-oak forest in central Japan. *Ecol. Res.* **2015**, *30*, 867–877. [CrossRef]

25. Tang, J.; Baldocchi, D.D. Spatial-temporal variation in soil respiration in an oak-grass savanna ecosystem in California and its partitioning into autotrophic and heterotrophic components. *Biogeochemistry* **2005**, *73*, 183–207. [CrossRef]

26. Geological Survey of Japan, AIST (Ed.) Catalog of Eruptive Events during the Last 10,000 Years in Japan, Version 2.2. Available online: https://gbank.gsj.jp/volcano/eruption/index.html (accessed on 29 December 2016).

27. Kato, S.; Komiyama, A. Distribution patterns of understory trees and diffuse light under the canopy of a beech forest. *Jpn. J. Ecol.* **1999**, *49*, 1–10.

28. Runkle, J.R. Gap regeneration in some old-growth forests of the eastern United States. *Ecology* **1981**, *62*, 1041–1051. [CrossRef]

29. Hirota, M.; Zhang, P.; Gu, S.; Shen, H.; Kuriyama, T.; Li, Y.; Tang, Y. Small-scale variation in ecosystem CO_2 fluxes in an alpine meadow depends on plant biomass and species richness. *J. Plant Res.* **2010**, *123*, 531–541. [CrossRef] [PubMed]

30. Abbas, F.; Fares, A.; Fares, S. Field calibrations of soil moisture sensors in a forested watershed. *Sensors* **2011**, *11*, 6354–6369. [CrossRef] [PubMed]

31. Yang, Y.S.; Chen, G.S.; Guo, J.F.; Xie, J.S.; Wang, X.G. Soil respiration and carbon balance in a subtropical native forest and two managed plantations. *Plant Ecol.* **2007**, *193*, 71–84. [CrossRef]

32. R Development Core Team. *R: A Language and Environment for Statistical Computing*; R Foundation for Statistical Computing: Vienna, Austria, 2014.

33. Bates, D.; Maechler, M.; Bolker, B.; Walker, S. lme4: Linear Mixed-Effects Models Using Eigen and S4. R Package Version 1.1-7. 2014. Available online: http://CRAN.R-project.org/package=lme4 (accessed on 22 April 2016).

34. Epron, D.; Farque, L.; Lucot, É.; Badot, P.M. Soil CO_2 efflux in a beech forest: Dependence on soil temperature and soil water content. *Ann. For. Sci.* **1999**, *56*, 221–226. [CrossRef]

35. Qi, Y.; Xu, M. Separating the effects of moisture and temperature on soil CO_2 efflux in a coniferous forest in the Sierra Nevada mountains. *Plant Soil* **2001**, *237*, 15–23. [CrossRef]

36. Søe, A.R.B.; Buchmann, N. Spatial and temporal variations in soil respiration in relation to stand structure and soil parameters in an unmanaged beech forest. *Tree Physiol.* **2005**, *25*, 1427–1436. [CrossRef] [PubMed]

37. Knohl, A.; Søe, A.R.B.; Kutsch, W.L.; Göckede, M.; Buchmann, N. Representative estimates of soil and ecosystem respiration in an old beech forest. *Plant Soil* **2008**, *302*, 189–202. [CrossRef]

38. Tomotsune, M.; Masuda, R.; Yoshitake, S.; Anzai, T.; Koizumi, H. Seasonal and inter-annual variations in contribution ratio of heterotrophic respiration to soil respiration in a cool-temperate deciduous forest. *J. Geogr.* **2013**, *122*, 745–754. [CrossRef]

39. Maier, C.A.; Kress, L. Soil CO_2 evolution and root respiration in 11 year-old loblolly pine (*Pinus taeda*) plantations as affected by moisture and nutrient. *Can. J. For. Res.* **2000**, *30*, 347–359. [CrossRef]

40. Dörr, H.; Münnich, K.O. Annual variation in soil respiration in selected areas of the temperate zone. *Tellus* **1987**, *39*, 114–121. [CrossRef]

41. Boone, R.D.; Nadelhoffer, K.J.; Canary, J.D.; Kaye, J.P. Roots exert a strong influence on the temperature sensitivity of soil respiration. *Nature* **1998**, *396*, 570–572. [CrossRef]

42. Liski, J.; Ilvesniemi, H.; Mäkelä, A.; Westman, C.J. CO_2 emissions from soil in response to climatic warming are overestimated: The decomposition of old soil organic matter is tolerant of temperature. *Ambio* **1999**, *28*, 171–174.

43. Liang, N.; Hirano, T.; Zheng, Z.-M.; Tang, J.; Fujinuma, Y. Soil CO_2 efflux of a larch forest in Northern Japan. *Biogeosciences* **2010**, *7*, 3447–3457. [CrossRef]

44. Vesterdal, L.; Elberling, B.; Christiansen, J.R.; Callesen, I.; Schmidt, I.K. Soil respiration and rates of soil carbon turnover differ among six common European tree species. *For. Ecol. Manag.* **2012**, *264*, 185–196. [CrossRef]

45. Luan, J.; Liu, S.; Wang, J.; Zhu, X. Factors affecting spatial variation of annual apparent Q_{10} of soil respiration in two warm temperate forests. *PLoS ONE* **2013**, *8*, e64167. [CrossRef] [PubMed]

46. Jia, S.; Akiyama, T.; Mo, W.; Inatomi, M.; Koizumi, H. Temporal and spatial variability of soil respiration in a cool-temperate broad-leaved forest 1. Measurement of spatial variance and factor analysis. *Jpn. J. Ecol.* **2003**, *53*, 13–22.

47. Scott-Denton, L.E.; Sparks, K.L.; Monson, R.K. Spatial and temporal controls of soil respiration rate in a high-elevation, subalpine forest. *Soil Biol. Biochem.* **2003**, *35*, 525–534. [CrossRef]

48. Saiz, G.; Green, C.; Butterbach-Bahl, K.; Kiese, R.; Avitabile, V.; Farrell, E.P. Seasonal and spatial variability of soil respiration in four Sitka spruce stands. *Plant Soil* **2006**, *287*, 161–176. [CrossRef]

49. Ngao, J.; Epron, D.; Delpierre, N.; Bréda, N.; Granier, A.; Longdoz, B. Spatial variability of soil CO_2 efflux linked to soil parameters and ecosystem characteristics in a temperate beech forest. *Agric. For. Meteorol.* **2012**, *154–155*, 136–146. [CrossRef]

50. Stoyan, H.; De-Polli, H.; Böhm, S.; Robertson, G.P.; Paul, E.A. Spatial heterogeneity of soil respiration and related properties at the plant scale. *Plant Soil* **2000**, *222*, 203–214. [CrossRef]

51. Adachi, M.; Bekku, Y.S.; Rashidah, W.; Okuda, T.; Koizumi, H. Differences in soil respiration between different tropical ecosystems. *Appl. Soil Ecol.* **2006**, *34*, 258–265. [CrossRef]

![forests logo] *forests*

MDPI

Article

Temporal Variability of Soil Respiration in Experimental Tree Plantations in Lowland Costa Rica

James W. Raich

Department of Ecology, Evolution & Organismal Biology, Iowa State University, Ames, IA 50011, USA; jraich@iastate.edu; Tel.: +1-515-294-5073

Academic Editors: Robert Jandl and Mirco Rodeghiero
Received: 10 January 2017; Accepted: 2 February 2017; Published: 8 February 2017

Abstract: The principal objective of this study was to determine if there is consistent temporal variability in soil respiration from different forest plantations in a lowland tropical rainforest environment. Soil respiration was measured regularly over 2004 to 2010 in replicated plantations of 15- to 20-year-old evergreen tropical trees in lowland Costa Rica. Statistically significant but small differences in soil respiration were observed among hours of the day; daytime measurements were suitable for determining mean fluxes in this study. Fluxes varied more substantially among months, with the highest average emissions (5.9 $\mu mol \cdot m^{-2} \cdot s^{-1}$) occurring in September and low emissions (3.7 $\mu mol \cdot m^{-2} \cdot s^{-1}$) occurring in January. Three of the six tree species had significantly increasing rates of soil respiration across 2004–2010, with fluxes increasing at an average of 0.09 $\mu mol \cdot m^{-2} \cdot s^{-1}$ per year: the three other species had no long-term trends. It was hypothesized that there would be a tradeoff between carbon allocation aboveground, to produce new leaves, and belowground, to sustain roots and mycorrhizae, but the relationship between canopy leaf fall—a surrogate for canopy leaf flushing—and soil respiration was significantly positive. The similarities observed among temporal trends across plantation types, and significant relationships between soil respiration, soil water content and soil temperature, suggest that the physical environment largely controlled the temporal variability of soil respiration, but differences in flux magnitude among tree species were substantial and consistent across years.

Keywords: climate change; climate warming; forest carbon cycle; plant-soil system; soil carbon cycle; tropical forest; tropical forest phenology; tropical rainforest

---

1. Introduction

The tropical forest biome covers only about 12% of Earth's land surface [1,2], yet it harbors much of the world's biological diversity [3–6] and disproportionally influences global biogeochemistry. Tropical forests are responsible for one-third of the world's terrestrial plant productivity [1,2,7], one-third of the world's terrestrial respiration [8] and one-third of terrestrial evapotranspiration [9]. Although they vary greatly in composition and structure, tropical rainforests often contain substantial living biomass, and large stocks of detrital carbon and organic nitrogen [10–13] that sustain high rates of microbial activity. In brief, the warm, moist conditions that prevail year-round in tropical forests support abundant life and rapid land-atmosphere exchanges of energy, water and carbon. Both warmer temperatures and an accelerated hydrologic cycle are expected in tropical forest regions in future years [14]. Those changes will alter tropical forests and soils and, thus, land-atmosphere exchanges of carbon, energy and water. We do not fully envision what the resulting impacts will be.

Large stocks of detritus, warm temperatures, and humid conditions together create a strong potential for warming (without drying) to stimulate decomposer activity and the mineralization of detritus to CO_2 and other trace gases. That could generate a positive feedback to global warming, because CO_2 is an important greenhouse gas [15–17]. The likelihood that this particular feedback will emerge has been extensively discussed [18–23] but has not been ruled out [24–29]. As a result, despite decades of discussion and debate, the question persists. Although greater warming is expected in northern latitudes than in the tropics [14], CO_2 fluxes from humid tropical forests are already large, and small changes in large fluxes can be important. For example, a 3% increase in a tropical soil respiration of 1500 $g \cdot C \cdot m^{-2} \cdot year^{-1}$ is greater than a 20% increase in a tundra soil-respiration rate of 200 $g \cdot C \cdot m^{-2} \cdot year^{-1}$. Based on studies of moist tropical forests, it has been proposed that decomposition rates increase faster at warmer temperatures than does net primary productivity (NPP) [30]. That would be consistent with a positive-feedback scenario for tropical forest soil respiration. A variety of evidence suggests that the tropics already influence interannual variations in atmospheric CO_2 concentrations [31]. Measurements of soil respiration are insufficient, by themselves, to address this debate, but tropical forests play an important role in the earth's carbon cycle; they will be impacted by climatic changes; they are likely to influence future atmospheric changes; and they remain understudied in relation to their potential to cause change.

Objectives

The over-riding objective of this study was to determine if there is consistent temporal variability in soil respiration (*Rsoil*) from evergreen forest plantations in a lowland tropical rainforest environment. To do so, *Rsoil* was measured over diel, monthly and multi-annual time frames. Second, I sought to identify the environmental variables underlying observed temporal variations in *Rsoil*, to inform the process-based understanding of tropical forest carbon dynamics that is needed to improve predictive modeling [32]. Despite an increasing body of empirical data from tropical forests, there is a lack of agreement on several issues. For instance, diel variations in *Rsoil* have been observed in some evergreen tropical forests, but not in others [33–38]. That issue underlies an important concern: most measurements of tropical forest *Rsoil* are made only in the daytime, especially in remote locations. If fluxes are significantly different at night than during daylight hours, then we have a poor grasp of what average tropical forest *Rsoil* truly is [35]. Additionally, daytime photosynthesis by a productive tree canopy may drive diel variations in *Rsoil*, via temporally variable canopy-to-root carbon translocation [39–41]. Over longer time frames, most studies indicate that temporal variability in tropical forest *Rsoil* is attributable to variations in soil moisture content or temperature [37,38,42], indicating that climatic variability is important. However, plant phenology may independently influence the temporal dynamics of *Rsoil*, via seasonally variable rates of aboveground litter production or root growth or turnover [43]. Furthermore, if *Rsoil* is limited by carbon (C) fluxes to roots, then trade-offs between aboveground and belowground C allocation may also cause seasonal variability in *Rsoil*. For instance, belowground fluxes may be lower at times when forest canopies are producing a new crop of leaves. If so, that might suggest that root-shoot allocation patterns vary seasonally within a forest, and that carbon limitation influences the magnitude of *Rsoil*, at least during some months.

I tested the following hypotheses, based on the measurements collected. First, Hypothesis 1: soil respiration varies on a diel basis. Such temporal variations in *Rsoil* have important implications for interpretation of daytime-only measurements, and might result from a variety of processes. For instance, lowland, humid tropical forests have a unique climatic feature: diel variations in temperature often exceed the variations in mean monthly temperatures. Warmer daytime temperatures may drive faster daytime CO_2 production in, and diffusion from, soils. Also, observed night-time warming is greater than observed daytime warming in the tropics, probably due to increased cloud cover [44–47], yet most data available on soil CO_2 emissions from tropical rainforests are derived from daytime measurements. In tropical forests, soil temperatures typically exceed air temperatures at night, and that could promote movement of CO_2-rich soil air into the atmosphere. I also tested Hypothesis 2, that *Rsoil* varies among months of the year. Even within relatively aseasonal wet tropical forests, there is climatic variability and there are phenological differences among tree species [48] that likely influence the magnitudes of aboveground and belowground C fluxes. However, seasonal patterns in belowground carbon fluxes in evergreen tropical forests are very poorly understood. To further evaluate seasonal variability, I also tested Hypothesis 3, that *Rsoil* and canopy leaf flushing are negatively correlated with one another. This hypothesis is based on the supposition that there is a trade-off between plant carbon use aboveground, to produce a new crop of canopy leaves, and belowground, to support root systems. Leaf flushing by tropical evergreen broad-leaved trees represents an important and seasonally variable investment of available carbon in a new crop of leaves. As new leaves expand out of their buds, old leaves are dropped. Thus, leaf flushing appears as seasonality of leaf fall by canopy trees. Finally, I tested Hypothesis 4, that *Rsoil* increases with forest age in developing forest stands, as suggested by [49]. This hypothesis is put forth to determine if there is a multi-annual trajectory in *Rsoil*, either positive or negative, in the 15- to 20-year old experimental tree plantations that were the focus of this research. It is widely accepted that forest productivity increases with age in young forests, to a point, but does not continue to increase continuously, and very often decreases as forests mature [50–52]. Such non-linear trends may be difficult to disprove, but tropical tree plantations can grow rapidly [53]; quantifying belowground dynamics in a variety of different forest types as they mature and age, within a single rainforest location, can provide valuable insights into the temporal dynamics of belowground processes, about which we know very little.

2. Materials and Methods

2.1. Study Site

This research was undertaken at the La Selva Biological Station of the Organization for Tropical Studies, in the Caribbean lowlands of northeastern Costa Rica (10°26′ N, 84°03′ W). Over 1997–2009, annual precipitation averaged 4537 mm and mean air temperature was 25.1 °C [54]. The driest season at La Selva generally extends from February through April, but rainfall averages > 150 mm·mol^{-1} in every month. Forest evapotranspiration averages > 2000 mm·year^{-1} [55]. The native vegetation at this site is species-rich broad-leaved evergreen rainforest that has a high abundance of *Pentaclethra macroloba* (Willd.) Kuntze (Fabaceae, Mimosoideae) trees, and many subcanopy palms. In primary forest on rolling terrain, there are an average of 528 trees·ha^{-1} having a basal area of 23 m^2·ha^{-1} [56]. Long-term studies of old-growth forest at La Selva show that litterfall averages 9.1 Mg·ha^{-1}·year^{-1}, 78% of which is leaves [54]. Tree growth averages half that and their sum, aboveground net primary productivity (ANPP), averages 13.7 Mg·ha^{-1}·year^{-1} [54]. More information about the physical and biological features of La Selva is provided in [57].

Field studies for this research were conducted in replicated experimental tree plantations that were established on formerly grazed pastures, to test the influence of tree species identity on soil properties after reforestation [58,59]. The pastures were established from mature forest in the mid-1950s, and trees were planted, after removal of the cattle, in the winter of 1988–1989 [53,58,60]. This study began 15 years later. The study plots were at 10°26′ N, 83°59′ W, on hilly terrain with elevations of 44 to 89 m. They were approximately 3.3 km from the biological station and weather station. The soils were Oxisols derived from volcanic parent materials, and belong to the Matabuey consociation [61]. In brief, they are deep, acidic, highly permeable clays with low base saturation and relatively high soil organic matter contents [61–63]. Surface-soil pH (in water) was <4.5 in all plots [63]. The available data indicate that total soil C and nitrogen (N) stocks were lower in the plantations than in nearby, undisturbed forest, on average; and that soil organic C mineralization rates (gCO$_2$-C gSoil^{-1}·day^{-1}) were lower in plantation soils than in mature forest soils [8,53,63]. Nevertheless, rates of N cycling in the plantations were very high [64]. Fine-root biomass was concentrated in the surface 15 cm of mineral soil, and declined rapidly with increasing soil depth to 1 m [65]. The forest floors were comprised primarily of leaves and particulate debris that decomposed rapidly: forest floor turnover rates among the plantations we studied ranged from 1.5 to 2.3 year^{-1} [66]. The diversity and abundance of litter-dwelling organisms were high.

The experimental design included four randomized blocks composed of twelve 50 m × 50 m plots, each of which originally contained a single tree species planted at 3 m × 3 m spacing, except that the control plots were not planted or otherwise managed in any way [58,67]. Each block was centered across a single hill and thus encompassed a range of slope and aspect positions. Plantations of the fastest growing species were thinned three years after planting, and additional thinning was done three years later [67,68]. Each of four replicate plots of six plantation types, a total 24 plots, were included in this study. These included plantations of *Pinus patula* subsp. *tecunumanii* (Eguiluz and J. P. Perry) Styles; *Hieronyma alchorneoides* Allemao; *Pentaclethra macroloba* (Willd.) Kuntze; *Virola koschnyi* Warb.; *Vochysia ferruginea* Mart.; and *Vochysia guatemalensis* Donn. Sm. *Pentaclethra* is a Mimosoid legume with occasionally nodulated roots; *Pinus patula* was the only conifer and the only non-native species included in this study. Site management included understory clearing over the first four years, and thinning, generally to 50% of the standing basal area (i.e., every other tree), but the timing of thinning varied among species [67]. *Hieronyma* and *Vochysia* spp. coppiced after being thinned, and *Pentaclethra* apparently was never thinned. In 2004, when this study began, these plantations had an average of 405 trees·ha^{-1} with a total basal area of 23 m^2·ha^{-1} and average tree heights that ranged from 14 to 26 m (Table 1). Over the course of this study, there was a decline in tree stem densities whereas overstory basal areas and tree heights increased, as is typical in growing forests. The exceptions were plantations of *Pinus patula* and *Vochysia ferruginea*, which lost basal area as a result, apparently, of fungal attacks. The understories of the experimental plantations developed dense and diverse understories [67,68] that continued to grow throughout this study.

Table 1. Characteristics of the experimental tree plantations in 2004, fifteen years after planting, and in 2010, at the end of this study. Values are means ± 1 S.E. based on measurements of all trees in four replicates of each plantation type.

Tree Species	Year of Measurement	Density (stems·ha^{-1})	Basal Area (m^2·ha^{-1})	Height [1] (m)	Diameter [2] (cm)
Hieronyma alchorneoides	2004	349 ± 13	15.2 ± 0.9	31.1 ± 1.1	23.6 ± 1.0
	2010	307 ± 8	17.4 ±1.0	32.0 ± 0.6	26.9 ± 1.1
Pentaclethra macroloba	2004	611 ± 87	21.2 ± 4.4	21.6 ± 1.7	21.2 ± 2.9
	2010	548 ± 50	25.3 ± 5.9	23.3 ± 1.7	24.0 ± 3.3
Pinus patula	2004	289 ± 9	19.8 ± 1.0	28.9 ± 0.8	29.6 ± 1.2
	2010	155 ± 18	14.0 ± 2.2	30.9 ± 2.0	33.8 ± 1.2
Virola koschnyi	2004	455 ± 48	19.8 ± 3.6	26.7 ± 1.4	23.1 ± 1.7
	2010	432 ± 50	22.2 ± 4.2	27.1 ± 1.4	25.1 ± 1.8
Vochysia ferruginea	2004	206 ± 44	20.3 ± 2.3	31.3 ± 0.7	36.4 ± 2.2
	2010	156 ± 45	19.1 ± 3.7	31.7 ± 0.7	41.0 ± 3.4
Vochysia guatemalensis	2004	519 ± 30	40.2 ± 3.6	33.9 ± 1.7	31.4 ± 1.3
	2010	476 ± 31	43.6 ± 1.7	36.2 ± 1.3	34.3 ± 1.5

[1] The mean of the tallest tree in each plot ($N = 4$); [2] the diameter at breast height of an average-sized tree (i.e., the quadratic mean diameter).

2.2. Measurements of Soil Respiration

Measurements of soil respiration were initiated in September 2004 in each of the four replicate plots of six plantation types and continued until February 2010 in four of the plantation types. All studies were based on LI-COR© automated soil gas flux instruments (LI-COR Biosciences, NE, USA) that were returned to the factory for cleaning and recalibration annually. The specific measurement chambers utilized varied, but they provided fully comparable measurements of in situ soil respiration, *Rsoil*. I herein report the results of two principal studies: diel study and survey measurements. The diel study characterized variations in *Rsoil* each hour across 2-day and longer periods, based on automated sampling with a single instrument that was moved to sample each of the 24 study plots at least twice. The Survey study involved sampling each of the experimental plots at one time, during daytime hours, so that the effects of the experimental treatments could be cleanly compared. It sometimes took two days to sample all the chambers in all plots. Measurements were conducted through time so that seasonal variations in *Rsoil* could be quantified. Summary results of the first two years of this study were reported earlier [53,65], but data from 2008 to 2010 have not been previously reported, nor have data from plantations of *Vochysia ferruginea* or from the diel study. Based on the data collected during both studies, I also tested for longer-term trends in fluxes in the maturing plantations (Hypothesis 4), which were 16 years old in 2004 and 21 years old in 2009. Thus, hourly, monthly, and inter-annual variations in *Rsoil* were assessed for this article.

2.2.1. Diel Measurements

Soil CO_2 emissions were measured every hour over a total of 52 continuous periods of >2 days between November 2004 and February 2010, with a single chamber in one plot at a time, to investigate diel variations in *Rsoil*. Measurements were made with a LI-COR 8100 soil CO_2 flux system attached to an 8100-101 long-term chamber (LI-COR© Environmental, Lincoln, NE, USA) and powered by a deep-charge marine battery, all of which were back-packed into an experimental plot and then left in place for the duration of measurements. The battery allowed for measurements to be collected every hour over 2–3 day periods. A day before measurements began, a single 20-cm diameter, 12-cm tall polyvinyl chloride collar was inserted approximately 2 cm deep into the surface soil of a plot by carefully cutting through the forest floor with a sharp knife. All measurements were at randomly located positions without regard to tree positions. On the measurement date, the automatic chamber was carefully situated such that it closed firmly on top of the collar, and lifted freely after each measurement. During measurements, the chamber remained tightly closed for 2 min, and within-chamber CO_2 concentrations were monitored as the chamber headspace was circulated through the IRGA (Infrared Gas Analyzer, total system volume averaged 7 L). The first 30 s of measurements were ignored to allow for internal mixing to complete. Over the remaining 90 s, the concentration of CO_2 within the chamber was measured every second, and the rate of change in headspace CO_2 concentrations was used to quantify *Rsoil*. The resulting time-series data were analyzed using LI-COR's embedded FV8100 file viewer software. In most cases, exponential fits were utilized because the rate at which within-chamber CO_2 concentrations increased typically declined through time. The R^2 values of curve fits averaged 0.99 ($N = 2650$) and the minimum R^2 was 0.75. The first hourly measurement collected at a new location was discarded because of consistently high ambient CO_2 concentrations that were attributable to human respiration. A summary of the 52 measurement periods is provided in Table S1. Due to time commitments to other studies, no diel measurements were made from May through September.

During diel measurements, near-surface air temperatures, relative humidity, atmospheric pressure, and CO_2 concentrations were monitored continuously as the measurement chamber closed and then within the closed chamber. Mineral-soil temperature at 10 cm depth was monitored with a soil thermistor probe (LI-COR 8150-203). Surface-soil volumetric moisture content was assessed at four locations surrounding the chamber, at the beginning and end of each measurement period, using a CS620 Hydrosense system with 12-cmrods (Campbell Scientific, Inc., Logan, UT, USA). Following successful completion of one measurement period (i.e., one continuous sequence of hourly measurements at one plot, typically lasting >50 h), the equipment was disassembled and relocated to a different plot. All 24 plots were measured twice between January 2008 and February 2010, to capture both dry and wet periods (Supplementary Table S1). Additional measurements collected in 2004, before sensor malfunctions halted measurements, are included in the analyses (Table S1).

Weather data encompassing 2008 through 2010 were obtained from the La Selva meteorological station, which was approximately 3.3 km from the experimental plots. These data included half-hourly air temperature (°C), relative humidity (RH, %), rainfall (mm), mean wind speed ($m \cdot s^{-1}$), solar radiation ($W \cdot m^{-2}$) and photosynthetic photon flux density (PPFD, $\mu mol \cdot m^{-2} \cdot s^{-1}$). To meld data into a single file for analyses of flux-$\times$-environment relationships, measurements were matched to the nearest half-hourly meteorological data. I also assigned each measurement to its nearest hour so that 'hour-of-day' could be used as a categorical variable for statistical analyses. Detailed weather data were not available for 2004.

2.2.2. Across-Site Surveys of Soil Respiration

The study site was on hilly terrain, and measurements were conducted with backpacked instruments and were limited to daytime hours. Altogether, 99.5% of the measurements were collected between 07:30 and 16:30 sunrise and sunset were at approximately 06:00 and 18:00 daily. The survey measurements were undertaken to test for significant influences of plantation type (tree species) on the magnitude of *Rsoil*. These data are utilized herein for three purposes: (a) to quantify monthly variability in *Rsoil* (Hypothesis 2); (b) to test Hypothesis 3, which posits a temporal offset between aboveground and belowground C fluxes; and (c) to identify longer-term (2004–2010) trends in *Rsoil* (Hypothesis 4). The survey measurements undertaken for this study are summarized in Table S2.

Soil respiration within the plantations was measured 66 times between August 2004 and March 2010 using a LI-COR® 8100 soil CO_2 flux system. From 2004 through 2008, measurements were made with an 8100-102 chamber placed on top of a 10-cm diameter, 5-cm tall plastic collar that was carefully inserted by cutting with a sharp knife through the forest floor and about 1 cm deep into the mineral soil. There were 3–4 collars per plot (12–16 collars per species each date), and they were moved to new, randomly selected locations every year or whenever they were disturbed. At each collar on each measurement date, soil CO_2 fluxes were monitored every second for 90 s. Fluxes were calculated over the final 70 s of chamber closure, using the embedded LI-COR file viewer software, with the first 20 s being excluded to allow for full mixing of the within-chamber atmosphere. The mean flux from the final 70 s was based on an exponential fit of the CO_2 concentration over time relationship. Initial measurements included six species, in one plot per block of each. Measurements in *Vochysia ferruginea* were discontinued after May 2006 because entire trees began falling as a result of butt rot, which seemed to spread among adjacent trees via root grafts. Measurements were discontinued in *Pinus palustris* after March 2008 because trees dropped all their needles and disintegrated. From February 2009 through February 2010 an 8100-103 (20-cm diameter) chamber was utilized, with four 20-cm diameter collars in each of the four plots of each of the remaining four tree species (Supplementary Table S2). Within-chamber CO_2 concentrations were monitored every second for four minutes, with the first 40 s being allowed for full mixing of the air within the chamber, and the subsequent 200 s being used to calculate CO_2 emissions. It was sometimes possible to collect data from all soil collars within all four blocks in one day, but it often took two. Each measurement cycle began at a randomly selected block to minimize the potential for temporal bias to influence comparisons among blocks and species.

At the time of soil respiration measurements, a thermistor probe (LI-COR 8100-201) was inserted 5 cm into the mineral soil approximately 20 cm distant from the measurement collar at three locations, to measure soil temperature. In 2009, we started measuring soil temperature (*Tsoil*) at 10 cm depth. After chamber closure, we measured surface-soil water content with a hand-held electronic soil moisture sensor that was inserted into three locations around but >20 cm from the chamber. Early measurements were made with an ECH_2O Dielectric Aquameter (Model EC-10, Decagon Devices, Inc., Pullman, WA, USA; LI-COR part 8100-202). In 2005, that instrument was replaced with a Campbell 620 Hydrosense system with 12-cm long probes (Campbell Scientific, Inc., Logan, UT, USA), and it proved more durable under the site conditions. We cross-calibrated the two water-content sensors such that they provided the similar readings across the breadth of observed soil moisture contents. Local weather data were obtained from the La Selva Biological Station, which was 3.3 km from the closest study plots.

2.3. Canopy Leaf Fall

Total fine litterfall was measured with four traps per plot that were emptied at the middle and end of each month. Traps had wooden frames that had 1.3×0.4 m internal dimensions and 2-mm mesh screen bottoms. They were supported about 30 cm above the soil on steel legs. The collected materials, which included branches $\leq$ 1 cm in diameter and all non-woody materials, were combined within months to generate a single sample per plot that was sorted into four fractions: branches and bark; canopy leaves; other leaves; and miscellaneous materials [65,66]. Traps were removed, repaired and repositioned annually. More than four years of litterfall measurements were made from October 2003 to December 2009. They are used herein to test for temporal offsets between leaf fall and soil respiration (i.e., Hypothesis 3).

2.4. Statistical Analyses

Statistical analyses were conducted using JMP® 11.2.0 (SAS Institute, Inc., Cary, NC, USA). The underlying experimental design was a randomized complete-block with four blocks and six plantation types, which are referred to by the single tree species that was planted into each plot. A combination of parametric and non-parametric statistical analyses was applied to test for significant differences among the test variable, typically time (hour or month or year); tree species; and their interaction. Tests of hourly and monthly data were based on binned values, 24 h of day and 12 months of year, rather than the continuous time variables that were collected. Several features characterize the datasets. First, the hypotheses were designed to test for landscape-level patterns but, commonly, results varied among the plantation types. Also, sample sizes differed among treatments. Therefore, hypotheses typically were tested at both the all-plot (landscape) and treatment (tree species) levels. Homogeneity of variances was specifically tested with the Brown-Forsythe and Levene tests and was very frequently rejected. In those cases, Welch's test was applied. In cases where data were not normally distributed, transformations were applied prior to statistical analysis. For instance, measurements of daytime *Rsoil* (N = 4962) were not normally distributed and included occasional very high fluxes (maximum = 46 μmol·m^{-2}·s^{-1}). Natural-log transformation of the data reduced kurtosis from 54.7 to 0.8, reduced skewness from 3.6 to -0.13, and changed the median-to-mean ratio from 0.94 to 1.01. Statistical tests of *Rsoil* therefore were based on ln(*Rsoil*), but results typically are presented as actual values to facilitate understanding. In all cases of comparisons among groups, a non-parametric Kruskal–Wallis test was applied to validate the results of parametric analyses of variance; reported results represent the more conservative result (i.e., lowest P value). Differences among factors were considered significant at P = 0.05.

The diel study comprised 52 independent observational studies that were designed to determine the effect of time-of-day on *Rsoil*. Each study included automated measurements of *Rsoil* that were made every hour over more than two continuous days, at one location in one plantation type. Measurements were made a minimum of two times in each plantation type in each block, at different times of year (Table S1). Thus, the measurements incorporated variability attributable to tree species, hour-of-day, and season-of-year, all of which were found to significantly affect fluxes, at least sometimes. Therefore, to quantify the hour-of-day effect, each of the 39 to 70 hourly measurements made during each of the 52 measurement periods was normalized to a measurement-period mean value of precisely 1.0. Within an individual measurement period, then, fluxes varied through time but had an overall average normalized flux rate of one. As a result, all measurement periods had fully comparable fluxes that varied only with time of day. The normalized data then were combined for statistical analyses. The resulting dataset had a sample size of 2650, a median = 0.99, a range of 0.025–1.798, a CV (coefficient of variation) of 0.14, a skewness of -0.12, and a kurtosis of 6.5. To test for diel variations in *Rsoil* (Hypothesis 1), data were assigned to hourly bins (categorical variable, hour 14 = 13:30–14:29).

The survey study similarly comprised individual observational comparative studies during which *Rsoil* was measured within plantations within all four blocks. This study included 66 such measurement periods: the species being measured varied through time (Table S2). To test for seasonal differences (Hypothesis 2), measurements were assigned to monthly bins. Block was treated as a random effect. Tree species and their interaction significantly influenced *Rsoil* (least squares ANOVA, N = 4858), so comparisons among months were done on a per-species basis. In most cases, variances among months were not homogeneous, so the results of Welch's test are reported. In cases where variables were not normally distributed, the non-parametric ranked-sum Kruskal–Wallis Test was applied.

To compare the seasonality of leaf flushing with that of *Rsoil*, to test Hypothesis 3, mean monthly soil respiration data were paired with mean monthly canopy leaf fall data from the same species and time frame. Pearson product–moment correlations were calculated on a single dataset including all months and species, i.e., mean monthly canopy leaf fall and mean monthly *Rsoil* for each month and species (N = 72).

To assess whether there were long-term trends in *Rsoil* over the duration of the study (Hypothesis 4), data from both the diel and survey studies were combined into a single dataset, and the hypothesis was tested independently for each species. Least-squares linear regression modeling was applied with ln(*Rsoil*) as the dependent variable. Results from simple correlations, non-linear models, or multiple linear models having additional time-related variables (e.g., time^2) added no meaningful information, and are not reported.

3. Results

This study ran from October 2003 through February 2010, during which largely normal weather patterns prevailed at La Selva. Rainfall averaged 4420 mm·year^{-1} and temperatures averaged 25.1 °C.; both are similar to longer-term means (Table S3). Dry-season precipitation (January through April) was below average in 2007 and 2008, but exceeded 65 mm in every month. Overall, both the diel and survey measurements included a broad range of the weather conditions that typify the climate of La Selva, including brief rain-free periods.

3.1. Diel Variability in Rsoil and the Environment

Soil respiration varied widely within and among days, without obvious regularity (Figure 1). Nevertheless, normalized fluxes did vary significantly among hours of the day (Kruskal–Wallis test, df = 23, χ^2 < 0.0001). This finding supports Hypothesis 1, that *Rsoil* varies on a diel basis. However, the R^2 was low (0.02) and variability within hours was high. Overall, *Rsoil* was lowest between 06:30 and 10:30, when it averaged 4.03 µmol·m^{-2}·s^{-1} (Figure 2a). There was no equivalent identifiable period of maximum emissions; all other hours of the day had statistically equivalent fluxes.

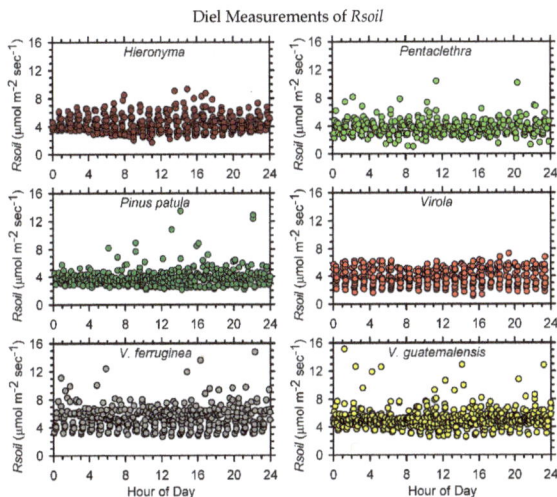

Figure 1. Soil respiration was highly variable at all hours of the day and night, within all plantation types. This figure includes all data collected during diel measurement periods (N = 2650).

Several meteorological variables that potentially influenced emissions also varied through time (Figure 2) and, thus, potentially provide insight into environmental controls over within-soil CO_2 production and its rate of escape into the atmosphere. Note that data in Figure 2 represent averages calculated from hours during which *Rsoil* was measured. Across all measuremement dates, soil temperature (*Tsoil*) at 10-cm depth averaged <24.2 °C from 07:30–12:30 but >24.5 between 16:30 and 23:30 (Figure 2c). Air temperatures increased from an average daybreak minimum of 21.6 °C to a post-noon maximum of 27.3 °C (Figure 2d). The temperature difference between the soil and atmosphere varied more widely (Figure 2b), ranging from −3.1 °C at 13:00 to 2.7 °C at 06:00. Atmospheric relative humidities were quite high, on average (Figure 2e), but they too varied across the day: they averaged >93% between 02:30 and 08:30 but <83% between 11:30 and 16:30. Among those variables, *Rsoil* most closely paralleled changes in soil temperature, and increased with *Tsoil* at 10 cm depth with a Q_{10} of 3.3 (Least-squares linear regression, $N = 2432$, $R^2 = 0.07$, $P < 0.0001$, temperature exponent = 0.1204 ± 0.0086 per °C).

Ambient CO_2 concentrations above the forest floor varied significantly with hour-of-day in plantations of each of the study species (GLM (General Linear Modeling procedure), $N = 398$ to 539, $\chi^2 < 0.0001$) and across all species combined ($N = 2650$, $\chi^2 < 0.0001$). Carbon dioxide, presumably from the soil, accumulated on-site when wind speeds (measured in the open) averaged <1 m·s^{-1}, and atmospheric CO_2 typically exceeded 500 μmol·mol^{-1} during the nighttime (Figure 2g). Ambient CO_2 and wind speed (Figure 2h) were strongly negatively correlated ($N = 2492$, $r = -0.61$, $P < 0.0001$), based on wind-speed measurements at the La Selva Biological Station, about 3.5 km from the field plots. Also measured at that weather station, over 2008 through 2010, were solar irradiance and photosynthetic photon flux density (PPFD). During diel measurements of *Rsoil*, irradiance averaged 551 W·m^{-2} at noon and had a slightly skewed distribution, with slightly more sunlight in the morning than afternoon (Figure 2f). Based on parallel measurements at the same location, photosynthetic photon flux density (PPFD, not shown) was linearly correlated with irradiance according to

$$\text{PPFD (μmol·m}^{-2}\text{·s}^{-1}) = 7.88 + 2.49 \times \text{Irradiance (W·m}^{-2}) \tag{1}$$

(Least-squares linear regression, $N = 1889$, $R^2 = 0.99$, $P < 0.0001$). This suggests that either variable would be equally suitable for predictive modeling. In the current study, stepwise linear regression indicated that the normalized soil CO_2 efflux was negatively related to incident sunlight received two days earlier, and positively related to sunlight received 5.5 days previously:

$$\text{Normalized[ln(}Rsoil\text{)]} = 1.00 - 0.0000204 \times \text{PPFD}(-2.0 \text{ days}) + 0.0000209 \times \text{PPFD}(-5.5 \text{ days}) \tag{2}$$

(stepwise linear regression, selected for minimum BIC (Bayesian Information Criterion), $N = 1889$, $R^2 = 0.02$, $P < 0.0001$). The three parameters in this equation were significant at $P < 0.0001$, 0.0015 and 0.0025, respectively. The predictive utility of this relationship is virtually nil, but a 5.5-day lag between photosynthesis and soil-CO_2 emissions is consistent with a 4 to 5 day lag for forest trees [41].

Diel variability in *Rsoil* and environment

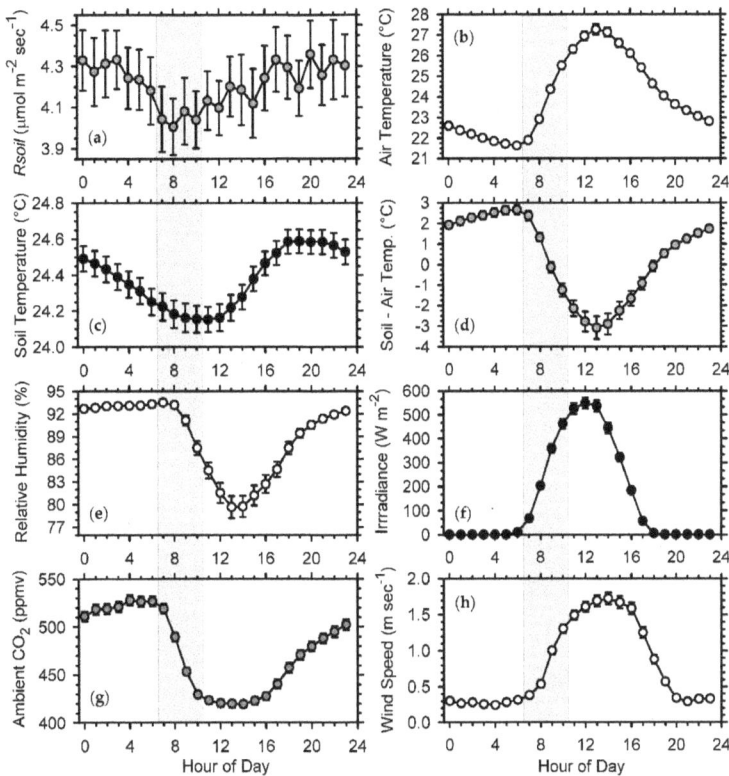

Figure 2. Observed diel variability in (**a**) soil respiration (*Rsoil*) was significant, but inconsistent ($R^2 = 0.02$), whereas the variability in potentially controlling environmental variables was pronounced. The shaded hours highlight the period between 06:30 and 10:30 when *Rsoil* was lowest on average; (**b**) Mean surface air temperatures; (**c**) soil temperatures at 10-cm depth; (**d**) the difference between air and soil temperatures; (**e**) atmospheric relative humidity and (**g**) ambient atmospheric CO_2 concentration were all measured in situ, with *Rsoil*. Shown are means (±S.E.) of each variable, based on all diel measurement dates and times; (**f**) Solar irradiance and (**h**) mean wind speeds were measured at the La Selva weather station, about 3.5 km from the field sites; the data included here were temporally synchronized with the diel measurements.

3.2. Seasonal Variability in Rsoil and Environment

Soil respiration during across-site surveys ranged from 0.85 to 46.1 μmol·m^{-2}·s^{-1} ($N = 4858$). Fluxes varied significantly among species (Kruskal–Wallis $\chi^2 < 0.0001$, Welch's $P < 0.0001$): they were lowest in plantations of *Pinus*, at 3.8 μmol·m^{-2}·s^{-1}; and highest in plantations of *Vochysia* spp., at 5.3 μmol·m^{-2}·s^{-1} (Table 2). Fluxes varied significantly among months in all six plantation types ($N = 274$–1040, Welch's $P < 0.0001$), consistent with Hypothesis 2. In all six plantation types, *Rsoil* was highest in either August or September, and lowest in January or February (Table 3, Table S4). That is, there were remarkably consistent seasonal patterns of soil respiration across the six different plantation types.

Table 2. Annual average (±1 S.E.) rates of daytime soil respiration (*Rsoil*) and litterfall in experimental plantations of evergreen trees at La Selva, Costa Rica during this study. Means are weighted by days per month; and standard errors reflect variation among months. Both fluxes varied significantly among the plantation types.

Tree Species	Rsoil (2004–2010) ($\mu mol \cdot m^{-2} \cdot s^{-1}$)	Litterfall (2003–2009) ($g \cdot m^{-2} \cdot day^{-1}$)
Hieronyma alchorneoides	4.91 ± 0.64	2.88 ± 0.21
Pentaclethra macroloba	4.15 ± 0.42	2.65 ± 0.17
Pinus patula	3.81 ± 0.84	2.63 ± 0.14
Virola koschnyi	4.08 ± 0.40	2.13 ± 0.09
Vochysia ferruginea	5.26 ± 0.97	3.12 ± 0.17
Vochysia guatemalensis	5.26 ± 0.54	2.69 ± 0.09

Soils were wettest, on average, in November, when 53% of the total soil volume was water, and driest (28%) in April, at the end of La Selva's drier season (Table 3). Soil moisture content varied significantly among months within all six plantation types (Welch's tests, $P < 0.0001$), and the correlation between mean monthly *Rsoil* and mean monthly soil moisture content was significantly negative ($N = 72$, Pearson's $r = -0.44$, $P < 0.0001$). Soil temperature also varied significantly among months, based on measurements at both 5-cm and 10-cm depth (Welch's tests, $P < 0.0001$). Soil temperatures were coolest in January or February, and were warmest in August or September, depending upon the depth of measurement. Thus, the seasonality observed in *Tsoil* matched that observed in *Rsoil*, and soil temperature and *Rsoil* correlated positively for most species (Table S6). Incident solar radiation is bimodally distributed at the study site, peaking in April and September each year, with a wintertime minimum (Table 3). Such seasonal variations in climate and microclimate potentially drive seasonal variability in *Rsoil*.

Table 3. Mean monthly soil respiration (*Rsoil*), soil temperatures (*Tsoil*), soil water contents, canopy leaf fall and incident solar radiation across 24 monodominant plantations of six tree species in the Atlantic lowlands of Costa Rica. Not all species were measured over the duration of the study; the mean values for each month were determined from the mean monthly values determined for each species. Soil temperatures refer to 2008–2010, when they were measured at 10 cm depth. Error intervals are ±1 S.E.

Month	Rsoil ($\mu mol \cdot m^{-2} \cdot s^{-1}$)	Tsoil [1] (°C)	Soil H_2O ($cm \cdot m^{-1}$)	Canopy Leaf Fall ($g \cdot m^{-2} \cdot day^{-1}$)	Mean Daily Irradiance [2] ($W \cdot m^{-2}$)
January	3.67 ± 0.18	21.6 ± 0.04	51.0 ± 1.8	1.41 ± 0.14	177 ± 43
February	3.80 ± 0.27	24.0 ± 0.09	52.0 ± 1.7	1.61 ± 0.17	154 ± 24
March	4.50 ± 0.30	22.8 ± 0.04	38.1 ± 1.0	2.01 ± 0.39	214 ± 24
April	4.73 ± 0.29	24.7 ± 0.04	31.2 ± 2.6	1.24 ± 0.19	272 ± 23
May	4.84 ± 0.30	25.2 ± 0.04	36.3 ± 1.8	1.24 ± 0.15	232 ± 15
June	4.68 ± 0.31	25.4 ± 0.05	38.8 ± 1.6	1.27 ± 0.15	178 ± 16
July	4.31 ± 0.32	25.4 ± 0.03	44.8 ± 2.0	1.29 ± 0.17	187 ± 22
August	4.58 ± 0.49	25.5 ± 0.03	47.6 ± 1.6	1.21 ± 0.15	236 ± 29
September	5.85 ± 0.33	26.1 ± 0.03	38.6 ± 0.9	1.60 ± 0.23	214 ± 26
October	4.97 ± 0.38	25.3 ± 0.03	50.6± 1.4	1.78 ± 0.25	193 ± 22
November	4.12 ± 0.33	24.3 ± 0.03	53.5 ± 1.0	1.34 ± 0.22	176 ± 43
December	4.25 ± 0.31	24.0 ± 0.05	52.4 ± 2.3	1.02 ± 0.12	159 ± 22

[1] January 2008 through February 2010; [2] on dates of *Rsoil* measurements; data courtesy of La Selva Biological Station.

3.3. Litterfall and Leaf Fall

Total fine litterfall over the course of this study ranged from an average of $2.13 \pm 0.09 \ g \cdot m^{-2} \cdot day^{-1}$ in plantations of *Virola* to $3.12 \pm 0.17 \ g \cdot m^{-2} \cdot day^{-1}$ in plantations of *V. ferruginea*. Across all plantation types and years, both total and canopy leaf litter production varied by species, by month, and with a significant species × month interaction (GLM, $N = 1072$, $\chi^2 < 0.0001$). Canopy leaf fall varied significantly through the year in all species except *Vochysia ferruginea*. Canopy leaves made up an average of 53% of the total fine litterfall and largely defined the seasonality of leaf fall; other litterfall components were not strongly seasonal. Across all species and plots, canopy leaf fall had a bimodal distribution that enveloped the vernal and autumnal equinoxes (Table 3, Table S7). Contrary to Hypothesis 3, which posited that canopy leaf fall and *Rsoil* would be negatively correlated, mean monthly canopy leaf fall and mean monthly *Rsoil* were positively correlated (Figure 3, $N = 72$, Pearson's $r = 0.42$, $P < 0.001$).

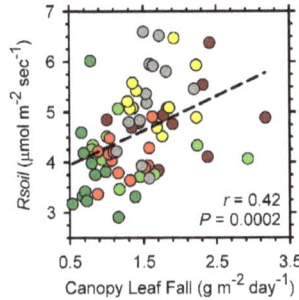

Figure 3. Mean monthly canopy leaf fall and soil respiration (*Rsoil*) correlated positively across all plantations studied ($N = 72$, $P = 0.0002$, $r = 0.42$). Symbol colors are as in Figure 1. Each symbol is the mean of all measurements from a particular plantation type and month, from 2004 to 2010.

3.4. Longer-Term Trends in Rsoil

Soil respiration in plantations of *Heironyma* ($N = 1486$, $\chi^2 < 0.0001$), *Pinus* ($N = 846$, $\chi^2 < 0.001$) and *Virola* ($N = 1493$, $\chi^2 < 0.0004$) increased significantly over the five-year duration of this study, with slopes ($\pm$ SE) of 0.13 ± 0.026, 0.079 ± 0.028 and 0.054 ± 0.031 $\mu mol \cdot m^{-2} \cdot s^{-1} \cdot year^{-1}$, respectively (Figure 4). There were no discernible influences of time, across years, on *Rsoil* in any of the other plantation types.

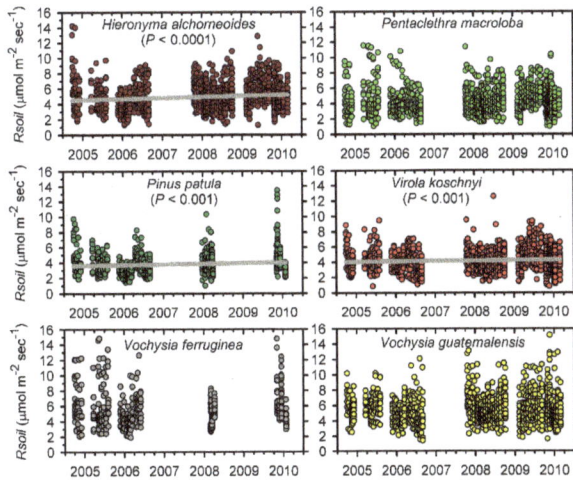

Figure 4. From late 2004 to early 2010, soil respiration (*Rsoil*) increased significantly in plantations of *Hieronyma*, *Pinus* and *Virola*, but not in other plantation types. Lines show the least-square regressions of significant relationships; among those three species the rate of increase averaged 0.09 $\mu mol \cdot m^{-2} \cdot s^{-1} \cdot year^{-1}$.

4. Discussion

The spatial variability of *Rsoil* at the study site was high: coefficients of variation (CV) among fluxes gathered over individual across-site surveys averaged 37% and, in some cases, exceeded 100% (Table S2). Previous studies at this site demonstrated differences in *Rsoil* among tree species. Those differences have persisted through time. For instance, over 2004–2010, *Rsoil* in plantations of *Vochysia* spp. averaged 30% greater than emissions in *Virola* plantations (Table 2), as they did after years one [65] and two [53]. Thus, some of the spatial variability encountered during this study resulted directly from the experimental treatment. Temporal variability was significant but somewhat less important quantitatively: CVs of flux measurements from individual chambers that were sequentially monitored every hour at a single location averaged 18% (Table S1). It was apparent in the field, while measurements were being collected, that the activities of fauna, particularly ants, also generated substantial spatial flux variability, as demonstrated elsewhere [69]. Despite this variability, demonstrable differences were apparent.

The principal objective of this study was to characterize temporal variability in *Rsoil* within evergreen tree plantations in a lowland tropical rainforest environment. Soil respiration varied significantly on diel, monthly and, in three of six species, multi-annual time frames (Figure 2, Table 3 and Figure 4, respectively). Temporal variations in soil respiration have been documented at other sites [38,49,70–72]. They commonly are attributed to temporal variations in plant productivity, or to environmental characteristics such as prevailing temperatures and water availability.

4.1. Diel Variability

The diel variability of *Rsoil* in this study was not pronounced (Figure 1), but significant differences among hours of the day were observed ($\chi^2 < 0.0001$). Specifically, lower fluxes were observed between 06:30 and 10:30, with sunrise occurring at about 06:00. One important reason for measuring hourly variations in *Rsoil* is to determine if daytime-only measurements cause bias in the estimation of annual fluxes. In this study, more than 99% of the across-site survey measurements of *Rsoil* were made between 07:30 and 16:30, over which *Rsoil* averaged 4.12 μmol·m^{-2}·s^{-1}. That value is 2% greater than the morning minima, and 2% lower than the 24-h average flux of 4.21 μmol·m^{-2}·s^{-1}, based on hourly measurements. In short, there was no evidence that daytime measurements of soil respiration generated biased estimates of monthly or seasonal soil respiration from the plantations studied, as a result of diel flux variability.

The notable differences between daytime and nighttime ambient conditions in tropical rainforest environments (e.g., Figure 2) provide the opportunity to apply diel measurements to identify environmental controls over *Rsoil*. In this study, hourly variations in *Rsoil* correlated with hourly variations in soil temperatures (Figure 5a, $N = 24$, Pearson's $r = 0.82$, $P < 0.0001$), with high probability ($\chi^2 < 0.0001$) but low predictability ($R^2 = 0.02$). This also was true at the monthly scale (Figure 5b, $N = 72$, $r = 0.50$, $P < 0.0001$). Warmer soils had greater *Rsoil*. This is consistent with our understanding that rates of metabolism and gas diffusion both increase with temperature. Soil respiration also varied with soil water content (Table S5). However, soil moisture content was not monitored continuously during the diel studies, so sample sizes were too low to evaluate meaningfully. Other considered factors did not correlate with *Rsoil*, including concurrent atmospheric humidity, ambient CO_2 concentrations, the temperature difference between the air and soil, or solar radiation, all of which varied consistently across days (Figure 2).

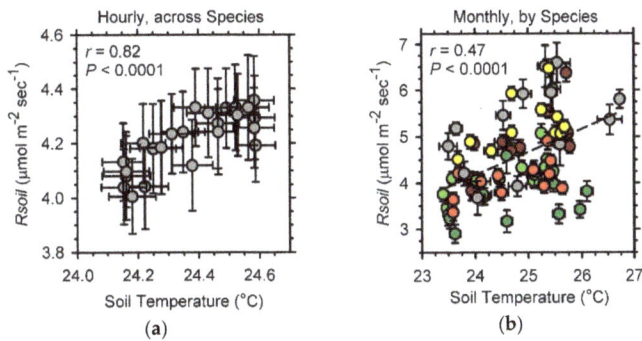

Figure 5. (**a**) Mean hourly soil respiration (*Rsoil*) in relation to mean hourly soil temperature over the course of the diel study (*N* = 24, Pearson *r* = 0.82). (**b**) Mean monthly soil respiration (*Rsoil*) in relation to mean monthly daytime soil temperature for all six species over the course of the across-site survey study (*N* = 72, Pearson's *r* = 0.50). Symbol colors follow Figure 1; error bars show ±1 S.E.

In contrast to the relatively predictable diel dynamics of many physical factors (Figure 2), precipitation at La Selva occurs at all hours of the day and night (Figure 6). That simple observation may explain the great within-hour variability in CO_2 emissions that was observed (Figure 1). Rainfall during this study averaged 4420 mm·year^{-1}. Annual canopy interception of rain may be as high as 710 mm [55], and transpiration is perhaps 1500 mm. Every time that 1 cm of rain falls, 442 times a year, about 8 mm of water enters the soil, where it advectively displaces 8 liters per m^{-2} of CO_2-rich soil atmosphere. The soil atmosphere, in other words, is awash with rainwater, more frequently than it is connected by air-filled pores through which soil gases might continuously diffuse. Every time that 1 cm of soil water is taken up by vegetation and transpired, 150 times each year, 10 liters of atmosphere are pulled into each m^2 of soil, refreshing it with oxygen. Water movemet into and out of the soil at La Selva occurs all all hours of the day and, I submit, obfuscates the temperature- and CO_2-concentration gradient-driven processes that typically control gas diffusion from soils. Carbon dioxide still escapes from the soil, at a very high rate, but likely through advection as frequently as diffusion, and when it can rather than when it is produced. Animal tunnels offer routes of escape.

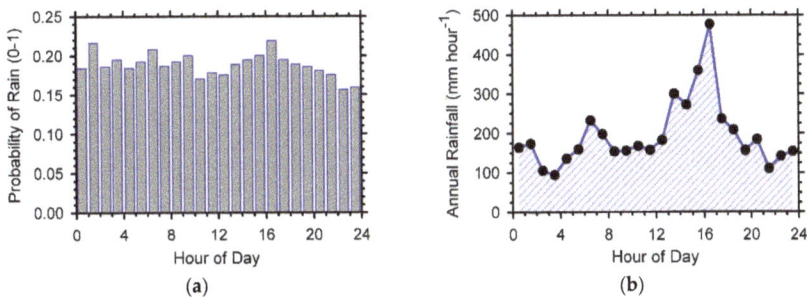

Figure 6. (**a**) The likelihood that rain will occur within any given hour at La Selva is quite uniform across the day. (**b**) The amount of rain that falls each hour of the day at La Selva is variable: afternoon showers bring more rain. These figures are based on 2009, a relatively typical year.

Data from the diel study contained evidence of a time lag between PPFD and *Rsoil* (Equation (2)). Such a lag is expected [73], if PPFD is a meaningful surrogate for canopy photosynthesis. The estimated time lag between canopy fixation of CO_2 and CO_2 release from soils is 4–5 days for forest trees [41], but there are few data from tropical forests. Additional studies are required to verify that this result was not a Type 1 statistical error: the model included 30 different time lags for selection; two were found to be highly significant, and of opposite direction.

The time series that emerged from each diel measurement period further provide insight into processes. Across days without rain (Figure 7a,b) there are distinct diel variations in soil temperature that are not necessarily associated with parallel changes in *Rsoil*. Figure 7a shows the best example of what predictable diel variability in *Rsoil* would look like: peak emissions were offset from peak soil temperatures by about 13 h, and were either 4 h before or 20 h after mid-day. This pattern was not again observed, however. Moderate diel variations in soil temperature in *Pentaclethra* (Figure 7b) did not apparently influence *Rsoil* on 28 March. A sharp decline in emissions at 14:00 that coincided with a rainstorm was observed in the time series retrieved from *Virola* (Figure 7c). The time series from *Vochysia guatemalensis* (Figure 7d) shows a typically rainy day. Rain seems to temporarily seal the soil surface with a skin of wet soil, impeding the outward diffusion of CO_2, but when the rain stops, emissions slowly increase, as the water penetrates the soil. A rainforest is defined by rain and, it seems, so are its soil-CO_2 emission rates.

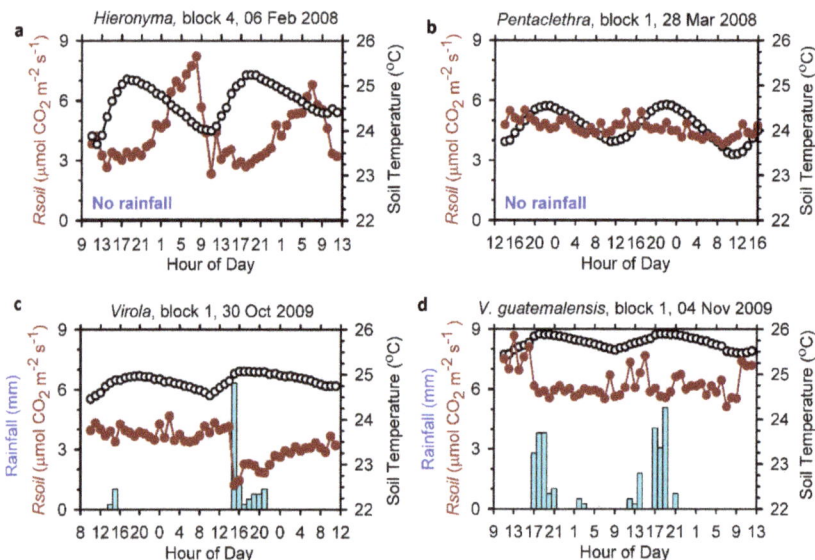

Figure 7. Soil temperatures at 10-cm depth (open symbols), soil respiration (*Rsoil*, dark red symbols and lines) and hourly rainfall (blue bars) over 52 continuous hours on different dates in plantations of four different tree species at La Selva. Dates refer to the beginning of the measurement period. (**a**) *Hieronyma alchorneoides*, in this case, exhibited increasing rates of *Rsoil* to morning peaks that were distinctly offset from maximum soil temperatures, on two consecutive rain-free days; (**b**) *Pentaclethra macroloba* showed no obvious trends in *Rsoil* across two consecutive rain-free days, despite obvious variations in soil temperatures; (**c**) A rainstorm in the early afternoon on day two resulted in a dramatic decline in *Rsoil* in a plantation of *Virola koschnyi*. Emissions slowly increased following the rain; (**d**) A typical day in a plantation of *Vochysia guatemalensis*: rainfall impedes CO_2 escape from the soil, and emissions increase after the rain stops.

4.2. Monthly Variability

Monthly variability in *Rsoil* was pronounced: mean soil respiration in September was 60% greater than it was in January (Table 3). Mean monthly fluxes correlated negatively with mean monthly soil moisture content ($N = 71$, Pearson's $r = -0.45$, $P < 0.0001$), demonstrating that *Rsoil* was greater in drier months. Emissions correlated positively with both mean monthly soil temperature (Figure 5b $N = 72$, $r = 0.50$, $P < 0.0001$) and irradiance $(W \cdot m^{-2})$ $(N = 72, r = 0.29, P < 0.02)$. Those findings indicate that monthly weather data could be usefully applied to estimate monthly *Rsoil* at this site. They do not, however, include any direct information about plant-to-soil carbon fluxes, which fundamentally control the amount of carbon available for respiration.

A specific objective of this study was to seek a plant-growth variable that would provide insight into aboveground–belowground interactions. Hypothesis 3 posited that *Rsoil* and canopy leaf flushing would be negatively correlated, if carbon allocation aboveground, for the production of a new crop of leaves, was temporally offset from carbon allocation belowground, to produce fine roots and mycorrhizae. The fall of canopy tree leaves is an indirect measure of the timing of leaf flushing. Across all six plantation types, maximum canopy leaf fall occurred in March, in the dry season, before the coming of more dependable rains in May (Table 3 and Table S7). They also shared a secondary peak of canopy leaf fall in October, before the colder, darker, and wetter winter months set in. Over this study, including all species and blocks, canopy leaf fall and soil respiration were not offset from one another, they paralleled one another ($N = 72$, Pearson's $r = 0.42$, $P = 0.0002$). Fortunately, the phenological characteristics of La Selva's forest trees have been monitored. The number of tree species flushing many new leaves is maximum in February and March, with a secondary peak in September; and the number of species producing mature fruits is maximum in September [74]. Those observations support the inference that March and September are periods of high allocation of carbon to aboveground processes. They are also months having high soil respiration. Thus, the evidence gathered in this study strongly suggests that above- and belowground C use covaried temporally at this site, opposite of the a priori conjecture of Hypothesis 3. More evidence addressing this topic is warranted, but the alternative hypothesis, i.e., that roots and leaves are not competing with one another for photosynthate but, rather, together depend upon recently produced photosynthates, is reasonable and consistent with the data. Covariance between litterfall and soil respiration was previously found in three forest types in subtropical China [49].

4.3. Physical and Biological Factors

Evidence for a significant time lag between irradiance and soil-CO_2 emissions was found (Equation 2), which suggests the possibility of a time lag between canopy photosynthesis and *Rsoil*. This finding is uncertain because of the large number of independent variables included in the model, and so deserves to be investigated further. The hypothesis that there would be a trade-off in C use between aboveground and belowground C use was tested and rejected. Positive relationships between soil temperatures and *Rsoil* were observed at diel and monthly time steps (Figure 5), and are evident in the entire dataset derived from across-site surveys (Figure 8a). A significantly negative relationship between soil moisture content and *Rsoil* was observed at the monthly scale ($r = -0.44$, $P < 0.0001$), which indicates that *Rsoil* was greater in drier months. Across all measurements, *Rsoil* $(\mu mol \cdot m^{-2} \cdot s^{-1})$ varied non-linearly with surface-soil moisture content (SoilH$_2$O, cm·m^{-1}) based on a quadratic polynomial fit:

$$\ln(Rsoil) = 2.036 - 0.01075 \times SoilH_2O - 0.00037758 \times (SoilH_2O - 44.12)^2 \tag{3}$$

(Figure 8b, $N = 4486$, $F_{2,4483} = 415$, $R^2 = 0.16$) with all three parameters being significant at $P < 0.0001$. Relationships between *Rsoil* and soil water content were observed for each of the six species, and were non-linear in all cases except for *V. ferruginea* (Table S5). At La Selva, soil respiration was limited by a lack of drier soils. Overall, these data support the conclusion that physical variables—soil moisture content, temperatures, solar radiation—can serve as useful inputs into models that are designed to evaluate the likely directional changes in *Rsoil* that would accompany changing climatic conditions. The absolute magnitude of fluxes was species-dependent, however.

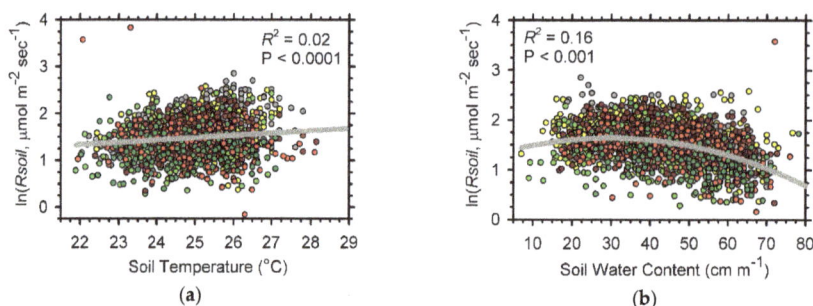

Figure 8. (a) The relationship between log (soil respiration), i.e., ln(*Rsoil*), and soil temperature at 5 cm depth was highly significant, signifying an exponential relationship between CO_2 emissions and temperature. Colors follow Figure 1; (b) The relationship between soil moisture content and soil respiration was non-linear (Equation (3)). Both figures include data from all across-site surveys; every other datum is shown to improve legibility.

5. Conclusions

In this study, the different tree plantations varied from 3.8 to 5.3 $\mu mol \cdot m^{-2} \cdot s^{-1}$ in mean annual emissions, i.e., by 40% (Table 2) at one site, with all plots exposed to the same weather and on the same soils. It is frightening to consider the global ramifications. We need to improve our capacity to translate plant (and animal) species effects to broader spatial scales in a way that could inform physical climate models. This study did not provide new insights into that issue.

It is sometimes put forth that modeling *Rsoil* is not particularly worthwhile (but see [75]), because *Rsoil* derives from different processes that respond differently to environmental conditions [76], and often is considered to be the sum of soil heterotrophic and autotrophic respiration [77]. There is merit in modeling those components, and perhaps others, such as roots as distinct from mycorrhizae, and soil arthropods and annelids as distinct from microbes, but we require all the information we can muster, to broadly advance understanding. The value of soil respiration measurements and models is that they reflect an empirically measurable and important soil-to-atmosphere CO_2 flux. At a minimum, they provide useful information for testing models of subcomponents. The problem with *Rsoil* measurements is that they reflect the composite emissions of multiple within-soil CO_2-producing sources that are individually influenced by multiple environmental factors.

Supplementary Materials: The following are available online at www.mdpi.com/1999-4907/8/2/40/s1, Table S1: Summary of diel measurements included in this study, Table S2: Summary of across-site survey measurements included in this study, Table S3: Annual climate data for the study site during this study, Table S4: Average soil respiration differed among months within species but in parallel among species, Table S5: Statistical relationships between observed *Rsoil* and soil water content based on data from the across-site survey study, Table S6: Linear relationships between observed *Rsoil* and soil temperature at 5 cm depth, Table S7: Mean monthly litterfall across 24 monodominant plantations of six tree species in the Atlantic lowlands of Costa Rica [78].

Acknowledgments: This work was funded by the U.S. National Science Foundation grants DEB-0343766 and DEB-0703561. The work benefitted substantially by the contributions of many, most notably Ricardo Bedoya Arrieta, Oscar Valverde-Barrantes, Dennes Chavarría, and Marlón Hernández, and from the staff and facilities of the La Selva Biological Station.

Conflicts of Interest: The author declares no conflict of interest. The funding agency had no role in the design of the study; in the collection, analyses, or interpretation of data; or in the preparation of the manuscript.

Abbreviations

The following abbreviations are used in this manuscript:

ANOVA	analysis of variance
BIC	Bayesian Information Criterion
CV	coefficient of variation
GLM	general linear modeling statistical procedure
IRGA	infrared gas analyzer
PPFD	photosynthetic photon flux density
Rsoil	soil respiration
Tsoil	temperature of the soil

References

1. Lieth, H. Primary production: Terrestrial ecosystems. *Hum. Ecol.* **1973**, *1*, 303–332. [CrossRef]
2. Whittaker, R.H. *Communities and Ecosystems*, 2nd ed.; Macmillan Publishing Co., Inc.: New York, NY, USA, 1975; p. 385.
3. Erwin, T.L. Tropical forests: Their richness in Coleoptera and other arthropod species. *Coleopt. Bull.* **1982**, *36*, 74–75.
4. Dirzo, R.; Raven, P.H. Global state of biodiversity and loss. *Annu. Rev. Environ. Resour.* **2003**, *28*, 137–167. [CrossRef]
5. Willig, M.R.; Kaufman, D.M.; Stevens, R.D. Latitudinal lgradients of biodiversity: Pattern, process, scale, and synthesis. *Annu. Rev. Ecol. Evol. Syst.* **2003**, *34*, 273–309. [CrossRef]
6. Brown, J.H. Why are there so many species in the tropics. *J. Biogeogr.* **2014**, *41*, 8–22. [CrossRef] [PubMed]
7. Saugier, B.; Roy, J.; Mooney, H.A. Estimations of global terrestrial productivity: Converging towards a single number? In *Terrestrial Global Productivity*; Roy, J., Saugier, B., Mooney, H.A., Eds.; Academic Press: San Diego, CA, USA, 2001; pp. 543–557.
8. Raich, J.W.; Lambers, H.; Oliver, D.J. Respiration in Terrestrial Ecosystems. In *Treatise on Geochemistry*, 2nd ed.; Holland, H.D., Turekian, K.K., Eds.; Elsevier: Oxford, UK, 2014; pp. 613–649.
9. Schlesinger, W.H.; Jasechko, S. Transpiration in the global water cycle. *Agric. For. Meteorol.* **2014**, *189–190*, 115–117. [CrossRef]
10. Dixon, R.K.; Brown, S.; Houghton, R.A.; Solomon, A.M.; Trexler, M.C.; Wisniewski, J. Carbon pools and flux of global forest ecosystems. *Science* **1994**, *263*, 185–190. [CrossRef] [PubMed]
11. Jobbágy, E.G.; Jackson, R.B. The vertical distribution of soil organic carbon and its relation to climate and vegetation. *Ecol. Appl.* **2000**, *10*, 423–436. [CrossRef]
12. Raich, J.W.; Russell, A.E.; Kitayama, K.; Parton, W.J.; Vitousek, P.M. Temperature influences carbon accumulation in moist tropical forests. *Ecology* **2006**, *87*, 76–87. [CrossRef] [PubMed]
13. Hansen, M.C.; Potapov, P.V.; Moore, R.; Hancher, M.; Turubanova, S.A.; Tyukavina, A.; Thau, D.; Stehman, S.V.; Goetz, S.J.; Loveland, T.R.; et al. High-resolution global maps of 21st-century forest cover change. *Science* **2013**, *342*, 850–853. [CrossRef] [PubMed]
14. Intergovernmental Panel on Climate Change (IPCC). *Climate Change 2013: The Physical Science Basis. Contribution of Working Group I to the Fifth Assessment Report of the Intergovernmental Panel on Climate Change*; Stocker, T.F., Qin, D., Plattner, G.-K., Tignor, M., Allen, S.K., Boschung, J., Nauels, A., Xia, Y., Bex, V., Midgley, P.M., Eds.; Cambridge University Press: Cambridge, UK; New York, NY, USA, 2013; p. 1535.
15. Arrhenius, S. On the Influence of Carbonic Acid in the Air upon the Temperature of the Ground. *Philos. Mag.* **1896**, *41*, 237–276. [CrossRef]
16. Schleser, G.H. The response of CO_2 evolution from soils to global temperature changes. *Z. Naturforsch.* **1982**, *37*, 287–291. [CrossRef]
17. Jenkinson, D.S.; Adams, D.E.; Wild, A. Model estimates of CO_2 emissions from soil in response to global warming. *Nature* **1991**, *351*, 304–306. [CrossRef]
18. Kirschbaum, M.U.F. The temperature dependence of soil organic matter decomposition, and the effect of global warming on soil organic C storage. *Soil Biol. Biochem.* **1995**, *27*, 753–760. [CrossRef]
19. Kirschbaum, M.U.F. Soil respiration under prolonged soil warming: Are rate reductions caused by acclimation or substrate loss? *Glob. Chang. Biol.* **2004**, *10*, 1870–1877. [CrossRef]

20. Davidson, E.A.; Janssens, I.A. Temperature sensitivity of soil carbon decomposition and feedbacks to climate change. *Nature* **2006**, *440*, 165–173. [CrossRef] [PubMed]
21. Hartley, I.P.; Heinemeyer, A.; Ineson, P. Effects of three years of soil warming and shading on the rate of soil respiration: Substrate availability and not thermal acclimation mediates observed response. *Glob. Chang. Biol.* **2007**, *13*, 1761–1770. [CrossRef]
22. Hopkins, F.M.; Torn, M.S.; Trumbore, S.E. Warming accelerates decomposition of decades-old carbon in forest soils. *Proc. Natl. Acad. Sci. USA* **2012**, *109*, E1753–E1761. [CrossRef] [PubMed]
23. Lu, M.; Zhou, X.; Yang, Q.; Li, H.; Luo, Y.; Fang, C.; Chen, J.; Yang, X.; Li, B. Responses of ecosystem carbon cycle to experimental warming: A meta-analysis. *Ecology* **2013**, *94*, 726–738. [CrossRef] [PubMed]
24. Knorr, W.; Prentice, I.C.; House, J.I.; Holland, E.A. Long-term sensitivity of soil carbon turnover to warming. *Nature* **2005**, *433*, 298–301. [CrossRef] [PubMed]
25. Schindlbacher, A.; Wunderlich, S.; Borken, W.; Kitzler, B.; Zechmeister-Boltenstern, S.; Jandl, R. Soil respiration under climate change: Prolonged summer drought offsets soil warming effects. *Glob. Chang. Biol.* **2012**, *18*, 2270–2279. [CrossRef]
26. Cox, P.M.; Pearson, D.; Booth, B.B.; Friedlingstein, P.; Huntingford, C.; Jones, C.D.; Luke, C.M. Sensitivity of tropical carbon to climate change constrained by carbon dioxide variability. *Nature* **2013**, *494*, 341–344. [CrossRef] [PubMed]
27. Huntingford, C.; Zelazowski, P.; Galbraith, D.; Mercado, L.M.; Sitch, S.; Fisher, R.; Lomas, M. Simulated resilience of tropical rainforests to CO_2-induced climate change. *Nat. Geosci.* **2013**, *6*, 268–273. [CrossRef]
28. Bradford, M.A.; Wieder, W.R.; Bonan, G.B.; Fierer, N.; Raymond, P.A.; Crowther, T.W. Managing uncertainty in soil carbon feedbacks to climate change. *Nat. Clim. Chang.* **2016**, *6*, 751–758. [CrossRef]
29. Carey, J.C.; Tang, J.; Templer, P.H.; Kroeger, K.D.; Crowther, T.W.; Burton, A.J.; Dukes, J.S.; Emmett, B.; Frey, S.D.; Heskel, M.A.; et al. Temperature response of soil respiration largely unaltered with experimental warming. *Proc. Natl. Acad. Sci. USA* **2016**, *113*, 13797–13802. [CrossRef] [PubMed]
30. Vitousek, P.M.; Aplet, G.H.; Raich, J.W.; Lockwood, J.P. Biological perspectives on Mauna Loa Volcano: A model system for ecological research. In *Mauna Loa Revealed: Structure, Composition, History, and Hazards*; Rhodes, J.M., Lockwood, J.P., Eds.; American Geophysical Union: Washington, DC, USA, 1995; pp. 117–126.
31. Wang, X.; Piao, S.; Ciais, P.; Friedlingstein, P.; Myneni, R.B.; Cox, P.; Heimann, M.; Miller, J.; Peng, S.; Wang, T.; et al. A two-fold increase of carbon cycle sensitivity to tropical temperature variations. *Nature* **2014**, *506*, 212–215. [CrossRef] [PubMed]
32. Bahn, M.; Janssens, I.A.; Reichstein, M.; Smith, P.; Trumbore, S.E. Soil respiration across scales: Towards an integration of patterns and processes. *New Phytol.* **2010**, *186*, 292–296. [CrossRef] [PubMed]
33. Vargas, R.; Allen, M.F. Environmental controls and the influence of vegetation type, fine roots and rhizomorphs on diel and seasonal variation in soil respiration. *New Phytol.* **2008**, *179*, 460–471. [CrossRef] [PubMed]
34. Zimmermann, M.; Meir, P.; Bird, M.; Malhi, Y.; Cahuana, A. Litter contribution to diurnal and annual soil respiration in a tropical montane cloud forest. *Soil Biol. Biochem.* **2009**, *41*, 1338–1340. [CrossRef]
35. Zimmermann, M.; Meir, P.; Bird, M.I.; Malhi, Y.; Ccahuana, A.J.Q. Temporal variation and climate dependence of soil respiration and its components along a 3000 m altitudinal tropical forest gradient. *Glob. Biogeochem. Cycles* **2010**, *24*, GB4012. [CrossRef]
36. Tan, Z.-H.; Zhang, Y.-P.; Liang, N.; Song, Q.-H.; Liu, Y.-H.; You, G.-Y.; Li, L.-H.; Yu, L.; Wu, C.-S.; Lu, Z.-Y.; et al. Soil respiration in an old-growth subtropical forest: Patterns, components, and controls. *J. Geophys. Res. Atmos.* **2013**, *118*, 2981–2990. [CrossRef]
37. Wood, T.E.; Detto, M.; Silver, W.L. Sensitivity of soil respiration to variability in soil moisture and temperature in a humid tropical forest. *PLoS ONE* **2013**, *8*, e80965. [CrossRef] [PubMed]
38. Hanpattanakit, P.; Leclerc, M.Y.; Mcmillan, A.M.S.; Limtong, P.; Maeght, J.-L.; Panuthai, S.; Inubushi, K.; Chidthaisong, A. Multiple timescale variations and controls of soil respiration in a tropical dry dipterocarp forest, western Thailand. *Plant Soil* **2015**, *390*, 167–181. [CrossRef]
39. Kuzyakov, Y.; Cheng, W. Photosynthesis controls of rhizosphere respiration and organic matter decomposition. *Soil Biol. Biochem.* **2001**, *33*, 1915–1925. [CrossRef]
40. Moyano, F.E.; Kutsch, W.L.; Rebmann, C. Soil respiration fluxes in relation to photosynthetic activity in broad-leaf and needle-leaf forest stands. *Agric. For. Meteorol.* **2008**, *148*, 135–143. [CrossRef]

41. Kuzyakov, Y.; Gavrichkova, O. Time lag between photosynthesis and carbon dioxide efflux from soil: A review of mechanisms and controls. *Glob. Chang. Biol.* **2010**, *16*, 3386–3406. [CrossRef]

42. Ohashi, M.; Kumagai, T.; Kume, T.; Gyokusen, K.; Saitoh, T.M.; Suzuki, M. Characteristics of soil CO_2 efflux variability in an aseasonal tropical rainforest in Borneo Island. *Biogeochemistry* **2008**, *90*, 275–289. [CrossRef]

43. Curiel Yuste, J.; Janssens, I.A.; Carrara, A.; Ceulemans, R. Annual Q_{10} of soil respiration reflects plant phenological patterns as well as temperature sensitivity. *Glob. Chang. Biol.* **2004**, *10*, 161–169. [CrossRef]

44. Easterling, D.R.; Horton, B.; Jones, P.D.; Peterson, T.C.; Karl, T.R.; Parker, D.E.; Salinger, M.J.; Razuvayev, V.; Plummer, N.; Jamason, P.; et al. Maximum and minimum temperature trends for the globe. *Science* **1997**, *277*, 364–367. [CrossRef]

45. Clark, D.A.; Piper, S.C.; Keeling, C.D.; Clark, D.B. Tropical rain forest tree growth and atmospheric carbon dynamics linked to interannual temperature variation during 1984–2000. *Proc. Natl. Acad. Sci. USA* **2003**, *100*, 5852–5857. [CrossRef] [PubMed]

46. Vincent, L.A.; Peterson, T.C.; Barros, V.R.; Marino, M.B.; Rusticucci, M.; Carrasco, G.; Ramirez, E.; Alves, L.M.; Ambrizzi, T.; Berlato, M.A.; et al. Observed trends in indices of daily temperature extremes in South America 1960–2000. *J. Clim.* **2005**, *18*, 5011–5023. [CrossRef]

47. Donat, M.G.; Alexander, L.V. The shifting probability distribution of global daytime and night-time temperatures. *Geophys. Res. Lett.* **2012**, *39*, L14707. [CrossRef]

48. Newstrom, L.E.; Frankie, G.W.; Baker, H.G. A new classification for plant phenology based on flowering patterns in lowland tropical rain forest trees at La Selva, CostaRica. *Biotropica* **1994**, *26*, 141–159. [CrossRef]

49. Yan, J.; Wang, Y.; Zhou, G.; Zhang, D. Estimates of soil respiration and net primary production of three forests at different succession stages in South China. *Glob. Chang. Biol.* **2006**, *12*, 810–821. [CrossRef]

50. Kira, T.; Shidei, T. Primary production and turnover of organic matter in different forest ecosystems of the western Pacific. *Jpn. J. Ecol.* **1967**, *17*, 70–87.

51. Ryan, M.G.; Binkley, D.; Fownes, J.H. Age-related decline in forest productivity: Pattern and process. *Adv. Ecol. Res.* **1997**, *27*, 213–262.

52. Ryan, M.G.; Binkley, D.; Fownes, J.H.; Giardina, C.P.; Senock, R.S. An experimental test of the causes of forest growth decline with stand age. *Ecol. Monogr.* **2004**, *74*, 393–414. [CrossRef]

53. Russell, A.E.; Raich, J.W.; Bedoya, R.; Valverde-Barrantes, O.; González, E. Impacts of individual tree species on carbon dynamics in a moist tropical forest environment. *Ecol. Appl.* **2010**, *20*, 1087–1100. [CrossRef] [PubMed]

54. Clark, D.A.; Clark, D.B.; Oberbauer, S.F. Field-quantified responses of tropical rainforest aboveground productivity to increasing CO_2 and climatic stress, 1997–2009. *J. Geophys. Res. Biogeosci.* **2013**, *118*, 783–794. [CrossRef]

55. Loescher, H.W.; Gholz, H.L.; Jacobs, J.M.; Oberbauer, S.F. Energy dynamics and modeled evapotranspiration from a wet tropical forest in Costa Rica. *J. Hydrol.* **2005**, *315*, 274–294. [CrossRef]

56. Hartshorn, G.S.; Hammel, B.E. Vegetation types and floristic patterns. In *La Selva Ecology and Natural History of a Neotropical Rain Forest*; McDade, L.A., Bawa, K.S., Hespenheide, H.A., Hartshorn, G.S., Eds.; University of Chicago Press: Chicago, IL, USA, 1994; pp. 73–89.

57. McDade, L.A.; Bawa, K.S.; Hespenheide, H.A.; Hartshorn, G.S. *La Selva Ecology and Natural History of a Neotropical Rain Forest*; The University of Chicago Press: Chicago, IL, USA, 1994; p. 486.

58. Fisher, R.F. Amelioration of degraded rain forest soils by plantations of native trees. *Soil Sci. Soc. Am. J.* **1995**, *59*, 544–549. [CrossRef]

59. Haggar, J.P.; Briscoe, C.B.; Butterfield, R.P. Native species: A resource for the diversification of forestry production in the lowland humid tropics. *For. Ecol. Manag.* **1998**, *106*, 195–203. [CrossRef]

60. González, J.E.; Fisher, R.F. Growth of native forest species planted on abandoned pasture land in Costa Rica. *For. Ecol. Manag.* **1994**, *70*, 159–167. [CrossRef]

61. Sollins, P.; Sancho, F.M.; Mata, R.C.; Sanford, R.L., Jr. Soils and soil process research. In *La Selva Ecology and Natural History of a Neotropical Rain Forest*; McDade, L.A., Bawa, K.S., Hespenheide, H.A., Hartshorn, G.S., Eds.; University of Chicago Press: Chicago, IL, USA, 1994; pp. 34–53.

62. Kleber, M.; Schwendenmann, L.; Veldkamp, E.; Rößner, J.; Jahn, R. Halloysite versus gibbsite: Silicon cycling as a pedogenetic process in two lowland Neotropical rain forest soils of La Selva, Costa Rica. *Geoderma* **2007**, *138*, 1–11. [CrossRef]

63. Russell, A.E.; Raich, J.W.; Valverde-Barrantes, O.J.; Fisher, R.F. Tree species effects on soil properties in experimental plantations in tropical moist forest. *Soil Sci. Soc. Am. J.* **2007**, *71*, 1389–1397. [CrossRef]
64. Russell, A.E.; Raich, J.W. Rapidly growing tropical trees mobilize remarkable amounts of nitrogen, in ways that differ surprisingly among species. *Proc. Natl. Acad. Sci. USA* **2012**, *109*, 10398–10402. [CrossRef] [PubMed]
65. Valverde-Barrantes, O.J. Relationships among litterfall, fine root growth, and soil respiration for five tropical tree species. *Can. J. For. Res.* **2007**, *37*, 1954–1965. [CrossRef]
66. Raich, J.W.; Russell, A.E.; Bedoya-Arrieta, R. Lignin and enhanced litter turnover in tree plantations of lowland Costa Rica. *For. Ecol. Manag.* **2007**, *239*, 128–135. [CrossRef]
67. Haggar, J.P.; Wightman, K.; Fisher, R.F. The potential of plantations to foster woody regeneration within a deforested landscape in lowland Costa Rica. *For. Ecol. Manag.* **1997**, *99*, 55–64. [CrossRef]
68. Powers, J.S.; Haggar, J.P.; Fisher, R.F. The effect of overstory composition on understory woody regeneration and species richness in 7-year-old plantations in Costa Rica. *For. Ecol. Manag.* **1997**, *99*, 43–54. [CrossRef]
69. Ohashi, M.; Kume, T.; Yamane, S.; Suzuki, M. Hot spots of soil respiration in an Asian tropical rainforest. *Geophys. Res. Lett.* **2007**, *34*, L08705. [CrossRef]
70. Oishi, A.C.; Palmroth, S.; Butnor, J.R.; Johsen, K.H.; Oren, R. Spatial and temporal variability of soil CO_2 efflux in three proximate temperate forest ecosystems. *Agric. For. Meteorol.* **2013**, *171*, 256–269. [CrossRef]
71. Wang, H.; Liu, S.; Wang, J.; Shi, Z.; Lu, L.; Zeng, J.; Ming, A.; Tang, J.; Yu, H. Effects of tree species mixture on soil organic carbon stocks and greenhouse gas fluxes in subtropical plantations in China. *For. Ecol. Manag.* **2013**, *300*, 4–13. [CrossRef]
72. Huang, Z.; Yu, Z.; Wang, M. Environmental controls and the influence of tree species on temporal variation in soil respiration in subtropical China. *Plant Soil* **2014**, *382*, 75–87. [CrossRef]
73. Högberg, P.; Read, D.J. Towards a more plant physiological perspective on soil ecology. *Trends Ecol. Evol.* **2006**, *21*, 548–554. [CrossRef] [PubMed]
74. Frankie, G.W.; Baker, H.G.; Opler, P.A. Comparative phenological studies of trees in tropical wet and dry forests in the lowlands of Costa Rica. *J. Ecol.* **1974**, *62*, 881–919. [CrossRef]
75. Phillips, C.L.; Bond-Lamberty, B.; Desai, A.R.; Lavoie, M.; Risk, D.; Tang, J.; Todd-Brown, K.; Vargas, R. The value of soil respiration measurements for interpreting and modeling terrestrial carbon cycling. *Plant Soil* **2016**. [CrossRef]
76. Raich, J.W.; Mora, G. Estimating root plus rhizosphere contributions to soil respiration in annual croplands. *Soil Sci. Soc. Am. J.* **2005**, *69*, 634–639. [CrossRef]
77. Savage, K.; Davidson, E.A.; Tang, J. Diel patterns of autotrophic and heterotrophic respiration among phenological stages. *Glob. Chang. Biol.* **2013**, *19*, 1151–1159. [CrossRef] [PubMed]
78. Raich, J.W.; Valverde-Barrantes, O.J. Soil CO_2 Flux, Moisture, Temperature, and Litterfall at La Selva, Costa Rica, 2003–2010. Available online: http://dx.doi.org/10.3334/ORNLDAAC/1373 (accessed on 7 February 2017).

forests

MDPI

Erratum

Erratum: Spatial Upscaling of Soil Respiration under a Complex Canopy Structure in an Old-Growth Deciduous Forest, Central Japan; *Forests* 2017, *8*, 36

Forests Editorial Office

MDPI AG, St. Alban-Anlage 66, 4052 Basel, Switzerland; forests@mdpi.com

Received: 27 February 2017; Accepted: 27 February 2017; Published: 6 March 2017

Due to a mistake during the production process, there was a spelling error in the Academic Editors' names in the original published version [1]. The *Forests* Editorial Office therefore wishes to make this correction to the paper:

Academic Editors: Robert Jandl and Mirco Rodeghiero

We would like to apologize for any inconvenience caused to the authors and readers by this mistake. We will update the article and the original version will remain available on the article webpage.

1.	Suchewaboripont, V.; Ando, M.; Yoshitake, S.; Iimura, Y.; Hirota, M.; Ohtsuka, T. Spatial Upscaling of Soil Respiration under a Complex Canopy Structure in an Old-Growth Deciduous Forest, Central Japan. *Forests* **2017**, *8*, 36. [CrossRef]

forests

MDPI

Article

Partitioning Forest-Floor Respiration into Source Based Emissions in a Boreal Forested Bog: Responses to Experimental Drought

Tariq Muhammad Munir [1,*], Bhupesh Khadka [1], Bin Xu [1] and Maria Strack [2]

[1] Department of Geography, University of Calgary, Calgary, AB T2N 1N4, Canada; bhupeshk@nait.ca (B.K.); binx@nait.ca (B.X.)

[2] Department of Geography and Environmental Management, University of Waterloo, Waterloo, ON N2L 3G1, Canada; mstrack@uwaterloo.ca

* Correspondence: tmmunir@ucalgary.ca; Tel.: +1-403-971-5693

Academic Editors: Robert Jandl and Mirco Rodeghiero
Received: 1 February 2017; Accepted: 7 March 2017; Published: 10 March 2017

Abstract: Northern peatlands store globally significant amounts of soil carbon that could be released to the atmosphere under drier conditions induced by climate change. We measured forest floor respiration (R_{FF}) at hummocks and hollows in a treed boreal bog in Alberta, Canada and partitioned the flux into aboveground forest floor autotrophic, belowground forest floor autotrophic, belowground tree respiration, and heterotrophic respiration using a series of clipping and trenching experiments. These fluxes were compared to those measured at sites within the same bog where water-table (WT) was drawn down for 2 and 12 years. Experimental WT drawdown significantly increased R_{FF} with greater increases at hummocks than hollows. Greater R_{FF} was largely driven by increased autotrophic respiration driven by increased growth of trees and shrubs in response to drier conditions; heterotrophic respiration accounted for a declining proportion of R_{FF} with time since drainage. Heterotrophic respiration was increased at hollows, suggesting that soil carbon may be lost from these sites in response to climate change induced drying. Overall, although WT drawdown increased R_{FF}, the substantial contribution of autotrophic respiration to R_{FF} suggests that peat carbon stocks are unlikely to be rapidly destabilized by drying conditions.

Keywords: forest floor respiration; root respiration; autotrophic respiration; heterotrophic respiration; disturbance; water table; drought; climate change; modeling; soil temperature

1. Introduction

Peatlands contain one of the largest terrestrial carbon (C) stocks, estimated at ~600 Gt C [1], with northern peatland C storage accounting for ~390–440 Gt [1,2]. The large C stock has been accumulated as a result of only a marginal difference, over millennia, between photosynthetic C uptake and loss of C as ecosystem respiration, methane (CH_4) emissions, and water-borne outflows [3]. The stored C is present in the form of highly mineralizable organic C [4,5] protected in water-saturated, anoxic conditions and is highly sensitive to warmer and drier climate [5–7]. Therefore, any increase in carbon dioxide (CO_2) emissions in response to the expected changes in climate has the potential to provide a positive feedback to global warming [4,7–9]. In general, many northern peatlands are expected to be drier under future climates [10,11], and while the response of peatland ecosystem respiration to water-table drawdown has been extensively studied [12–17], controlled field experimentation for partitioning ecosystem respiration into its source-based major components remains largely unexplored [18]. Research is needed to investigate source-based respiration fluxes in relation to potential changes in environmental conditions to improve our understanding of changes in ecosystem C storage or emissions to the atmosphere under climate change scenarios [19,20].

Ecosystem respiration includes the emission of CO_2 to the atmosphere from above and belowground parts of vegetation (autotrophic respiration; R_A), and from microbial decomposition of the soil organic matter including litter (heterotrophic respiration; R_H). The aboveground parts mainly include plant leaves and stems while the belowground parts comprise living roots with their associated mycorrhizal fungi and microbial populations [21,22]. Many northern peatlands are treed [23,24]. When respiration is measured at the ground layer of a treed peatland, this forest floor respiration (R_{FF}) includes respiration associated with tree roots, while respiration of the aboveground tree biomass is excluded. This research partitioned the bulk R_{FF} into source-based above and belowground respiration flux components as: (1) forest floor aboveground autotrophic respiration ($R_{FF\_A\_ag}$) and, (2) belowground shrub + herb (roots) autotrophic respiration ($R_{A\_SH\_bg}$) and belowground tree (roots) autotrophic respiration ($R_{A\_T\_bg}$). Separating the $R_{A\_SH\_bg}$ from $R_{A\_T\_bg}$ is important as specific vascular plant functional types may respire at different rates [25,26] due to the difference in their respective above and belowground productivities [27], and therefore modify the response of R_{FF} to changes in water-table (WT) level and soil temperature at 5 cm depth (T_5) [4,28].

Peatland respiration flux components are highly responsive to environmental changes such as WT level [26,29–32], T_5 [15,33,34], and plant functional (shrubs + herbs, trees) type [35–38]. Therefore, increased atmospheric or soil temperatures and subsequent WT lowering [39,40] may increase soil respiration rates [22] and ultimately alter the peatland C sink or source strength ([37] resulting from increasing atmospheric CO_2 [41]. However, increases in R_{FF} alone do not necessarily indicate a loss of soil C if only autotrophic respiration increases, illustrating the importance of determining the source of respiration and the relative response of each component to changing environmental conditions [42].

Ericaceous shrubs (e.g., *Rhododendron groenlandicum*) are important contributors to ecosystem productivity in many northern bogs [43,44], also making important contributions to ecosystem respiration. The contribution of shrub autotrophic respiration to R_{FF} in a shrub-dominated bog in Patuanak, Saskatchewan, Canada was estimated to be ~75% [45]. A median value of root: shoot ratios estimated from 14 sites in boreal forest was found to be 0.39 [46], suggesting that root respiration is likely to make a significant contribution to measured autotrophic respiration. Overall, root/rhizospheric respiration has been found to account for between 10% and 90% of R_{FF} depending upon vegetation type and season of year [28,47]. As similar controls apply to the above and belowground productivities of shrubs [48,49], therefore, the response of above and belowground respiration to environmental change should be similar [46].

Experimental partitioning of soil autotrophic and heterotrophic respiration components has been attempted using different methods with varying results. For example, methods include application of stable isotopes [50,51], root biomass regression with R_{FF} to determine belowground root respiration [52], and comparison of soils with and without root exclusion to determine tree root respiration [17,28]; however, evaluating the responses of various respiration sources ($R_{FF\_A\_ag}$, $R_{A\_SH\_bg}$, $R_{A\_T\_bg}$, and R_H) to environmental change has not been completed. Moreover, responses may vary between microforms (hummocks and hollows) due to initial differences in WT level, soil properties, and vegetation community [16,17,32]. Therefore, this study focused on partitioning R_{FF} emissions along a microtopographic gradients in order to evaluate responses of each respiration component to short (2 years) and longer term (12 years) water-table drawdown.

Our specific objectives were to:

1. Partition R_{FF} between the aboveground ground-layer autotrophic respiration ($R_{FF\_A\_ag}$), belowground autotrophic respiration of shrubs + herbs ($R_{A\_SH\_bg}$) and trees ($R_{A\_T\_bg}$), and heterotrophic respiration (R_H) across water-table (WT) treatments (control, experimental, drained),

2. Evaluate differences in source contributions to R_{FF} along a microtopographic (hummock, hollow) gradient in a boreal forested bog and how these contributions changed in response to WT treatments, and

3. Assess the respiration components' responses to the WT and soil temperature at 5 cm depth (T_5) over one growing season.

We hypothesized that experimental WT lowering would lead to increases in all respiration components, with greatest increases at hollows. We also hypothesized that increases in R_H would be greatest at hollows, while hummocks would have greater increases in autotrophic respiration (both above and belowground) and that all components of R_{FF} would have significant positive correlations with depth to WT and T_5.

2. Materials and Methods

2.1. Sites Description and Experimental Design

During the growing season (1 May to 31 October) of 2012, this research was conducted in a forested bog within the southern boreal forest and near the town of Wandering River, Alberta, Canada. Based on 30-year (1981–2010) averages, the mean growing season (May to October) temperature and precipitation for this region are 11.7 °C and 382 mm, respectively [53]. The mean growing season air temperature and precipitation measured during the 2012 study using a meteorological station installed at the study sites were 13.2 °C and 282 mm, respectively.

Within the dry ombrotrophic forested bog, three research sites, CONTROL (55°21′ N, 112°31′ W), EXPERIMENTAL (55°21′ N, 112°31′ W), and DRAINED (55°16′ N, 112°28′ W), were chosen or created (Figure 1). The control was an undisturbed site with a mean WT level of −38 cm whereas the experimental site was created adjacent to the control by ditching around, lowering the mean WT level to 78 cm below surface. One year prior to this study, WT level (± Standard Deviation (SD)) at the control (−56 ± 22 cm) and experimental (−57 ± 20 cm) sites were not different (negative values denote belowground WT; ANOVA, $F_{1,5} = 0.55$, $p = 0.492$). The drained site (part of the same bog and located 9 km to southwest) was drained inadvertently 12 years prior to the study as a result of a peat harvesting preparation on an adjacent section, and had a mean WT level of ~−120 cm in 2012 (Figure 2).

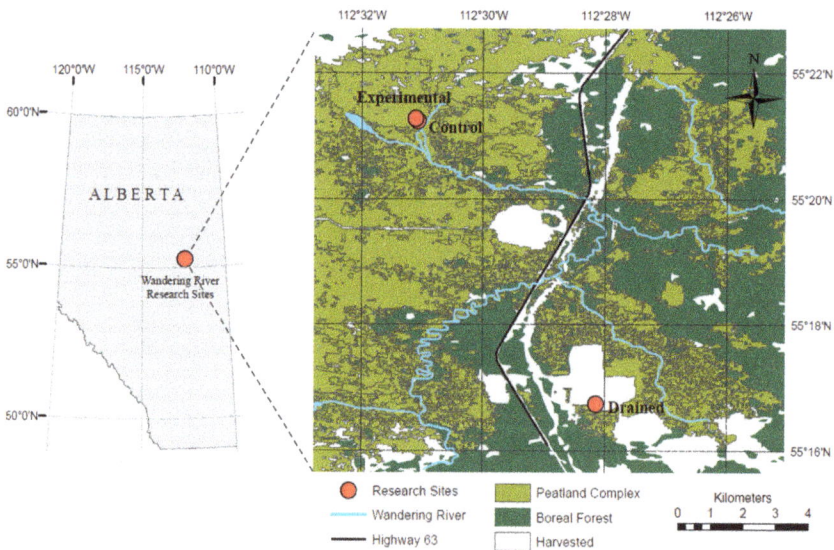

Figure 1. Geographical map of the Wandering River study sites located in a forested peatland complex within boreal forest in Alberta, Canada [54].

Based on plant species indicators, the studied bog was classified as a forested low shrub bog [55] with two distinct microtopographic features: hummock and hollow. One year prior to this study, the control and experimental site microforms had equal coverage of mosses with sparse shrubs, whereas the drained hummocks had the highest coverage of shrubs and the drained hollows had the greatest coverage of lichens [44]. At all sites, black spruce (*Picea mariana* (Mill.) B.S.P.) was the most abundant type of tree constituting >99% of the tree stand, with 25,766 stems·ha^{-1} consisting of 37% taller trees (>137 cm height) up to 769 cm high [44]. The black spruce stand had an average canopy height of 168 cm, projection coverage of 42%, and basal area of 73.5 m^2·ha^{-1} [17]. Trees were generally evenly distributed across the study plots (i.e., not clustered).

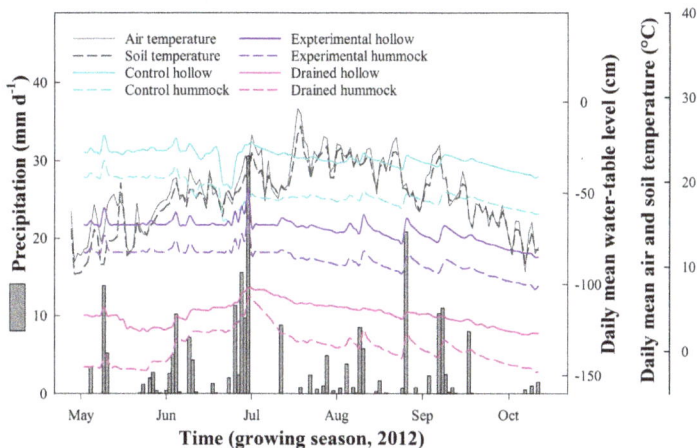

Figure 2. Daily mean site air temperature (°C), soil temperature (°C) at 5 cm depth, total precipitation (mm), representative hollow and hummock water-table (WT) level (cm) over the 2012 growing season (May to October). Note the two *y*-axes on the right side: daily mean WT level, and air and soil temperatures.

Mean (±SD) pH and electrical conductivity (μS·cm^{-1}) of pore water in the control (4.1 ± 0.1 and 16.6 ± 0.7, respectively) and experimental (4.4 ± 0.3 and 15.2 ± 2.5, respectively) sites were similar (ANOVA, pH: $F_{1,5}$ = 2.6, *p* = 0.166; EC: $F_{1,5}$ = 0.84, *p* = 0.401) prior to any manipulation. All the study sites had an average peat depth exceeding 4 m, and were underlain by a sandy clay substrate. Because all the study sites were part of the same bog and their initial primary characteristics of vegetation composition, WT levels, and chemistry were similar, their secondary attributes (e.g., respiration rates) were also assumed to be similar prior to WT manipulation.

From the available microtopography, we chose eight hummock and eight hollow microforms at each of the control and drained sites, and four of each microform type at the experimental site (due to its smaller area). In early May 2012, each of the chosen microform plots was fitted with a 60 cm × 60 cm collar having a groove at the top for placing the CO$_2$ flux chamber. The collar was carefully inserted into the peat surface to a depth of ~5 cm to keep the disturbance minimal [56]. Outside each collar, a Polyvinyl Chloride (PVC) well (length = 200 cm, diameter = 3.5 cm) with perforation and a nylon cloth covering on the lower 150 cm was inserted into a bore-hole drilled to a depth of 150 cm. WT level was manually measured at all the water wells every time CO$_2$ flux was measured during the growing season of 2012. The WT levels were also monitored during the study period at 20-min intervals using automatic, temperature compensating pressure transducers (Levelogger Junior 3001, Solinst, Georgetown, Ontario, Canada) installed in two randomly selected water wells at each site: one at a hummock and the other at a hollow plot.

2.2. CO_2 Flux Measurements

All CO_2 flux measurements were made during the day time of the growing season (May to October, 2012) using the same equipment. We used a closed chamber with dimensions 60 cm × 60 cm × 30 cm (width × length × height), made of opaque acrylic and fitted with a low-speed battery-operated fan to circulate air within the chamber headspace during and between CO_2 concentration measurements. The chamber had no pressure equilibrium port installed. A portable infrared gas analyzer (EGM-4, PP Systems, Amesbury, MA, USA) with a built-in CO_2 sampling pump operating at a flow rate of 350 mL·min^{-1} and compensating for temperature fluctuations within the chamber headspace was used to measure the instantaneous CO_2 concentration inside the chamber headspace. The chamber headspace temperature was measured using a thermocouple thermometer (VWR International, Edmonton, AB, Canada). The CO_2 concentration and temperature measurements were made every 15 s during a short chamber deployment period [57,58] of 1.75 min. Immediately after the CO_2 concentration measurements at a plot, soil temperature at a depth of 5 cm (T_5) was measured using a thermocouple thermometer, and the WT level relative to moss surface was manually measured from a permanently installed water well adjacent to the plot. The CO_2 flux was calculated from the linear change in CO_2 concentration in chamber headspace over time [17], as a function of air temperature, pressure, and volume within the chamber headspace, following the ideal gas law.

2.2.1. Forest Floor Respiration (R_{FF})

Prior to any manipulation at a plot, a CO_2 efflux measurement represented forest floor respiration (R_{FF}). During a 5-day long R_{FF} measurement campaign in May 2012, we measured the fluxes on four to five occasions in each plot. The measured R_{FF} is divided into major source-based respiration flux components as:

$$R_{FF} = R_{FF\_A\_ag} + R_{A\_SH\_bg} + R_{A\_T\_bg} + R_H, \tag{1}$$

where R_{FF} accounts for forest floor aboveground autotrophic respiration ($R_{FF\_A\_ag}$), belowground shrub + herb and tree autotrophic (rhizospheric) respiration ($R_{A\_SH\_bg} + R_{A\_T\_bg}$, respectively), and soil heterotrophic respiration (R_H). At the end of the R_{FF} measurement campaign, we clipped all plots using sharp scissors at the base of capitulum at 1 cm below the moss surface [27,59]. All the plots had their surface carefully cleared of any plant litter. The clipped shrubs + herbs were placed in labelled paper bags, taken to the Ecohydrology Lab, University of Calgary, AB, oven dried at 60 °C for 48 hours and weighed to calculate mean biomass (g·m^{-2}) at each plot at each site.

2.2.2. Partitioning Forest Floor Respiration

Following the R_{FF} measurement campaign, a 5-day long campaign for measuring the post-clipping CO_2 emissions from every plot was performed. During the campaign, we measured the emissions on four to five occasions in each plot. At plot level, the CO_2 emissions (g·CO_2·m^{-2}·day^{-1}) at clipped plots subtracted from R_{FF} (g·CO_2·m^{-2}·day^{-1}) represented $R_{FF\_A\_ag}$ (g·CO_2·m^{-2}·day^{-1}).

To estimate $R_{A\_T\_bg}$ at hummock or hollow microform at each site, we used a trenching method [17,52]. One half of the instrumented hummock or hollow plots (60 cm × 60 cm) were chosen at each site for the trenching procedure, whereas the other half of the plots were left untrenched. The trenched hummock or hollow plot R_{FF} was not significantly different from those of the untrenched plots at control (ANOVA, Hummock: $F_{1,25} = 0.667$, $p = 0.422$; Hollow: $F_{1,23} = 0.316$, $p = 0.580$), experimental (ANOVA, Hummock: $F_{1,7} = 0.000$, $p = 0.990$; Hollow: $F_{1,7} = 0.605$, $p = 0.466$), or drained (ANOVA, Hummock: $F_{1,23} = 0.041$, $p = 0.841$; Hollow: $F_{1,23} = 0.070$, $p = 0.790$) site. Therefore, in June 2012, we incised around the trenched plots to a depth of 30 cm and installed a thick polythene sheet to prevent root ingrowth while keeping the disturbance minimal. All the trenched and intact, untrenched plots were measured for CO_2 emission between July and September, 2012 to quantify the difference in the respiration rate for estimation of $R_{A\_T\_bg}$. The trenching method at this site had already been used at this bog [17].

We did not measure $R_{A\_SH\_bg}$ and calculated it by using regression equations ($y = a + bx$) generated by regressing the aboveground biomass of shrubs + herbs (x) with $R_{FF\_A\_ag}$ (y) following a previous study [52]. We did not sample/measure shrub + herb root biomass (B_{SH_bg}) to avoid disturbance to plots used for ongoing research and instead calculated B_{SH_bg} as: $B_{SH\_ag} \times 0.39$ [46], as previous research found the median root:shoot ratio of shrubs + herbs for 14 data points in boreal forest to be 0.39. Similar factors control net production both above and belowground, and the two rates were found to be related with each other [47]. The generated regression equations were used to determine the ($R_{A\_SH\_bg}$) by substituting B_{SH_ag} with B_{SH_bg} in the equation. The R_H was calculated from Equation (1) once all other components were estimated.

2.3. Seasonal Modeling

Only the measured respiration components (R_{FF}, $R_{FF\_A\_ag}$, $R_{A\_T\_bg}$) during the growing season (May to October) of 2012 were modeled using a multiple linear regression model with T_5 and WT level as:

$$R_{FF} = a \times T_5 + b \times WT\ level + c, \qquad (2)$$

where a, b, and c are regression coefficients (Table 1). Seasonal R_{FF}, $R_{FF\_A\_ag}$, and $R_{A\_T\_bg}$ were estimated for each 20-min period between 1 May to 31 October 2012, averaged daily, and summed separately for the growing season using T_5 (Onset®, HOBO®, Bourne, MA, USA) and WT level (Levelogger Junior, Solinst Canada Ltd., Georgetown, ON, Canada) measurements made on site. As the environmental variable logs were missing for the first 21 days of May 2012, they were filled by assuming that the first measured value was representative of the whole missing period. The field measured values of R_{FF}, $R_{FF\_A\_ag}$, and $R_{A\_T\_bg}$ were plotted against the model predicted values obtained using SPSS 24.0.0.1. Validation of the models showed excellent agreement between the measured and the modeled values (Appendix A: Figure A1).

The seasonal respiration rates (R_{FF}, $R_{FF\_A\_ag}$, $R_{A\_T\_bg}$) at hummock and hollow microforms were up-scaled by multiplying mean estimated growing season respiration by their corresponding coverage of 56% and 44% at the control, 55% and 45% at the experimental, and 52% and 48% at the drained site, respectively [60]. The seasonal value of $R_{A\_SH\_bg}$ was calculated separately for each microform/site combination by determining it as a proportion of corresponding instantaneous R_{FF} value and then estimating it as this proportion of the modeled seasonal R_{FF}. Seasonal R_H was determined by difference according to Equation (1).

2.4. Statistical Analyses

To test the effects of WT level and microform on R_{FF}, $R_{FF\_A\_ag}$, $R_{A\_SH\_bg}$, and $R_{A\_T\_bg}$, we conducted a repeated, linear mixed-effects model analysis (LMEM; IBM SPSS Statistics 24.0.0.1, IBM corporation, Armonk, New York, USA) with WT level (control, experimental, drained) and microform (hummock, hollow) as predictor variables and R_{FF}, $R_{FF\_A\_ag}$, $R_{A\_SH\_bg}$, or $R_{A\_T\_bg}$ as response variables, including the random effect of plot and repeated effect of time (as the same plots were used sequentially for all the measurements). We also tested the effect of B_{SH_ag} in combination with site and microform on $R_{FF\_A\_ag}$ with random effect of plot using a repeated, LMEM. In this case, non-significant terms were removed from the model one at a time, starting with the highest p-value and the model was re-run until only significant terms remained. In all the LMEMs used in this study, a combined symmetry covariance structure was used. The relationships of WT level and T_5 with R_{FF}, $R_{FF\_A\_ag}$, $R_{A\_SH\_bg}$, and $R_{A\_T\_bg}$ were also tested for their significance using linear regression model fitting where applicable. Differences between regression slopes were tested [61] where applicable.

3. Results

The microclimate of the study sites was monitored over the growing season (May to October) of 2012 and was warmer by 1.4 °C and drier by 79 mm than the 30-year (1981–2010) regional averages measured at Athabasca, Alberta, Canada. The WT levels at the experimental and drained sites were as much as 36 cm and 82 cm lower than at the control site (Figure 2). In general, as a result of 12 years of drainage, the mosses at the drained hummocks were replaced by shrubs and mosses at the drained hollows were replaced by lichens. Detailed site hydrological responses to the warmer and drier climate at these sites have been reported [17].

Table 1. Fitted model parameters, their values (±SE), residual standard error (RSE), *p* values, adjusted r^2, and the number of values (*n*) included in the regression analyses for the forest floor respiration (R_{FF}), forest floor aboveground autotrophic respiration ($R_{FF\_A\_ag}$), and belowground autotrophic respiration of tree roots ($R_{A\_T\_bg}$) models (Equation (2)) *.

Site/Microform	Flux	a	b	c	RSE	*p*	r^2	*n*
		Dimensionless			$g \cdot CO_2 \cdot m^{-2} \cdot day^{-1}$			
Control Hummock	R_{FF}	0.52 ± 0.14	-0.47 ± 0.15	-19.90 ± 5.1	1.09	<0.001	0.80	24
	$R_{FF\_A\_ag}$	0.36 ± 0.10	-0.45 ± 0.13	-22.51 ± 5.3	1.16	<0.001	0.67	24
	$R_{A\_T\_bg}$	0.43 ± 0.17	0.25 ± 0.17	7.88 ± 5.1	1.46	0.044	0.17	26
Hollow	R_{FF}	1.08 ± 0.21	-0.29 ± 0.14	-15.74 ± 5.1	2.45	<0.001	0.61	24
	$R_{FF\_A\_ag}$	0.85 ± 0.19	-0.54 ± 0.18	-24.59 ± 5.5	1.55	<0.001	0.75	24
	$R_{A\_T\_bg}$	0.86 ± 0.23	-0.21 ± 0.14	14.80 ± 5.7	2.34	0.002	0.43	22
Experimental Hummock	R_{FF}	0.05 ± 0.10	-0.24 ± 0.11	-15.67 ± 11.8	0.24	0.010	0.77	8
	$R_{FF\_A\_ag}$	0.96 ± 0.50	-0.83 ± 0.57	-86.98 ± 55.9	2.02	0.049	0.58	8
	$R_{A\_T\_bg}$	0.91 ± 0.52	-0.72 ± 0.63	-75.24 ± 61.5	1.24	0.032	0.83	6
Hollow	R_{FF}	2.59 ± 0.80	0.70 ± 0.09	-15.09 ± 13.2	1.59	0.058	0.55	8
	$R_{FF\_A\_ag}$	0.82 ± 1.60	-0.27 ± 0.18	-29.44 ± 10.0	2.56	0.061	0.54	8
	$R_{A\_T\_bg}$	0.94 ± 0.39	-0.69 ± 0.87	-61.08 ± 70.1	2.06	0.021	0.79	7
Drained Hummock	R_{FF}	-0.32 ± 0.16	-1.40 ± 0.20	-167.53 ± 30.9	3.27	<0.001	0.76	24
	$R_{FF\_A\_ag}$	0.93 ± 0.31	-0.49 ± 0.18	-70.88 ± 19.3	2.43	<0.001	0.75	24
	$R_{A\_T\_bg}$	1.69 ± 0.35	0.39 ± 0.64	34.86 ± 89.9	2.41	<0.001	0.79	24
Hollow	R_{FF}	0.79 ± 0.33	-0.58 ± 0.17	-61.70 ± 16.9	2.29	<0.001	0.70	20
	$R_{FF\_A\_ag}$	0.70 ± 0.23	-0.32 ± 0.10	-40.53 ± 10.6	2.15	<0.001	0.56	24
	$R_{A\_T\_bg}$	0.26 ± 0.19	-1.08 ± 0.42	-113.87 ± 46.4	1.28	<0.001	0.77	25

* R_{FF}, $R_{FF\_A\_ag}$, and $R_{A\_T\_bg}$ models were developed for each microform type (*n* = 3) at the control, experimental, and drained sites for the growing season of 2012. a and b are soil temperatures at 5 cm depth and WT level (below-ground) coefficients, respectively, and c is a regression constant. All modeled parameters are significant at $\alpha = 0.05$.

3.1. Controls on Respiration Flux Components

All the measured component fluxes (R_{FF}, $R_{FF\_A\_ag}$, $R_{A\_SH\_bg}$, $R_{A\_T\_bg}$) were well correlated to T_5 and WT (Figure 3) and therefore the modeled values matched the measured values well (Appendix A: Figure A1). Generally, the highest respiration occurred with warm temperatures and deep WT position.

3.2. Measured Respiration Components

3.2.1. Mean Forest Floor Respiration Rate (R_{FF})

There were significant effects of each of the WT (control, experimental, drained) and microform (hummock, hollow) types on R_{FF} measured in May, 2012; however, their interaction term did not significantly affect R_{FF} (Table 2). The drained site had significantly higher R_{FF} value (±SE) of 21.0 ± 2.1 $g \cdot CO_2 \cdot m^{-2} \cdot day^{-1}$ compared with those measured at the experimental (13.2 ± 2.5 $g \cdot CO_2 \cdot m^{-2} \cdot day^{-1}$; *p* = 0.042) and control (9.3 ± 2.1 $g \cdot CO_2 \cdot m^{-2} \cdot day^{-1}$; *p* = 0.006) sites that were not different (*p* = 0.455) from each other. Bonferroni pairwise comparisons revealed that the hummocks had overall significantly higher R_{FF} (17.9 ± 1.8 $g \cdot CO_2 \cdot m^{-2} \cdot day^{-1}$) than hollows ($11.2 \pm 1.8$ $g \cdot CO_2 \cdot m^{-2} \cdot day^{-1}$). Comparing microforms across sites (Figure 4) revealed that R_{FF} at hummocks was significantly different among all sites with highest rates at drained, followed by experimental, and then control, while hollows were not different ($p = 0.526$) across sites.

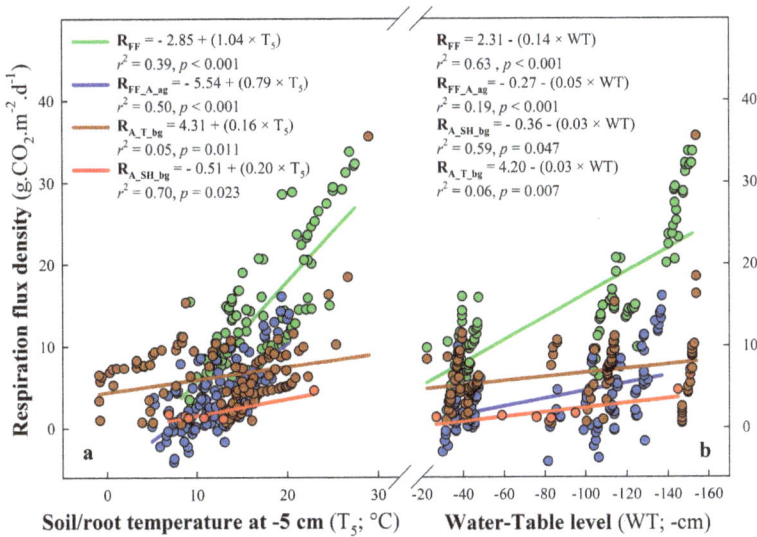

Figure 3. (a) Forest floor respiration flux (R_{FF}; $g \cdot CO_2 \cdot m^{-2} \cdot day^{-1}$), aboveground autotrophic respiration flux of forest floor ($R_{FF\_A\_ag}$; $g \cdot CO_2 \cdot m^{-2} \cdot day^{-1}$), belowground autotrophic respiration of shrubs + herbs ($R_{A\_SH\_bg}$; $g \cdot CO_2 \cdot m^{-2} \cdot day^{-1}$), and belowground autotrophic respiration of tree roots ($R_{A\_T\_bg}$; $g \cdot CO_2 \cdot m^{-2} \cdot day^{-1}$) versus soil temperature at -5 cm (T_5) and (b) R_{FF}, $R_{FF\_A\_ag}$, $R_{A\_SH\_bg}$, and $R_{A\_T\_bg}$ versus water-table (WT) level, at all sites/microforms. All the respiration components were statistically related to both the T_5 and WT level.

Table 2. Statistical results of repeated, linear mixed effects models. The models tested the fixed effects of site and microform on respiration of forest floor (R_{FF}), aboveground autotrophic respiration of forest floor ($R_{FF\_A\_ag}$), and belowground autotrophic respiration of shrubs + herbs and trees ($R_{A\_SH\_bg}$ and $R_{A\_T\_bg}$, respectively), separately *.

Effect	Flux Component							
	R_{FF}		$R_{FF\_A\_ag}$		$R_{A\_SH\_bg}$		$R_{A\_T\_bg}$	
	F	p	F	p	F	p	F	p
Site	$F_{2,7} = 8.1$	0.015	$F_{2,15} = 4.9$	0.023	$F_{2,12} = 25.3$	<0.001	$F_{2,12} = 6.4$	0.012
Microform	$F_{1,7} = 6.9$	0.033	$F_{1,15} = 10.3$	0.005	$F_{2,12} = 44.4$	<0.001	$F_{2,27} = 2.5$	0.123
Site × Microform	$F_{2,7} = 3.5$	0.090	$F_{2,15} = 4.1$	0.037	$F_{2,12} = 35.6$	<0.001	$F_{2,12} = 0.9$	0.437

* All models included a random effect of plot at sites to account for repeated measurements made at each site.

3.2.2. Mean Forest Floor Aboveground Autotrophic Respiration Rate ($R_{FF\_A\_ag}$) and Belowground Autotrophic Respiration of Shrubs + Herbs ($R_{A\_SH\_bg}$)

There were significant effects of WT and microform treatments individually and interactively on both the $R_{FF\_A\_ag}$ and $R_{A\_SH\_bg}$ (Table 2). Similar to R_{FF}, the $R_{FF\_A\_ag}$ and $R_{A\_SH\_bg}$ values were highest at the drained site ($5.2 \pm 0.6 \, g \cdot CO_2 \cdot m^{-2} \cdot day^{-1}$; $2.8 \pm 0.6 \, g \cdot CO_2 \cdot m^{-2} \cdot day^{-1}$, respectively), followed by values at the experimental ($3.1 \pm 0.9 \, g \cdot CO_2 \cdot m^{-2} \cdot day^{-1}$, $1.5 \pm 0.2 \, g \cdot CO_2 \cdot m^{-2} \cdot day^{-1}$, respectively) and control ($2.4 \pm 0.6 \, g \cdot CO_2 \cdot m^{-2} \cdot day^{-1}$; $1.4 \pm 0.2 \, g \cdot CO_2 \cdot m^{-2} \cdot day^{-1}$, respectively) sites. The $R_{FF\_A\_ag}$ and $R_{A\_SH\_bg}$ fluxes were also overall higher at the hummocks ($4.9 \pm 0.6 \, g \cdot CO_2 \cdot m^{-2} \cdot day^{-1}$; $2.5 \pm 0.1 \, g \cdot CO_2 \cdot m^{-2} \cdot day^{-1}$, respectively) than that at the hollows ($2.1 \pm 0.6 \, g \cdot CO_2 \cdot m^{-2} \cdot day^{-1}$;

$1.3 \pm 0.1 \, \text{g·CO}_2 \cdot \text{m}^{-2} \cdot \text{day}^{-1}$, respectively). Comparing microforms across sites, drained hummocks had significantly higher $R_{FF\_A\_ag}$ emissions than all other plots ($p = 0.005$; Figure 4) while $R_{A\_SH\_bg}$ at hummocks was significantly different between all three sites. There were no significant differences at hollows across the WT treatment sites for either $R_{FF\_A\_ag}$ or $R_{A\_SH\_bg}$.

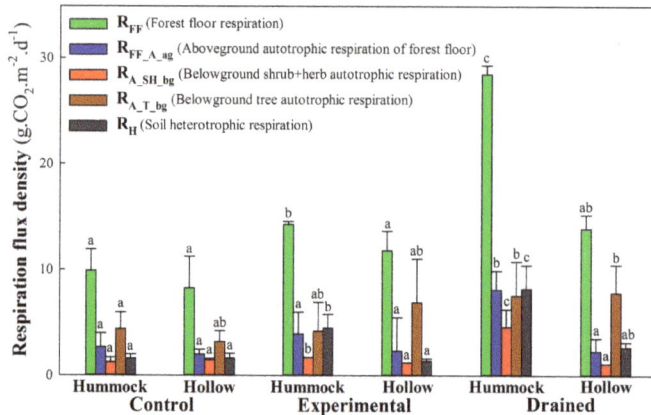

Figure 4. Mean ($\pm$SD) R_{FF}, $R_{FF\_A\_ag}$, $R_{A\_SH\_bg}$, and $R_{A\_T\_bg}$ measured during the growing season (May to October) of 2012 at all sites/microforms. Mean $R_{A\_SH\_bg}$ was determined by using a biomass regression method (Tables 3 and 4; Figure 5) [46,52]. Bars having no letters in common are significantly different ($p < 0.05$) while bars with same letters indicate no significant difference ($p > 0.05$); letters should be compared only within one flux component across all microforms.

We used B_{SH_ag} to calculate B_{SH_bg} ($B_{SH\_bg} = B_{SH\_ag} \times 0.39$; Table 3) and determine $R_{A\_SH\_bg}$ (Figures 4 and 5) using the regression equations we generated by regressing B_{SH_ag} with $R_{FF\_A\_ag}$ (explained in detail in Methods section). The B_{SH_ag} was not different among sites or microforms due to large variation between plots; however, the drained site had the highest, while experimental had the lowest B_{SH_ag} of all sites, and drained hummocks had higher B_{SH_ag} than those of control and experimental hummocks in that order (Table 4, Figure 4). The B_{SH_ag} and microform type significantly explained $R_{FF\_A\_ag}$ emissions individually and interactively (Table 4). Also, the $R_{FF\_A\_ag}$ was significantly related to an interaction between B_{SH_ag}, site and microform, where B_{SH_ag} was significantly related to the $R_{FF\_A\_ag}$ at all sites and microforms except at the experimental hollows which had the lowest or inconsistent B_{SH_ag} values (Table 4; Figure 5). The overall regression lines' slopes differed significantly at the hummocks and hollows ($z = 4.43$; 3.12, respectively). Regarding hollows, the slope was steeper at the drained site compared to control.

Table 3. Estimated aboveground biomass of shrubs + herbs (B_{SH_ag}) and trees (B_{T_bg}), and belowground biomass of shrubs + herbs and trees (B_{SH_bg}), at control, experimental, and drained sites *. All values ($\pm$SD) are in g·m^{-2}.

Vascular vegetation	Control			Experimental			Drained		
	Site	Hummock	Hollow	Site	Hummock	Hollow	Site	Hummock	Hollow
Shrubs + herbs									
B_{SH_ag}	90 ± 41	110 ± 54	65 ± 27	62 ± 34	70 ± 32	52 ± 36	152 ± 103	245 ± 180	52 ± 27
B_{SH_bg}	35 ± 16	43 ± 21	25 ± 10	24 ± 1	27 ± 0.6	20 ± 1	59 ± 35	95 ± 70	20 ± 1
Tree									
B_{T_ag}		2142 ± 376			1986			1964 ± 381	

* all values are mean $\pm$ SD ($n = 6$ for each of the B_{SH_ag} and B_{SH_bg}, and $n = 3$ (except experimental site with $n = 1$) for B_{T_ag}) [17]. Site biomass was determined by weighting the forest floor by the proportion of hummock and hollow microforms at each site (hummocks: control = 56%, experimental = 55%, drained = 52%), where applicable.

Table 4. Statistical results of a repeated, linear mixed effects model with fixed effects of site (control, experimental, drained), microform (hummock, hollow), and aboveground biomass of shrub + herb (B_{SH_ag}; covariate), random effect of plot, and an outcome variable of aboveground autotrophic respiration of shrubs + herbs at the forest floor ($R_{FF\_A\_ag}$) *.

Respiration	Effect	F	p
	B_{SH_ag}	$F_{1, 34} = 10.63$	0.003
	Site	$F_{2, 34} = 4.09$	0.026
$R_{FF\_A\_ag}$	Microform	$F_{1, 34} = 6.00$	0.020
	$B_{SH\_ag} \times$ Microform	$F_{1, 34} = 9.20$	0.005
	Site $\times$ Microform	$F_{2, 34} = 9.57$	0.001
	$B_{SH\_ag} \times$ Site $\times$ Microform	$F_{4, 34} = 3.93$	0.010

* Random effect of plot was included in the model to account for the repeated measurements made at each site.

3.3. Modeled Respiration Components

Overall, all modeled seasonal (May to October) respiration components ($g \cdot CO_2 \cdot m^{-2} \cdot$growing season^{-1}) were in the order of drained > experimental > control site, with greater overall increases at hollows than at hummocks at all the sites (Table 5). R_H accounted for approximately 48%, 43%, and 37% of R_{FF} at control, experimental, and drained sites, respectively. The R_{FF} and $R_{A\_T\_bg}$ fluxes at the drained site were significantly higher than those at the control site, while $R_{FF\_A\_ag}$ and $R_{A\_SH\_bg}$ were not different among sites, but were different between drained hummocks and drained hollows. The seasonal $R_{A\_T\_bg}$ and R_H values were highest at the drained hollows.

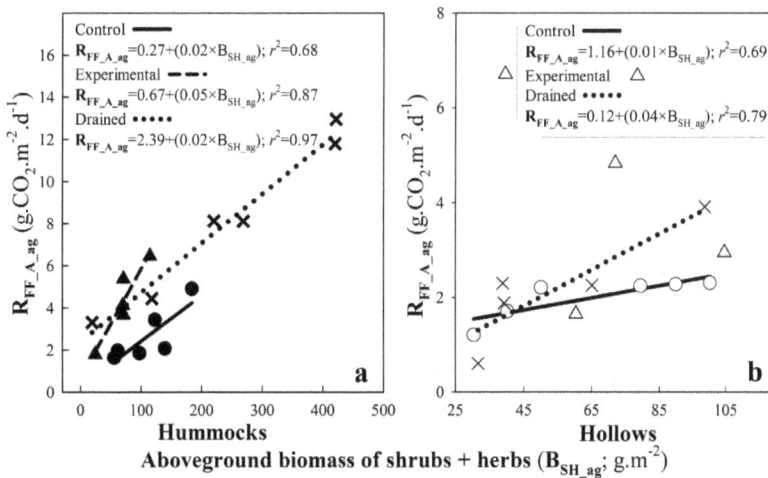

Figure 5. Aboveground autotrophic respiration of forest floor ($R_{FF\_A\_ag}$; $g \cdot CO_2 \cdot m^{-2} \cdot day^{-1}$) versus aboveground biomass of shrubs + herbs (B_{SH_ag}; $g \cdot m^{-2}$) at (**a**) hummock and at (**b**) hollow microforms at each site. Regression lines were plotted for each microform type at each site when statistically significant at $p < 0.05$.

4. Discussion

In agreement with the previous seasonal (2012) R_{FF} estimates made at these sites [17], this research found the greatest growing season R_{FF} values at the drained site ($422 \pm 22 \ g \cdot CO_2 \cdot m^{-2}$), smaller values at the experimental site ($354 \pm 16 \ g \cdot CO_2 \cdot m^{-2}$), and the smallest values at the control site ($255 \pm 10 \ g \cdot CO_2 \cdot m^{-2}$; Table 2); however, in general, the values at hollows across sites were slightly lower in the present study, as these R_{FF} values were modeled using measurements made over a shorter duration and from different plots (located in an area adjacent to the previous study plots [17]). The increased losses of CO_2 at the short- and long-term drained sites that we observed compare well with those reported by others from experimentally drained boreal peatlands [9,32,44,62,63]. Declining WT level promotes desiccation of aquatic vegetation and soil that progresses over time.

Table 5. Partitioning of the growing season (May to October, 2012) forest floor respiration ($\pm$SE; $g \cdot CO_2\text{-}C \cdot m^{-2}$) into major flux components *.

Site	Microform	R_{FF}	$R_{FF\_A\_ag}$	$R_{A\_SH\_bg}$	$R_{A\_T\_bg}$	R_H **
Control		255 ± 10.3 [a]	38 ± 6.7 [a]	37 ± 8.0 [a]	58 ± 8.9 [a]	122 ± 33.9 [a]
	Hummock	282 ± 4.6 [a,b]	42 ± 2.1 [a,b]	36 ± 6.2 [a,b]	80 ± 3.9 [a]	124 ± 16.8 [a]
	Hollow	221 ± 5.8 [a]	33 ± 4.6 [a,b]	39 ± 1.8 [a,b]	30 ± 5.1 [b]	119 ± 17.3 [a]
Experimental		354 ± 15.7 [a,b]	45 ± 11.8 [a]	40 ± 21.9 [a]	117 ± 11.1 [a,b]	152 ± 60.5 [a]
	Hummock	361 ± 2.2 [a,b]	57 ± 6.5 [a,b]	43 ± 6.8 [a,b]	101 ± 17.4 [a]	160 ± 22.9 [a]
	Hollow	346 ± 13.6 [a,b]	31 ± 5.3 [a,b]	36 ± 15.1 [a,b]	137 ± 13.8 [a]	142 ± 37.8 [a]
Drained		422 ± 21.9 [b]	66 ± 23.4 [a]	51 ± 22.7 [a]	150 ± 9.0 [b]	155 ± 77.0 [a]
	Hummock	429 ± 15.7 [b]	108 ± 11.2 [a]	69 ± 19.9 [a]	147 ± 5.2 [c]	105 ± 52.0 [a]
	Hollow	415 ± 6.2 [b]	21 ± 12.2 [b,c]	33 ± 2.8 [b,c]	154 ± 3.8 [c]	207 ± 25.0 [a]

* R_{FF}, $R_{FF\_A\_ag}$, and $R_{A\_T\_bg}$ denote forest floor respiration, above-ground autotrophic respiration of forest floor, and belowground autotrophic respiration of tree, respectively. $R_{A\_SH\_bg}$ represents belowground shrub + herb autotrophic respiration and was determined using biomass regression method (Tables 2 and 3). The seasonal value of $R_{A\_SH\_bg}$ is calculated by determining it as a proportion of instantaneous R_{FF} value and then estimating it as this proportion of the modeled seasonal R_{FF}. ** R_H was determined by difference using Equation (2). Values are significantly different if they have no letter in common; letters should be compared only within one flux component in a column between sites or microforms.

The R_{FF} was partitioned into major respiration components ($R_{FF\_A\_ag}$, $R_{A\_SH\_bg}$, $R_{A\_T\_bg}$) which were then modeled for seasonal estimates with WT level and T_5 as covariates. Model validations across sites/microforms showed excellent agreements within RSE of 0.24–2.41 $g \cdot CO_2 \cdot m^{-2}$ growing season^{-1} (Table 1, Figure A1). Many investigations on peatland or forest respiration components have shown that warm and dry conditions enhance autotrophic [9,30,64,65] and heterotrophic [25,26,39] respiration emissions with greater impact over longer time scales [17,32,66]. Warmer air and soil temperatures stimulate microbial activity, resulting in increased respiration fluxes; however, T_5 response of R_H depends on substrate type and availability of nutrients and moisture [67,68]. Water-table lowering in peatlands prompts increased respiration emissions that are enhanced with increase in peat surface temperature, as we noticed that R_{FF} values at our sites were well correlated to both WT level and T_5 (Tables 2 and 3; Figure 3). In contrast, a few studies suggest that the vascular vegetation (shrubs + herbs and trees) are less sensitive to WT lowering, as they can increase their rooting depth with deeper WT [15,30]. Peatland microforms have been observed to respond to changes in WT level and T_5 with different magnitudes and in different directions [16,60,69] mainly due to differences in vegetation coverage and composition. Counter to our hypothesis, we observed a significant increase in R_{FF} in response to WT lowering at hummocks, while R_{FF} at hollows was not significantly affected. Hummocks were drier and had significantly higher coverage of shrubs + herbs compared to hollows that were dominated by *Sphagnum* mosses at the research sites prior to WT manipulation. Therefore, differences in R_{FF} response to WT drawdown may be driven by changes in either autotrophic or heterotrophic respiration, or both.

At the control site, autotrophic respiration components ($R_{FF\_A\_ag}$, $R_{A\_SH\_bg}$, $R_{A\_T\_bg}$) were similar at hummocks and hollows. As for R_{FF}, drainage increased autotrophic respiration at hummocks, but only resulted in increased $R_{A\_T\_bg}$ at hollows. The greater $R_{FF\_A\_ag}$ and $R_{A\_SH\_bg}$ at hummocks (dominated by shrubs + herbs) was found to be related to B_{SH_ag} and B_{SH_bg} (Tables 4 and 5), respectively, indicating that drainage-induced increase in shrubs + herbs biomass had a strong control on above and belowground autotrophic respiration components at hummock microforms, which is similar to the findings of several studies [44,70,71]. Minimal change in B_{SH_ag} and B_{SH_bg} at hollows

(Table 3) resulted in the non-significant change in $R_{FF\_A\_ag}$ and $R_{A\_SH\_bg}$. As tree roots likely extend across all microforms, particularly as the depth of aerated peat is thickened by WT drawdown, the overall increase in tree productivity in response to drainage [17] resulted in increased $R_{A\_T\_bg}$ across both hummocks and hollows. Although $R_{FF\_A\_ag}$ did not increase significantly at hollows following drainage, the slope of the $R_{FF\_A\_ag}$ versus B_{SH_ag} became steeper, indicating increasing autotrophic respiration rates per unit increase in biomass. This may reflect the shift from moss dominated vegetation at the control site to more herbs and shrubs at the drained site, as vascular plants tend to have higher respiration rates than bryophytes [72,73].

Although R_H increased at some microforms (i.e. drained hollows) in response to WT drawdown, there was no significant difference across the WT treatments sites (Table 5). We hypothesized that R_{FF} increase would be greatest at hollows, as the WT drawdown would shift this microform position from largely anoxic to oxic conditions, resulting in large increases in R_H that would drive shifts in R_{FF}. We did observe substantial increases in R_H at hollows following WT lowering; however, these were masked by autotrophic respiration. Previous studies have also reported that $R_{FF\_A\_ag}$ can account for the majority of peatland respiration [45]. As WT drawdown enhanced plant productivity at the study site [17], R_H accounted for a declining proportion of growing season R_{FF} from 48% at control to 36% at the drained site. Hummock R_H did not change substantially in response to WT drawdown. As WT was initially deep below hummocks at the control site, due to the continental climate at the study site, the surface peat was well-aerated initially and R_H was likely rarely limited by saturated conditions. Drainage may actually result in desiccation of the surface peat that could result in conditions too dry for optimal rates of R_H [74]. The limited change in R_H between the WT treatments is therefore partially driven by the differential microform response where R_H is enhanced at hollows, with little change, or slight reduction at hummocks.

Overall, the substantial contribution of autotrophic respiration to R_{FF} suggests that the large increase in R_{FF} observed in response to WT drawdown will result in only slow loss of the C accumulated in peat, with these losses likely greatest at hollows [17]. However, comparison of R_H is partially complicated by the fact that it was calculated by difference once all the other partitioned flux components were estimated, and thus also contains error associated with the estimation of individual components. Isolated partitioning of the source-based respiration components remains to be developed, although the few manipulative field experiments that have investigated how climate change factors interact with one another to alter soil respiration [41,75] were not able to separate soil respiration into its components without significantly disrupting the soil [28]. Separating peat soil C into various major components is an important challenge for improving our understanding of peatland C cycling response to climate change [76].

5. Conclusions

Experimental water-table (WT) drawdown in a treed boreal bog increased forest floor respiration (R_{FF}). While all measured and estimated respiration components also increased following WT lowering, these were generally only significantly increased at hummock microforms. Increases in R_{FF} at hummocks were largely driven by increases in autotrophic respiration as shrub biomass increased. Drainage increased heterotrophic respiration at hollows, with less response at hummocks, suggesting that carbon is more likely to be released from stored peat at hollows. Overall, shifts in R_{FF} were largely driven by autotrophic respiration, indicating that rapid destabilization of peat carbon stocks under drying conditions are unlikely. Partitioning R_{FF} into its subcomponents accurately and without substantial disturbance to the soil is difficult and the development of partitioning methods are needed to better understand the fate of peat carbon stocks under various disturbances.

Acknowledgments: This research was funded by Alberta Innovated Technology Futures to MS, and University of Calgary to TM. Supplementary awards "Queen Elizabeth II, John D. Petrie Award and Karl C. Iverson Award" to TM helped complete the research project. Tak Fung provided valuable input on statistical analyses. We thank Mendel Perkins for help with site set up and collection of data, and two anonymous reviewers for their helpful comments on an earlier draft of this manuscript.

Author Contributions: M.S. and T.M. conceived and designed the experiments; T.M., B.K. and B.X. performed the experiments; T.M. and M.S. analyzed the data; T.M. and M.S. wrote the paper.

Conflicts of Interest: The authors declare no conflict of interest.

Appendix A

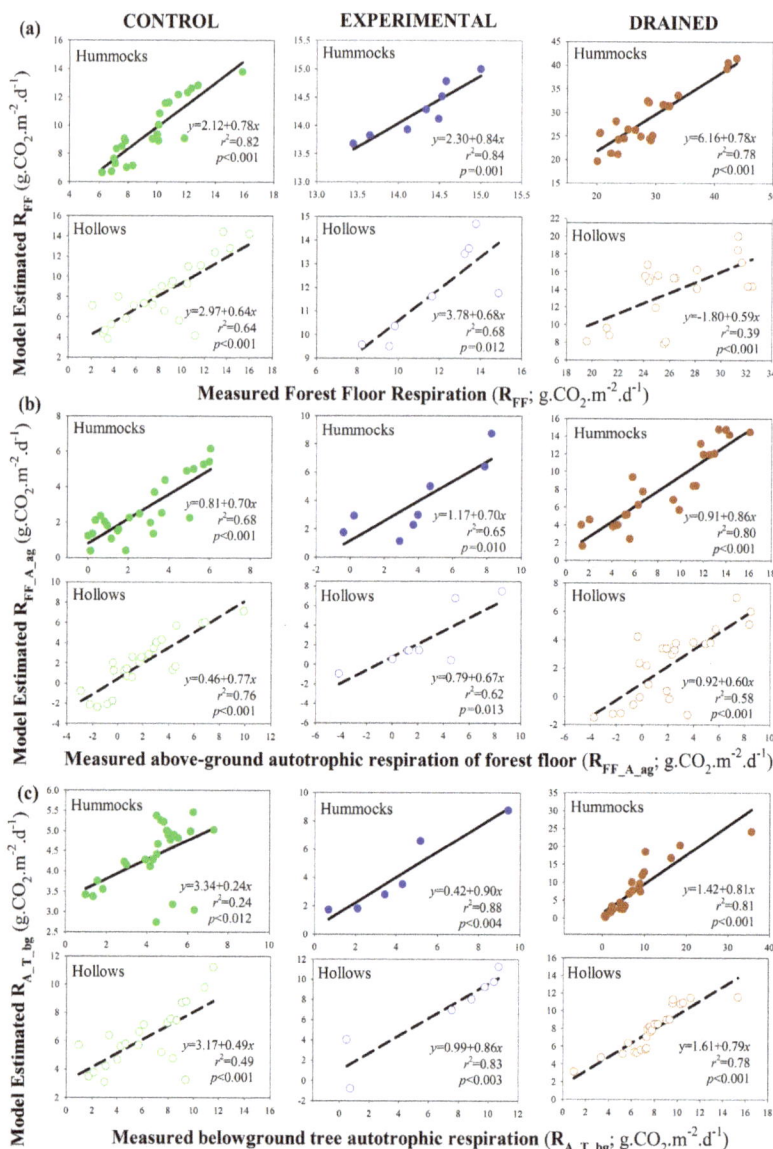

Figure A1. Goodness of fit between measured (year 2012) and model estimated values of (**a**) forest floor respiration (RFF), (**b**) aboveground autotrophic respiration of forest floor (RFF_A_ag), and (**c**) belowground autotrophic (rhizospheric) respiration of tree roots, across sites and microforms. Lines represent 1:1 fit.

References

1. Page, S.E.; Rieley, J.O.; Banks, C.J. Global and regional importance of the tropical peatland carbon pool. *Glob. Chang. Biol.* **2011**, *17*, 798–818. [CrossRef]
2. Loisel, J.; van Bellen, S.; Pelletier, L.; Talbot, J.; Hugelius, G.; Karran, D.; Yu, Z.; Nichols, J.; Holmquist, J. Insights and issues with estimating northern peatland carbon stocks and fluxes since the last glacial maximum. *Earth-Sci. Rev.* **2016**. [CrossRef]
3. Strack, M.; Cagampan, J.; Fard, G.H.; Keith, A.; Nugent, K.; Rankin, T.; Robinson, C.; Strachan, I.; Waddington, J.; Xu, B. Controls on plot-scale growing season CO_2 and CH_4 fluxes in restored peatlands: Do they differ from unrestored and natural sites? *Mires Peat* **2016**, *17*. [CrossRef]
4. Raich, J.W.; Tufekcioglu, A. Vegetation and soil respiration: Correlations and controls. *Biogeochemistry* **2000**, *48*, 71–90. [CrossRef]
5. Tarnocai, C. The effect of climate change on carbon in Canadian peatlands. *Glob. Planet. Chang.* **2006**, *53*, 222–232. [CrossRef]
6. Peng, S.; Piao, S.; Wang, T.; Sun, J.; Shen, Z. Temperature sensitivity of soil respiration in different ecosystems in China. *Soil Biol. Biochem.* **2009**, *41*, 1008–1014. [CrossRef]
7. Hiraishi, T.; Krug, T.; Tanabe, K.; Srivastava, N.; Baasansuren, J.; Fukuda, M.; Troxler, T. *IPCC (2014) 2013 Supplement to the 2006 IPCC Guidelines for National Greenhouse Gas Inventories: Wetlands*; Intergovernmental Panel on Climate Change: Geneva, Switzerland, 2014.
8. Ise, T.; Dunn, A.L.; Wofsy, S.C.; Moorcroft, P.R. High sensitivity of peat decomposition to climate change through water-table feedback. *Nat. Geosci.* **2008**, *1*, 763–766. [CrossRef]
9. Cai, T.; Flanagan, L.B.; Syed, K.H. Warmer and drier conditions stimulate respiration more than photosynthesis in a boreal peatland ecosystem: Analysis of automatic chambers and eddy covariance measurements. *Plant Cell Environ.* **2010**, *33*, 394–407. [CrossRef] [PubMed]
10. Roulet, N.; Moore, T.I.M.; Bubier, J.; Lafleur, P. Northern fens: Methane flux and climatic change. *Tellus B* **1992**, *44*, 100–105. [CrossRef]
11. Erwin, K.L. Wetlands and global climate change: The role of wetland restoration in a changing world. *Wetl. Ecol. Manag.* **2008**, *17*, 71. [CrossRef]
12. Clymo, R.S.; Turunen, J.; Tolonen, K. Carbon accumulation in peatland. *Oikos* **1998**, *81*, 368–388. [CrossRef]
13. Nykänen, H.; Alm, J.; Silvola, J.; Tolonen, K.; Martikainen, P.J. Methane fluxes on boreal peatlands of different fertility and the effect of long-term experimental lowering of the water table on flux rates. *Glob. Biogeochem. Cycles* **1998**, *12*, 53–69. [CrossRef]
14. Bubier, J.L.; Bhatia, G.; Moore, T.R.; Roulet, N.T.; Lafleur, P.M. Spatial and temporal variability in growing-season net ecosystem carbon dioxide exchange at a large peatland in Ontario, Canada. *Ecosystems* **2003**, *6*, 353–367.
15. Lafleur, P.M.; Moore, T.R.; Roulet, N.T.; Frolking, S. Ecosystem respiration in a cool temperate bog depends on peat temperature but not water table. *Ecosystems* **2005**, *8*, 619–629. [CrossRef]
16. Strack, M.; Waddington, J.M.; Rochefort, L.; Tuittila, E.S. Response of vegetation and net ecosystem carbon dioxide exchange at different peatland microforms following water table drawdown. *J. Geophys. Res. Biogeosci.* **2006**, *111*. [CrossRef]
17. Munir, T.M.; Perkins, M.; Kaing, E.; Strack, M. Carbon dioxide flux and net primary production of a boreal treed bog: Responses to warming and water-table-lowering simulations of climate change. *Biogeosciences* **2015**, *12*, 1091–1111. [CrossRef]
18. Lee, K.-H.; Jose, S. Soil respiration, fine root production, and microbial biomass in cottonwood and loblolly pine plantations along a nitrogen fertilization gradient. *For. Ecol. Manag.* **2003**, *185*, 263–273. [CrossRef]
19. Bond-Lamberty, B.; Bronson, D.; Bladyka, E.; Gower, S.T. A comparison of trenched plot techniques for partitioning soil respiration. *Soil Biol. Biochem.* **2011**, *43*, 2108–2114. [CrossRef]
20. Shaw, C.; Bona, K.; Thompson, D.; Dimitrov, D.; Bhatti, J.; Hilger, A.; Webster, K.; Kurz, W. *Canadian Model for Peatlands Version 1.0: A model Design Document*; Northern Forestry Centre: Edmonton, AB, Canada, 2016.
21. Ekblad, A.; Hogberg, P. Natural abundance of ^{13}C in CO_2 respired from forest soils reveals speed of link between tree photosynthesis and root respiration. *Oecologia* **2001**, *127*, 305–308. [CrossRef]
22. Lalonde, R.G.; Prescott, C.E. Partitioning heterotrophic and rhizospheric soil respiration in a mature douglas-fir (pseudotsuga menziesii) forest. *Can. J. For. Res.* **2007**, *37*, 1287–1297. [CrossRef]

23. Vitt, D.H.; Halsey, L.A.; Zoltai, S.C. The bog landforms of continental western Canada in relation to climate and permafrost patterns. *Arct. Alp. Res.* **1994**, *26*, 1–13. [CrossRef]
24. Roulet, N.T.; Moore, T.R. The effect of forestry drainage practices on the emission of methane from northern peatlands. *Can. J. For. Res.* **1995**, *25*, 491–499. [CrossRef]
25. Mäkiranta, P.; Minkkinen, K.; Hytönen, J.; Laine, J. Factors causing temporal and spatial variation in heterotrophic and rhizospheric components of soil respiration in afforested organic soil croplands in Finland. *Soil Biol. Biochem.* **2008**, *40*, 1592–1600. [CrossRef]
26. Mäkiranta, P.; Laiho, R.; Fritze, H.; Hytönen, J.; Laine, J.; Minkkinen, K. Indirect regulation of heterotrophic peat soil respiration by water level via microbial community structure and temperature sensitivity. *Soil Biol. Biochem.* **2009**, *41*, 695–703. [CrossRef]
27. Moore, T.R.; Bubier, J.L.; Frolking, S.E.; Lafleur, P.M.; Roulet, N.T. Plant biomass and production and CO_2 exchange in an ombrotrophic bog. *J. Ecol.* **2002**, *90*, 25–36. [CrossRef]
28. Hanson, P.J.; Edwards, N.T.; Garten, C.T.; Andrews, J.A. Separating root and soil microbial contributions to soil respiration: A review of methods and observations. *Biogeochemistry* **2000**, *48*, 115–146. [CrossRef]
29. Murphy, M.; Laiho, R.; Moore, T.R. Effects of water table drawdown on root production and aboveground biomass in a boreal bog. *Ecosystems* **2009**, *12*, 1268–1282. [CrossRef]
30. Murphy, M.T.; Moore, T.R. Linking root production to aboveground plant characteristics and water table in a temperate bog. *Plant Soil* **2010**, *336*, 219–231. [CrossRef]
31. Davidson, E.A.; Janssens, I.A. Temperature sensitivity of soil carbon decomposition and feedbacks to climate change. *Nature* **2006**, *440*, 165–173. [CrossRef] [PubMed]
32. Chimner, R.A.; Pypker, T.G.; Hribljan, J.A.; Moore, P.A.; Waddington, J.M. Multi-decadal changes in water table levels alter peatland carbon cycling. *Ecosystems* **2016**, 1–16. [CrossRef]
33. Rustad, L.; Campbell, J.; Marion, G.; Norby, R.; Mitchell, M.; Hartley, A.; Cornelissen, J.; Gurevitch, J.; GCTE-NEWS. A meta-analysis of the response of soil respiration, net nitrogen mineralization, and aboveground plant growth to experimental ecosystem warming. *Oecologia* **2001**, *126*, 543–562.
34. Kunkel, V.; Wells, T.; Hancock, G. Soil temperature dynamics at the catchment scale. *Geoderma* **2016**, *273*, 32–44. [CrossRef]
35. Alm, J.; Shurpali, N.J.; Minkkinen, K.; Aro, L.; Hytanen, J.; Lauriia, T.; Lohila, A.; Maijanen, M.; Martikainen, P.J.; Mäkiranta, P. Emission factors and their uncertainty for the exchange of CO_2, CH_4 and N_2O in Finnish managed peatlands. *Boreal Environ. Res.* **2007**, *12*, 191–209.
36. Ojanen, P.; Minkkinen, K.; Penttilä, T. The current greenhouse gas impact of forestry-drained boreal peatlands. *For. Ecol. Manag.* **2013**, *289*, 201–208. [CrossRef]
37. Lemprière, T.C.; Kurz, W.A.; Hogg, E.H.; Schmoll, C.; Rampley, G.J.; Yemshanov, D.; McKenney, D.W.; Gilsenan, R.; Beatch, A.; Blain, D.; et al. Canadian boreal forests and climate change mitigation. *Environ. Rev.* **2013**, *21*, 293–321. [CrossRef]
38. Jandl, R.; Bauhus, J.; Bolte, A.; Schindlbacher, A.; Schüler, S. Effect of climate-adapted forest management on carbon pools and greenhouse gas emissions. *Curr. For. Rep.* **2015**, *1*, 1–7. [CrossRef]
39. Minkkinen, K.; Lame, J.; Shurpali, N.J.; Mäkiranta, P.; Alm, J.; Penttilä, T. Heterotrophic soil respiration in forestry-drained peatlands. *Boreal Environ. Res.* **2007**, *12*, 115–126.
40. Strack, M.; Waddington, J.M.; Lucchese, M.C.; Cagampan, J.P. Moisture controls on CO_2 exchange in a *sphagnum*-dominated peatland: Results from an extreme drought field experiment. *Ecohydrology* **2009**, *2*, 454–461. [CrossRef]
41. Chen, X.; Post, W.M.; Norby, R.J.; Classen, A.T. Modeling soil respiration and variations in source components using a multi-factor global climate change experiment. *Clim. Chang.* **2011**, *107*, 459–480. [CrossRef]
42. Jandl, R.; Rodeghiero, M.; Martinez, C.; Cotrufo, M.F.; Bampa, F.; van Wesemael, B.; Harrison, R.B.; Guerrini, I.A.; deB Richter, D.; Rustad, L. Current status, uncertainty and future needs in soil organic carbon monitoring. *Sci. Total Environ.* **2014**, *468*, 376–383. [CrossRef] [PubMed]
43. Lafleur, P.M.; Roulet, N.T.; Bubier, J.L.; Frolking, S.; Moore, T.R. Interannual variability in the peatland-atmosphere carbon dioxide exchange at an ombrotrophic bog. *Glob. Biogeochem. Cycles* **2003**, *17*. [CrossRef]
44. Munir, T.M.; Xu, B.; Perkins, M.; Strack, M. Responses of carbon dioxide flux and plant biomass to water table drawdown in a treed peatland in northern Alberta: A climate change perspective. *Biogeosciences* **2014**, *11*, 807–820. [CrossRef]

45. Crow, S.E.; Wieder, R.K. Sources of CO$_2$ emission from a northern peatland: Root respiration, exudation, and decomposition. *Ecology* **2005**, *86*, 1825–1834. [CrossRef]
46. Mokany, K.; Raison, R.J.; Prokushkin, A.S. Critical analysis of root: Shoot ratios in terrestrial biomes. *Glob. Chang. Biol.* **2006**, *12*, 84–96. [CrossRef]
47. Nadelhoffer, K.J.; Raich, J.W. Fine root production estimates and belowground carbon allocation in forest ecosystems. *Ecology* **1992**, *73*, 1139–1147. [CrossRef]
48. Klepper, B. Root-shoot relationships. In *Plant Roots: The Hidden Half*; Dekker, M., Ed.; CRC Press: New York, NY, USA, 1991; pp. 265–286.
49. Korrensalo, A.; Hajek, T.; Vesala, T.; Mehtatalo, L.; Tuittila, E.S. Variation in photosynthetic properties among bog plants. *Botany* **2016**, *94*, 1127–1139. [CrossRef]
50. Schuur, E.A.G.; Trumbore, S.E. Partitioning sources of soil respiration in boreal black spruce forest using radiocarbon. *Glob. Chang. Biol.* **2006**, *12*, 165–176. [CrossRef]
51. Biasi, C.; Pitkamaki, A.S.; Tavi, N.M.; Koponen, H.T.; Martikainen, P.J. An isotope approach based on ^{13}C pulse-chase labelling vs. The root trenching method to separate heterotrophic and autotrophic respiration in cultivated peatlands. *Boreal Environ. Res.* **2012**, *17*, 184–193.
52. Wang, X.G.; Zhu, B.; Wang, Y.Q.; Zheng, X.H. Field measures of the contribution of root respiration to soil respiration in an alder and cypress mixed plantation by two methods: Trenching method and root biomass regression method. *Eur. J. For. Res.* **2008**, *127*, 285–291. [CrossRef]
53. Environment Canada. *Canadian Climate Normals and Averages: 1971–2000. National Climate Data and Information Archives*; Environment Canada: Fredericton, NB, Canada, 1 February 2010.
54. DMTI Spatial. *DMTI Canmap Postal Geograph (Retrieved from University of Calgary). DMTI Spatial [Producer(s)]: DMTI Spatial Mapping Academic Research Tools (SMART) Program*; DMTI Spatial: Markham, ON, Canada, 2014.
55. John, R. *Flora of the Hudson Bay Lowland and Its Postglacial Origins*; NRC Research Press: Ottawa, ON, Canada, 2003; p. 237.
56. Heinemeyer, A.; di Bene, C.; Lloyd, A.R.; Tortorella, D.; Baxter, R.; Huntley, B.; Gelsomino, A.; Ineson, P. Soil respiration: Implications of the plant-soil continuum and respiration chamber collar-insertion depth on measurement and modelling of soil CO$_2$. *Eur. J. Soil Sci.* **2011**, *62*, 82–94. [CrossRef]
57. Lai, D.; Roulet, N.; Humphreys, E.; Moore, T.; Dalva, M. The effect of atmospheric turbulence and chamber deployment period on autochamber CO$_2$ and CH$_4$ flux measurements in an ombrotrophic peatland. *Biogeosciences* **2012**, *9*, 3305. [CrossRef]
58. Koskinen, M.; Minkkinen, K.; Ojanen, P.; Kämäräinen, M.; Laurila, T.; Lohila, A. Measurements of CO$_2$ exchange with an automated chamber system throughout the year: Challenges in measuring night-time respiration on porous peat soil. *Biogeosciences* **2014**, *11*, 347. [CrossRef]
59. Loisel, J.; Gallego-Sala, A.V.; Yu, Z. Global-scale pattern of peatland *sphagnum* growth driven by photosynthetically active radiation and growing season length. *Biogeosciences* **2012**, *9*, 2737–2746. [CrossRef]
60. Munir, T.M.; Strack, M. Methane flux influenced by experimental water table drawdown and soil warming in a dry boreal continental bog. *Ecosystems* **2014**, *17*, 1271–1285. [CrossRef]
61. Clogg, C.C.; Petkova, E.; Haritou, A. Statistical methods for comparing regression coefficients between models. *Am. J. Sociol.* **1995**, *100*, 1261–1293. [CrossRef]
62. von Arnold, K.; Hånell, B.; Stendahl, J.; Klemedtsson, L. Greenhouse gas fluxes from drained organic forestland in Sweden. *Scand. J. For. Res.* **2005**, *20*, 400–411. [CrossRef]
63. Simola, H.; Pitkänen, A.; Turunen, J. Carbon loss in drained forestry peatlands in Finland, estimated by re-sampling peatlands surveyed in the 1980s. *Eur. J. Soil Sci.* **2012**, *63*, 798–807. [CrossRef]
64. Aurela, M.; Riutta, T.; Laurila, T.; Tuovinen, J.-P.; Vesala, T.; Tuittila, E.-S.; Rinne, J.; Haapanala, S.; Laine, J. CO$_2$ exchange of a sedge fen in southern Finland—The impact of a drought period. *Tellus B* **2007**, *59*, 826–837. [CrossRef]
65. Straková, P.; Anttila, J.; Spetz, P.; Kitunen, V.; Tapanila, T.; Laiho, R. Litter quality and its response to water level drawdown in boreal peatlands at plant species and community level. *Plant Soil* **2010**, *335*, 501–520. [CrossRef]
66. Strack, M.; Waddington, J.M.; Tuittila, E.S. Effect of water table drawdown on northern peatland methane dynamics: Implications for climate change. *Glob. Biogeochem. Cycles* **2004**, *18*. [CrossRef]
67. Blodau, C.; Basiliko, N.; Moore, T.R. Carbon turnover in peatland mesocosms exposed to different water table levels. *Biogeochemistry* **2004**, *67*, 331–351. [CrossRef]

68. Updegraff, K.; Bridgham, S.D.; Pastor, J.; Weishampel, P.; Harth, C. Response of CO_2 and CH_4 emissions from peatlands to warming and water table manipulation. *Ecol. Appl.* **2001**, *11*, 311–326.
69. Waddington, J.M.; Roulet, N.T. Carbon balance of a boreal patterned peatland. *Glob. Chang. Biol.* **2000**, *6*, 87–97. [CrossRef]
70. Laine, A.; Sottocornola, M.; Kiely, G.; Byrne, K.A.; Wilson, D.; Tuittila, E.-S. Estimating net ecosystem exchange in a patterned ecosystem: Example from blanket bog. *Agric. For. Meteorol.* **2006**, *138*, 231–243. [CrossRef]
71. Laiho, R. Decomposition in peatlands: Reconciling seemingly contrasting results on the impacts of lowered water levels. *Soil Biol. Biochem.* **2006**, *38*, 2011–2024. [CrossRef]
72. Riutta, T.; Laine, J.; Tuittila, E.-S. Sensitivity of CO_2 exchange of fen ecosystem components to water level variation. *Ecosystems* **2007**, *10*, 718–733. [CrossRef]
73. Ward, S.E.; Ostle, N.J.; Oakley, S.; Quirk, H.; Henrys, P.A.; Bardgett, R.D. Warming effects on greenhouse gas fluxes in peatlands are modulated by vegetation composition. *Ecol. Lett.* **2013**, *16*, 1285–1293. [CrossRef] [PubMed]
74. Tuittila, E.-S.; Vasander, H.; Laine, J. Sensitivity of c sequestration in reintroduced sphagnum to water-level variation in a cutaway peatland. *Restor. Ecol.* **2004**, *12*, 483–493. [CrossRef]
75. Wan, S.; Norby, R.J.; Ledford, J.; Weltzin, J.F. Responses of soil respiration to elevated CO_2, air warming, and changing soil water availability in a model old-field grassland. *Glob. Chang. Biol.* **2007**, *13*, 2411–2424. [CrossRef]
76. Fahey, T.J.; Tierney, G.L.; Fitzhugh, R.D.; Wilson, G.F.; Siccama, T.G. Soil respiration and soil carbon balance in a northern hardwood forest ecosystem. *Can. J. For. Res.* **2005**, *35*, 244–253. [CrossRef]

![forests logo] *forests*

MDPI

Article

Climate Impacts on Soil Carbon Processes along an Elevation Gradient in the Tropical Luquillo Experimental Forest

Dingfang Chen [1], Mei Yu [2,*], Grizelle González [3], Xiaoming Zou [2] and Qiong Gao [2]

[1] Department of Biology, University of Puerto Rico-Río Piedras, San Juan, PR 00936, USA;
    dingfang.chen@gmail.com (D.C.)
[2] Department of Environmental Sciences, University of Puerto Rico-Río Piedras, San Juan, PR 00936, USA;
    xzou2011@gmail.com (X.Z.); shiqun.gao@gmail.com (Q.G.)
[3] USDA Forest Service, International Institute of Tropical Forestry, 1201 Ceiba Street, Río Piedras, PR 00926,
    USA; ggonzalez@fs.fed.us
*   Correspondence: meiyu@ites.upr.edu; Tel.: +1-787-764-0000 (ext. 4375)

Academic Editors: Robert Jandl and Mirco Rodeghiero
Received: 14 January 2017; Accepted: 13 March 2017; Published: 19 March 2017

Abstract: Tropical forests play an important role in regulating the global climate and the carbon cycle. With the changing temperature and moisture along the elevation gradient, the Luquillo Experimental Forest in Northeastern Puerto Rico provides a natural approach to understand tropical forest ecosystems under climate change. In this study, we conducted a soil translocation experiment along an elevation gradient with decreasing temperature but increasing moisture to study the impacts of climate change on soil organic carbon (SOC) and soil respiration. As the results showed, both soil carbon and the respiration rate were impacted by microclimate changes. The soils translocated from low elevation to high elevation showed an increased respiration rate with decreased SOC content at the end of the experiment, which indicated that the increased soil moisture and altered soil microbes might affect respiration rates. The soils translocated from high elevation to low elevation also showed an increased respiration rate with reduced SOC at the end of the experiment, indicating that increased temperature at low elevation enhanced decomposition rates. Temperature and initial soil source quality impacted soil respiration significantly. With the predicted warming climate in the Caribbean, these tropical soils at high elevations are at risk of releasing sequestered carbon into the atmosphere.

Keywords: soil respiration; tropical forest; soil translocation experiment; elevation gradient; climate change

---

1. Introduction

Soil respiration, defined as CO_2 emission from the soil surface through the activities of soil microbes, plant roots, and other organisms, is one of the major pathways to release carbon fixed by vegetation into the atmosphere [1,2]. Because the quantity of carbon stored in soils is double that stored in either the atmosphere or the terrestrial vegetation, soil respiration is a critical component in the global carbon cycle [1,3]. Tropical biosphere stores 46% of the world's living terrestrial carbon and 11% of the world's soil carbon [4] and is sensitive to changes in climate [5–7], therefore, it plays an important role in global C dynamics [8]. At steady state, the carbon emission from soil, as the second largest carbon flux between the atmosphere and the terrestrial biomes, can be balanced by CO_2 net uptake by plants (net primary production, NPP) [1,9]. However, any small changes in soil respiration and carbon caused by climate change could have huge impacts on the global carbon cycle and future global climate [2,9]. With the ongoing and predicted warming/drought in the tropics [10,11], tropical ecosystems might change from a C sink to a C source.

Soil respiration involves the interactions among plant roots, rhizosphere, soil microbes, soil fauna, and physicochemical conditions [12]. Previous experiments have indicated that changes in environmental conditions would affect the soil respiration rate and might have a significant impact on the global carbon cycle [9,13–15]. Soil temperature, moisture, and the carbon substrate available for microorganisms (which is related to vegetation type) are major factors influencing the soil respiration rate at a given site [13,16–18].

Temperature is thought to be the primary driver of soil respiration [3,19–21]. Soil respiration generally increases exponentially with temperature [14,15,22] to maintain the increased metabolism of plant roots, soil microbes, and soil fauna [23–25]. When temperature reaches a certain maximum point, most of the enzymatic activity involved in respiration will be inhibited due to enzyme malfunction, and soil respiration stops increasing [26,27]. The optimal temperature for cryophiles is below 20 °C whereas that for mesophiles and thermophiles is 20 °C–40 °C and above 40 °C, respectively [28].

Soil moisture is generally considered to positively correlate with soil respiration [14,26]. Soil respiration, especially root respiration, is relatively low in dry conditions, and increases to a maximum in intermediate moisture conditions [14,17]. Although some microorganisms develop strategies to survive and grow under low soil moisture conditions [27,29], water deficiency prevails to limit activities of soil fauna and microbes. On the other hand, when soil is saturated, oxygen can be limited. Anaerobic conditions suppress aerobic microbial activities, resulting in limited soil respiration [15,26,27]. Soil moisture also affects the availability of soil nutrients [27,30]. Some of the nutrients available to plants and soil microorganisms need to be dissolved in water (e.g., N) [22,31,32]. The influence of soil moisture on soil respiration varies greatly among ecosystems. Changes in soil moisture can have a significant influence in semiarid ecosystems [16], but may not significantly alter soil respiration for humid ecosystems, except during warm and dry seasons [14]. Furthermore, the effects of moisture combined with temperature were suggested experimentally to be more reliable predictors [9,16,33].

The availability of soil carbon substrates is also an important factor controlling decomposition and heterotrophic respiration [34–36]. Heterotrophic respiration (i.e., CO_2 emissions produced through soil organic matter (SOM) decomposition), is primarily driven by the activities of soil microorganisms and soil fauna, and their richness and abundance primarily control the decomposition rate of SOM [23–25]. The spatial distribution of soil microbes is, to a great extent, affected by the availability of carbon substrates [2,32,35]. Moreover, soil microbes themselves have particular C:N balance needs [35,36]. Therefore, soil carbon substrates can affect soil respiration by controlling the distribution and activities of soil fauna and microbes.

Soil respiration varies with vegetation types [15,22,37]. Global mean soil respiration rates vary widely among major vegetation biomes, and the lowest rates occur in tundra and northern bogs, while the highest rates occur in tropical moist forests [22]. Although the distribution pattern of soil respiration may be partially affected by temperature and moisture, substrate quality and species composition of fauna and microbes may differ substantially among vegetation types, and could partly explain the difference in soil respiration [22,37,38].

Almost all the environmental conditions influencing soil respiration (e.g., forest type, solar radiation, temperature and moisture, and soil fauna such as earthworms) have distinct elevation patterns in the Luquillo Experimental Forest (LEF). In general, temperature, plant species richness and abundance, and NPP decrease with elevation, whereas SOM, soil organic carbon (SOC), precipitation, and soil moisture increase [39–42]. The number of earthworm species also significantly increased along the elevation from low to top [43]. The elevation gradient provides a natural *in situ* simulation of climate change [14,17,44]. Therefore, studying the variation in soil carbon along an elevation gradient in LEF is an ideal approach to investigate the impacts of climate change on tropical carbon processes. This paper describes a soil translocation experiment conducted along the elevation gradient in the LEF. We hypothesized that (1) soil respiration will increase but SOC will decrease with enhanced soil temperature and moisture; and (2) soils originated from high elevation with high SOC content will respire more than those from low elevation under similar climatic conditions. Warming in the tropics might boost soil respiration, especially in soils at high elevations with large SOC content, therefore this study may contribute to better understanding of C dynamics in tropical regions in the context of global change.

2. Materials and Methods

2.1. Study Area

All three experiment sites are located along an elevation gradient in LEF (18°20′ N, 65°49′ W), northeastern Puerto Rico (Figure 1), a tropical wet montane forest. Annual rainfall ranges from an average of 3537 mm at low elevation to 4849 mm at high elevation, and monthly temperatures in January and September change from 23.5 °C and 27 °C at low elevation to 17 °C and 20 °C at high elevation, respectively [45]. Soils are mainly derived from volcaniclastic sediments and are classified as "clay" based on their particle size distribution, except for one high-elevation area where the soils are derived from quartz diorite with lower clay content and are classified as "clay loam" [42]. The distribution of vegetation in LEF exhibits a distinct elevation pattern. Tabonuco forest, Palm forest, Palo Colorado, and Elfin woodland distribute along the elevation gradient from the foothill to the top [41,46]. The three sites are facing north to northwest with the slopes of 5.0, 26.0, and 9.7° at 350, 600, and 1000 m, respectively. We chose a relatively flat place at each site to implement the soil translocations. The soils are acidic with the pH of 4.2, 4.6, and 4.4 at the three sites. Soil clay contents are 53%, 56%, and 36%, and soil bulk densities are 0.8, 0.6, and 0.9 g·cm^{-3} at 350, 600, and 1000 m, respectively. The soil Ca contents are 0.6, 0.2, and 0.2 mg·g^{-1}, and soil P contents are 0.3, 0.2, and 0.4 mg·g^{-1} at the three sites, respectively. The lowest site (18°19′24″ N, 65°49′3″ W) is within the Tabonuco forest, a subtropical wet forest dominated by *Dacryodes excelsa* and *Prestoea montana*. The middle site (18°18′56″ N, 65°48′46″ W) is within the Palo Colorado and adjacent to the Tabonuco forest. Palo Colorado is a lower montane wet forest dominated by *Cyrilla racemiflora*, *Prestoea montana*, *Henriettea squamulosa*, and *Magnolia splendens*. The highest site (18°18′29″ N, 65°47′43″ W) is within the Elfin woodland characterized by stunted dwarf trees and dominated by *Cyathea bryophila*, *Eugenia borinquensis*, *Ocotea spathulata*, and *Tabebuia rigida* [47].

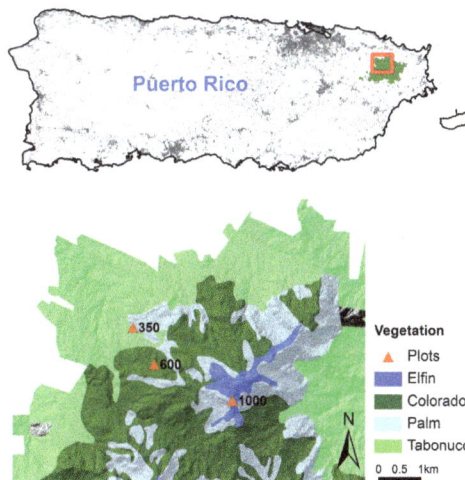

Figure 1. The soil translocation experiment sites were located along the elevation gradient (350, 600, and 1000 m) within different dominant vegetation types in the Luquillo Experimental Forest, northeastern Puerto Rico.

2.2. Soil Translocation Experiment

Nine soil cores were excavated from each site, six of them were translocated to the other two sites (three for each), and three soil cores remained at the source site but at different places. Before collecting the samples, all the aboveground litter was removed. The organic-rich soils (0–5 cm) were collected separately, and later put back on the top of soil cores after reinstallation in the translocated sites. The soil cores were taken using polyvinyl (PVC) tubes 10 cm in diameter and 15 cm in length. When excavating the soil cores, the tubes were inserted vertically into the subsoil to take an intact soil monolith with a depth of 15 cm. When soil cores were reinstalled, the soils were kept intact in the tubes and separated from the surrounding soils. The bottom of the coring tube was covered with iron mesh (63-µm) to further prevent large roots from growing and to balance the effects of temperature and moisture in the soil within and outside the tubes.

In each of the three soil translocation sites, one soil temperature and one moisture probe were installed at both surface and 15 cm in-depth in an undisturbed place and connected to a data logger (HOBO® Micro station, ONSET Computer Corporation, Bourne, MA, USA). Soil temperature and moisture were recorded every 30 min from July 2011 to March 2012. Due to a battery failure, data were missing for the site at 600 m in November 2011 and February–March 2012, and the site at 1000 m in November 2011 and March 2012. Temperature measurements at the 1000-m site in December 2011–March 2012 were recalibrated using a mobile temperature probe because of a disturbance to the installed probe in December 2011. The automatically measured soil temperature and moisture were only used to display the naturally occurred gradients in temperature and moisture along the elevation gradient.

Soil respiration rates (*Rs*) were measured monthly with a portable InfraRed Gas Analyzer (EMG4, PP Systems, Amesbury, MA, USA), with a cylindrical cuvette (CPY-2) inserted in soil cores. The soil respiration fluxes were recorded for 2–3 min after the CO_2 concentration in the closed chamber increased steadily. The real-time temperature and moisture for both soil and air were measured with external sensors or a thermometer. All the measurements were repeated three times.

At the end of the soil translocation experiment in May 2012, the SOC and SOM contents of all the soil cores translocated were measured at the International Institute of Tropical Forestry (IITF) of USDA Forest Service Chemistry Laboratory. The soil samples were oven dried to constant mass at 50 °C, and then ground and passed through a 20 mesh sieve (size of 0.84 mm) to screen coarse root detritus and other organic materials. SOC was determined using the LECO TruSpec CN Analyzer (LECO®, LECO Corporation, Saint Joseph, MI, USA). There are no carbonates found in these soils. Total carbon (TC) is equal to total organic carbon. Soil organic matter (SOM) was estimated using loss on ignition (LOI). LOI was determined using the LECO TGA-701 Analyzer at 490 °C for two hours (or until constant weight is reached at less than 2.5% variability) in an oxygen saturated environment. Soil total nitrogen content was determined using the LECO CN Analyzer.

2.3. Statistical Analyses

To test our hypotheses of impacts of soil source quality, translocation, and associated climate shift on soil respiration, we applied paired-*t*, ANOVA, and linear mixed-effects models on the measured soil respiration rates. Soil respiration rates were examined for normality (Shapiro–Wilk normality test), and log transformation was performed if needed. Multiple paired-*t* was used to test the impacts of soil source quality, such as SOC and SOM contents, on *Rs*. For example, in order to compare the *Rs* of the soils originating from 350 m with those from 1000 m, we took the paired *Rs* of $R_{350,350} \sim R_{1000,350}$, $R_{350,600} \sim R_{1000,600}$ and $R_{350,1000} \sim R_{1000,1000}$, for which the first subscript indicates the soil source and the second subscript is for the translocated site. By pairing the *Rs* of soils originating from two different sites but translocated at the same sites, we virtually excluded the impacts of microclimate. ANOVA was applied to test the impacts of translocation site on soil respiration. We also used *t*-tests to compare the respirations of the soils with the same source but translocated to different elevations. The linear mixed-effects model was finally used as a synthetic analysis to test the impacts of translocated sites and microclimate on *Rs*, with soil source set as a random effect. Statistical analyses were run in R software [48].

3. Results

3.1. Naturally Occurred Gradient in Soil Temperature and Moisture and Changes in Soil C and N Contents before and after the Translocation Experiment along the Elevation Gradient

Soil temperature and moisture at the three translocation sites, on average, followed the naturally occurring gradients in climate along the elevation gradient, i.e., decreased temperature but increased moisture from the foothill to the top (Figure 2). The averaged soil temperature within the depth of 0–15 cm during the period July 2011–March 2012 decreased from 22.2 °C at 350 m, to 21.8 °C at 600 m, and to 19.9 °C at 1000 m. On the other hand, the averaged soil moisture increased from 0.27 $m^3 \cdot m^{-3}$ at 350 m, to 0.31 $m^3 \cdot m^{-3}$ at 600 m, and to 0.4 $m^3 \cdot m^{-3}$ at 1000 m. The difference in soil moisture between the surface and the depth of 15 cm was larger at the site of 600 m than at the other two sites.

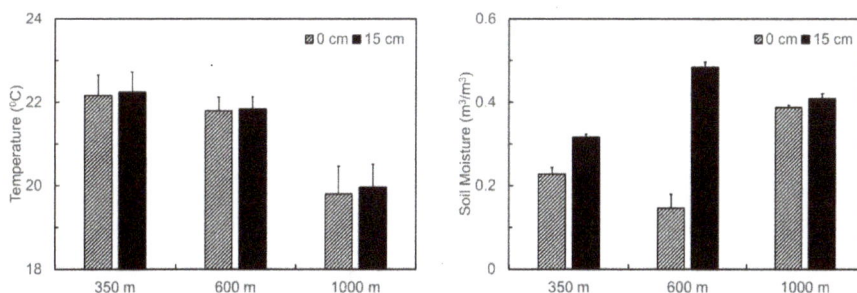

Figure 2. Averaged soil temperature and moisture at 0 and 15 cm in depth throughout the experiment period at the three soil translocation sites (with elevations at 350 m, 600 m, and 1000 m, respectively) in the Luquillo Experimental Forest, Puerto Rico. Bars stand for standard errors.

Initial patterns of SOC, SOM, and C:N all showed increased trends along the elevation gradient from low to top (Figure 3a–c). SOC increased from 5.4% at 350 m to 6.8% at 600 m, and to 20.5% at 1000 m, and SOM increased from 22.5% to 26.4% and to 48.2%, respectively. C:N increased from 13.6, to 17.2, and to 24.8 at the sites of 350, 600, and 1000 m, respectively. Soil total nitrogen was higher at 1000 m than at 350 or 600 m (Figure 3d). When measured 11 months after translocation, SOC increased to 6.2%, 7.5%, and 21.3% for the soils originating from 350, 600, and 1000 m, respectively. SOC originating from 350 m changed to 6.4% and 5.9% when translocated to 350 and 1000 m, respectively. For soils originating from 1000 m, SOC changed to 20.6% and 23.2% when translocated to 350 and 1000 m, respectively. Soil total nitrogen also increased after the soil translocation experiment. For the soils with the same source but translocated to different elevations, changes in soil total nitrogen follow the pattern of changes in SOC.

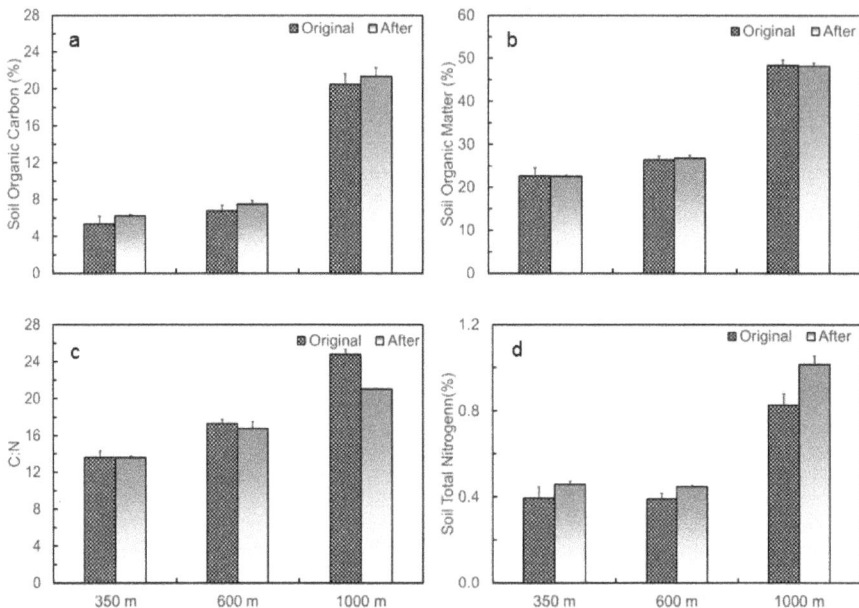

Figure 3. Changes in (**a**) soil organic carbon, (**b**) soil organic matter, (**c**) C to N ratio, and (**d**) soil total nitrogen of the soils at 0–15 cm at the three soil translocation sites (with elevations of 350, 600, and 1000 m), measured before and after the soil translocation experiment in the Luquillo Experimental Forest, Puerto Rico.

3.2. Impacts of Soil Source Quality on Soil Respiration

We took a natural logarithm transformation of the soil respiration rate before performing the multiple paired-*t* tests because the original *Rs* did not pass the normality test, i.e., Shapiro–Wilk test in R. Paired-*t* tests on the log-transformed *Rs* were only applied to the comparisons between the soil source at 1000 m and that at 350 m or 600 m, according to the result of the normality test. Soils originating from 1000 m have a significantly larger respiration rate than those from 600 m (estimated difference = 0.31, *t* = 2.3, *df* = 21, *p* = 0.03) and from 350 m (estimated difference = 0.23, *t* = 1.7, *df* = 24, *p* = 0.1).

3.3. Impacts of Soil Translocated Site and Associated Microclimate on Soil Respiration

An ANOVA test on the log-transformed *Rs* among the three translocated sites indicated that the treatment of the translocation site had significant effects on soil respiration ($F_{(2,71)}$ = 6.07, *p* = 0.004). Soils translocated to the site at 350 m had a higher respiration rate than those translocated to 600 or 1000 m.

As for the soils with the same source but translocated to different elevations (Figure 4), soils originating from 350 m have significantly higher respiration rates when translocated to 1000 m, as indicated by the comparison of the sample mean of 1.447 versus 1.024 $\mu mol \cdot m^{-2} \cdot s^{-1}$ (*t* = 2.17, *df* = 16, *p* = 0.05). Soils originating from 1000 m also have significantly higher respiration rates when translocated to 350 m, 2.890 versus 1.298 $\mu mol \cdot m^{-2} \cdot s^{-1}$ in mean (*t* = 4.2, *df* = 10, *p* = 0.002).

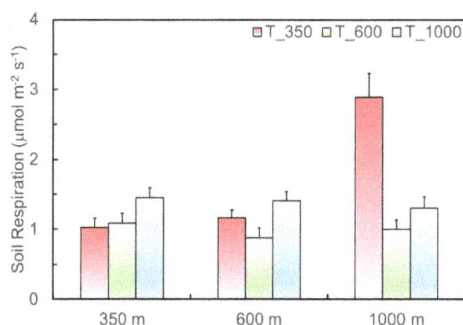

Figure 4. Soil respiration of the soils originating from 350, 600, and 1000 m (x axis), but translocated to 350 (T_350), 600 (T_600), and 1000 m (T_1000) in the Luquillo Experimental Forest, Puerto Rico, from August 2011 to May 2012.

The linear mixed effects model on the log-transformed Rs with translocation sites as a fixed effect and soil source as a random effect confirmed that the translocation site has a significant effect on soil respiration (upper part in Table 1). When temperature, moisture, and their interaction were added to the fixed model, the effects of both translocated site and temperature were significant. However, the effects of moisture and its interaction with temperature were not significant (lower part in Table 1). The AIC and BIC (Akaike or Bayesian Information Criterion) of the later model are smaller when temperature and moisture were added as additional explanatory variables (Table 1).

Table 1. The results of linear mixed effects models on the log-transformed soil respiration rate ($\mu\text{mol}\cdot\text{m}^{-2}\cdot\text{s}^{-1}$) with translocation sites (upper part) or translocation sites, soil temperature, moisture, and interaction of temperature and moisture (lower part) as fixed effects, and soil source as random effects. Temp. and M. stand for temperature and moisture, respectively. AIC and BIC stand for Akaike and Bayesian Information Criterion, respectively.

Fixed Effects Variable	Value	Standard Error	df	t-Value	p-Value
log(Rs)~Translocation I source			AIC = 105.6 BIC = 116.9 #observation = 74 #group = 3		
T_350	0.34	0.12	69	2.87	0.005
T_600	−0.08	0.12	69	−0.68	0.50
T_1000	0.26	0.12	69	2.15	0.04
log(Rs)~Translocation + Temp. + M. + Temp. × M. I source			AIC = 91.0 BIC = 107.4 #observation = 63 #group = 3		
T_350	−3.33	1.91	55	−1.75	0.09
T_600	−3.57	1.84	55	−1.95	0.06
T_1000	−3.09	1.92	55	−1.61	0.11
Temperature	0.17	0.09	55	1.98	0.05
Moisture	6.95	6.40	55	1.09	0.28
Temperature × Moisture	−0.35	0.30	55	−1.18	0.24

4. Discussion

Tropical forests play important roles in the interaction between terrestrial ecosystems and the global climate system, such as C dynamics. Warming and drought have the potential to change tropical forests from a C sink to a C source. Soil translocation experiments along altitudinal gradients provide an approach to assess the impacts of climate change on soil C dynamics. Our soil translocation experiment in tropical forests in northeastern Puerto Rico showed that both soil C and the soil respiration rates were altered by changes in temperature, moisture, and initial soil organic C.

Soil organic carbon, organic matter, total nitrogen, and C:N all increased with elevation (Figure 3), which conformed with other altitudinal studies involving naturally occurring gradients of decreasing temperature but increasing moisture [49]. Compared to the content before translocation, SOC increased slightly at the end of the experiment, which might be caused by the degradation of dead roots and litter in the soil cores. With the same source location, the changes in SOC after the translocation experiment followed the patterns of *Rs*. The greater the *Rs* at the translocated site, the greater the decrease in soil carbon.

Increasing SOC, SOM, and C:N along the altitudinal gradient (Figure 3) also implied improved substrate quantity/quality for soil respiration. By virtually excluding the effects of microclimate using the paired-*t* test, our study highlights that soil source has significant impacts on soil respiration. The soils originating from high elevation with high SOC, SOM, and C:N had a much larger respiration rate than those originating from low elevation with low SOC, SOM, and C:N. The results support our hypothesis on the important role of soil substrate in decomposition.

In addition to soil substrates, difference in microenvironment at the translocated sites along the elevation gradient could also differentiate the soil respiration. Our analysis on *Rs* via linear mixed effects models revealed that translocation significantly impacted soil respiration (Table 1). As one important factor influencing carbon processes, temperature is considered to have positive effects on SOM decomposition and soil respiration rates [33,50,51]. Our experiment supported the hypothesis of positive temperature effect on soil respiration. Soils translocated to the lowest site, thus with the highest temperature, had a higher mean respiration rate than those translocated to the middle and the highest elevations. This is particularly prominent for the soils originating from high elevation, thus having high SOC, SOM, and C:N (right columns in Figure 4 with soils from 1000 m). The linear mixed-effects model result confirmed a significantly positive effect of temperature on *Rs* (Table 1). Our estimated Q10 values also revealed high *Rs* sensitivity to the temperature of these soils, i.e., 3.1, 6.9, and 8.7 at 350, 600, and 1000 m, respectively.

However, environmental controls on soil carbon processes are complex. Temperature alone could not explain the pattern of soil respiration rates in LEF as signaled by the increases in *Rs* of the soils from 350 m but translocated to 1000 m with lowered temperature (left columns in Figure 4). Although not significant, the effect of soil moisture on *Rs* is indicated as being positive (Table 1). Soil moisture increases with elevation and reaches the highest at 1000 m (Figure 2), which might stimulate the decomposition rate during the dry season as mentioned above. Existing studies also suggested the effects of soil fauna and microbial biomass on *Rs*, and showed that soil fauna and microbes varied with microclimate conditions, soil properties (e.g., clay), and vegetation [34,52–54]. The soil fauna and microbes from high elevation might prefer the SOC and SOM with lower C:N originating from low elevation [26] and thus accelerate the decomposition. Particularly, soil microbial mass positively relates to SOM [34,52]. Since SOM increases with elevation in LEF (Figure 3), soil microbial mass might also increase along the elevation from low to top in LEF. Therefore, the increased *Rs* of the soils translocated from low elevation to high elevation might be related to the fact that the positive impacts of soil moisture, specific microbes, and microbial biomass at high elevation outweighed the limitation of decreased temperature.

Our conclusions of complex environmental controls on soil respiration along an elevation gradient are consistent with others. Kane et al. (2003) measured soil respiration rates along a gradient in the Olympic National Park, Washington, USA, where soil temperature at a high elevation site was 4.5 °C lower than that at low elevation. However, there was no significant relation detected between soil respiration and temperature [14]. Similarly, in a study on climate dependence of heterotrophic soil respiration along a 3000-m elevation gradient in a tropical forest in Peru, Zimmermann et al. (2009) also concluded that the soil respiration rate did not vary significantly along the elevation gradient with decreasing temperature, although SOC stocks increased linearly with increased elevation [17].

The ecosystem carbon balance primarily depends on the differences between the responses of productivity and those of respiration to climate change, especially warming [51,55,56]. Existing studies found resource quality could significantly affect the temperature sensitivity of SOM decomposition [26,57]. The labile SOM is sensitive to climate change and generally decomposes rapidly, which can largely contribute to increases in soil CO_2 fluxes [58,59]. Our estimated Q10 values, a widely-used parameter to describe the temperature sensitivity of soil C, increased with elevation. The Q10 value at high elevation, with low temperature, high moisture, SOC, SOM, and C:N, is more than double that at low elevation. Therefore, understanding the environment controls on the temperature sensitivity of soil respiration is critical to predict the responses of carbon processes to climate change [50,58,60–62].

With the ongoing and predicted warming climate in the Caribbean region [11], tropical soils are at severe risk of releasing large amounts of CO_2 into the atmosphere. Our soil translocation experiment improved the understanding of the impacts of environmental conditions on soil carbon processes by highlighting the significant effects of soil source and temperature, the nonsignificant positive effect of moisture, and overall complex environmental controls along the altitudinal gradient. Further long-term and multi-factor soil translocation experiments at the ecosystem level, incorporating, additionally, soil microbes, fauna, and litter inputs, should be established to study the impacts of climate change on tropical soil carbon balance, as well as their feedback to the climate.

5. Conclusions

Tropical forests play important roles in the interactions between terrestrial ecosystems and the global climate system, and have the potential to change from a C sink to a C source under climate change. Soil translocation experiments along the altitudinal gradient provide an approach to assess the impacts of climate change on soil C dynamics. Our soil translocation experiment at tropical forests in northeastern Puerto Rico showed that both soil C and soil respiration rates were altered by variations in temperature, moisture, and initial soil organic C. When soils were translocated to a lower elevation, the increased temperature enhanced soil respiration and therefore less soil organic C was left at the end of the experiment. Such an effect is more significant for those soils with an initial high content of soil organic C, i.e., soils originating from high elevations. On the other hand, soil respiration could also be enhanced when soils were translocated to higher elevation due to altered moisture conditions and soil microbes. Further comprehensive studies involving soil microbial composition and biomass and the quality of C substrates would improve the mechanistic understanding of soil respiration in response to climate change, and advance the earth system modeling.

Acknowledgments: This research was supported by the grant DEB-1546686 from the National Science Foundation to the Institute for Tropical Ecosystem Studies (ITES), Department of Environmental Science, University of Puerto Rico, and to the International Institute of Tropical Forestry (IITF), USDA Forest Service, as part of the Luquillo Long-Term Ecological Research Program. The USDA Forest Service and University of Puerto Rico gave additional field and lab support. Technicians at the El Verde Field Station helped with experiment set-up. Technicians (María M. Rivera, Maysaá Ittayem, and Mary Jane Sánchez) at IITF USDA Forest Service provided assistance on field work and laboratory analyses.

Author Contributions: M.Y., G.G., and X.Z. conceived, designed the experiments, and reviewed various versions of the manuscript; D.C. performed the experiments and field work; D.C., M.Y., and Q.G. analyzed the data; G.G. contributed lab analyses, equipment/materials, and field work; D.C., M.Y., and Q.G. wrote the paper.

Conflicts of Interest: The authors declare no conflict of interest. The founding sponsors had no role in the design of the study; in the collection, analyses, or interpretation of data; in the writing of the manuscript, and in the decision to publish the results.

References

1. Townsend, A.R.; Vitousek, P.M.; Trumbore, S.E. Soil organic-matter dynamics along gradients in temperature and land-use on the island of Hawaii. *Ecology* **1995**, *76*, 721–733. [CrossRef]
2. Schlesinger, W.H.; Andrews, J.A. Soil respiration and the global carbon cycle. *Biogeochemistry* **2000**, *48*, 7–20. [CrossRef]

3. Silver, W.L. The potential effects of elevated CO$_2$ and climate change on tropical forest soils and biogeochemical cycling. *Clim. Chang.* **1998**, *39*, 337–361. [CrossRef]

4. Pan, Y.; Birdsey, R.A.; Fang, J.; Houghton, R.; Kauppi, P.E.; Kurz, W.A.; Phillips, O.L.; Shvidenko, A.; Lewis, S.L.; Canadell, J.G.; et al. A large and persistent carbon sink in the world's forests. *Science* **2011**, *333*, 988–993. [CrossRef] [PubMed]

5. Meir, P.; Wood, T.E.; Galbraith, D.R.; Brando, P.M.; da Costa, A.C.L.; Rowland, L.; Ferreira, L.V. Threshold responses to soil moisture deficit by trees and soil in tropical rain forests: Insights from field experiments. *Bioscience* **2015**, *65*, 882–892. [CrossRef] [PubMed]

6. Rowland, L.; da Costa, A.C.L.; Galbraith, D.R.; Oliveira, R.S.; Binks, O.J.; Oliveira, A.A.R.; Pullen, A.M.; Doughty, C.E.; Metcalfe, D.B.; Vasconcelos, S.S.; et al. Death from drought in tropical forests is triggered by hydraulics not carbon starvation. *Nature* **2015**, *528*, 119–122. [CrossRef] [PubMed]

7. Gatti, L.V.; Gloor, M.; Miller, J.B.; Doughty, C.E.; Malhi, Y.; Domingues, L.G.; Basso, L.S.; Martinewski, A.; Correia, C.S.C.; Borges, V.F.; et al. Drought sensitivity of amazonian carbon balance revealed by atmospheric measurements. *Nature* **2014**, *506*, 76–80. [CrossRef] [PubMed]

8. Anderegg, W.R.L.; Ballantyne, A.P.; Smith, W.K.; Majkut, J.; Rabin, S.; Beaulieu, C.; Birdsey, R.; Dunne, J.P.; Houghton, R.A.; Myneni, R.B.; et al. Tropical nighttime warming as a dominant driver of variability in the terrestrial carbon sink. *Proc. Natl. Acad. Sci. USA* **2015**, *112*, 15591–15596. [CrossRef] [PubMed]

9. Wan, S.; Norby, R.J.; Ledford, J.; Weltzin, J.F. Responses of soil respiration to elevated CO$_2$, air warming, and changing soil water availability in a model old-field grassland. *Glob. Chang. Biol.* **2007**, *13*, 2411–2424. [CrossRef]

10. IPCC. *Climate Change 2013: The Physical Science Basis. Contribution of Working Group I to the Fifth Assessment Report of the Intergovernmental Panel on Climate Change*; Cambridge University Press: Cambridge, UK; New York, NY, USA, 2013; p. 1535.

11. Yu, M.; Gao, Q.; Gao, C.; Wang, C. Extent of night warming and spatially heterogeneous cloudiness differentiate temporal trend of greenness in mountainous tropics in the new century. *Sci. Rep.* **2017**, *7*, 41256. [CrossRef] [PubMed]

12. Baggs, E.M. Partitioning the components of soil respiration: A research challenge. *Plant Soil* **2006**, *284*, 1–5. [CrossRef]

13. Cramer, W.; Bondeau, A.; Woodward, F.I.; Prentice, I.C.; Betts, R.A.; Brovkin, V.; Cox, P.M.; Fisher, V.; Foley, J.A.; Friend, A.D.; et al. Global response of terrestrial ecosystem structure and function to CO$_2$ and climate change: Results from six dynamic global vegetation models. *Glob. Chang. Biol.* **2001**, *7*, 357–373. [CrossRef]

14. Kane, E.S.; Pregitzer, K.S.; Burton, A.J. Soil respiration along environmental gradients in olympic national park. *Ecosystems* **2003**, *6*, 326–335. [CrossRef]

15. Li, Y.; Xu, M.; Zou, X.; Xia, Y. Soil CO$_2$ efflux and fungal and bacterial biomass in a plantation and a secondary forest in wet tropics in Puerto Rico. *Plant Soil* **2005**, *268*, 151–160. [CrossRef]

16. Conant, R.T.; Dalla-Betta, P.; Klopatek, C.C.; Klopatek, J.M. Controls on soil respiration in semiarid soils. *Soil Biol. Biochem.* **2004**, *36*, 945–951. [CrossRef]

17. Zimmermann, M.; Meir, P.; Bird, M.I.; Malhi, Y.; Ccahuana, A.J.Q. Climate dependence of heterotrophic soil respiration from a soil-translocation experiment along a 3000 m tropical forest altitudinal gradient. *Eur. J. Soil Sci.* **2009**, *60*, 895–906. [CrossRef]

18. Nottingham, A.T.; Whitaker, J.; Turner, B.L.; Salinas, N.; Zimmermann, M.; Malhi, Y.; Meir, P. Climate warming and soil carbon in tropical forests: Insights from an elevation gradient in the Peruvian Andes. *Bioscience* **2015**, *65*, 906–921. [CrossRef] [PubMed]

19. Lloyd, J.; Taylor, J.A. On the temperature-dependence of soil respiration. *Funct. Ecol.* **1994**, *8*, 315–323. [CrossRef]

20. Kirschbaum, M.U.F. The temperature dependence of soil organic matter decomposition, and the effect of global warming on soil organic C storage. *Soil Biol. Biochem.* **1995**, *27*, 753–760. [CrossRef]

21. Knorr, W.; Prentice, I.C.; House, J.I.; Holland, E.A. Long-term sensitivity of soil carbon turnover to warming. *Nature* **2005**, *433*, 298–301. [CrossRef] [PubMed]

22. Raich, J.W.; Tufekcioglu, A. Vegetation and soil respiration: Correlations and controls. *Biogeochemistry* **2000**, *48*, 71–90. [CrossRef]

23. Heneghan, L.; Coleman, D.C.; Zou, X.; Crossley, D.A.; Haines, B.L. Soil microarthropod contributions to decomposition dynamics: Tropical-temperate comparisons of a single substrate. *Ecology* **1999**, *80*, 1873–1882.

24. González, G.; Seastedt, T.R. Soil fauna and plant litter decomposition in tropical and subalpine forests. *Ecology* **2001**, *82*, 955–964. [CrossRef]

25. Dechaine, J.; Ruan, H.; Sánchez-de León, Y.; Zou, X. Correlation between earthworms and plant litter decomposition in a tropical wet forest of Puerto Rico. *Pedobiologia* **2005**, *49*, 601–607. [CrossRef]

26. Sjögersten, S.; Wookey, P.A. Climatic and resource quality controls on soil respiration across a forest-tundra ecotone in Swedish Lapland. *Soil Biol. Biochem.* **2002**, *34*, 1633–1646. [CrossRef]

27. Lawrence, C.R.; Neff, J.C.; Schimel, J.P. Does adding microbial mechanisms of decomposition improve soil organic matter models? A comparison of four models using data from a pulsed rewetting experiment. *Soil Biol. Biochem.* **2009**, *41*, 1923–1934. [CrossRef]

28. Mikan, C.J.; Schimel, J.P.; Doyle, A.P. Temperature controls of microbial respiration in arctic tundra soils above and below freezing. *Soil Biol. Biochem.* **2002**, *34*, 1785–1795. [CrossRef]

29. Xu, L.; Baldocchi, D.D.; Tang, J. How soil moisture, rain pulses, and growth alter the response of ecosystem respiration to temperature. *Glob. Biogeochem. Cycles* **2004**, *18*. [CrossRef]

30. Qi, Y.; Xu, M.; Wu, J. Temperature sensitivity of soil respiration and its effects on ecosystem carbon budget: Nonlinearity begets surprises. *Ecol. Model.* **2002**, *153*, 131–142. [CrossRef]

31. Rastetter, E.B.; Ryan, M.G.; Shaver, G.R.; Melillo, J.M.; Nadelhoffer, K.J.; Hobbie, J.E.; Aber, J.D. A general biogeochemical model describing the responses of the C-cycle and N-cycle in terrestrial ecosystems to changes in CO_2, climate, and N-deposition. *Tree Physiol.* **1991**, *9*, 101–126. [CrossRef] [PubMed]

32. Gifford, R. The global carbon cycle: A viewpoint on the missing sink. *Funct. Plant Biol.* **1994**, *21*, 1–15. [CrossRef]

33. Li, Y.; Xu, M.; Zou, X. Heterotrophic soil respiration in relation to environmental factors and microbial biomass in two wet tropical forests. *Plant Soil* **2006**, *281*, 193–201. [CrossRef]

34. Wang, W.J.; Dalal, R.C.; Moody, P.W.; Smith, C.J. Relationships of soil respiration to microbial biomass, substrate availability and clay content. *Soil Biol. Biochem.* **2003**, *35*, 273–284. [CrossRef]

35. Vance, E.D.; Chapin Iii, F.S. Substrate limitations to microbial activity in taiga forest floors. *Soil Biol. Biochem.* **2001**, *33*, 173–188. [CrossRef]

36. Blagodatsky, S.; Blagodatskaya, E.; Yuyukina, T.; Kuzyakov, Y. Model of apparent and real priming effects: Linking microbial activity with soil organic matter decomposition. *Soil Biol. Biochem.* **2010**, *42*, 1275–1283. [CrossRef]

37. Smith, D.L.; Johnson, L. Vegetation-mediated changes in microclimate reduce soil respiration as woodlands expand into grasslands. *Ecology* **2004**, *85*, 3348–3361. [CrossRef]

38. Kleb, H.R.; Wilson, S.D. Vegetation effects on soil resource heterogeneity in prairie and forest. *Am. Nat.* **1997**, *150*, 283–298. [PubMed]

39. Bruijnzeel, L.A.; Veneklaas, E.J. Climatic conditions and tropical montane forest productivity: The fog has not lifted yet. *Ecology* **1998**, *79*, 3–9. [CrossRef]

40. Vázquez, J.A.; Givnish, T.J. Altitudinal gradients in tropical forest composition, structure, and diversity in the sierra de manantlán. *J. Ecol.* **1998**, *86*, 999–1020.

41. Gould, W.A.; González, G.; Carrero, R.G. Structure and composition of vegetation along an elevational gradient in Puerto Rico. *J. Veg. Sci.* **2006**, *17*, 653–664. [CrossRef]

42. Barone, J.A.; Thomlinson, J.; Cordero, P.A.; Zimmerman, J.K. Metacommunity structure of tropical forest along an elevation gradient in Puerto Rico. *J. Trop. Ecol.* **2008**, *24*, 525–534. [CrossRef]

43. González, G.; García, E.; Cruz, V.; Borges, S.; Zalamea, M.; Rivera, M.M. Earthworm communities along an elevation gradient in northeastern Puerto Rico. *Eur. J. Soil Biol.* **2007**, *43*, S24–S32. [CrossRef]

44. Zimmermann, M.; Meir, P.; Bird, M.I.; Malhi, Y.; Ccahuana, A.J.Q. Temporal variation and climate dependence of soil respiration and its components along a 3000 m altitudinal tropical forest gradient. *Glob. Biogeochem. Cycles* **2010**, *24*. [CrossRef]

45. Garcia-Martinó, A.G.; Warner, G.S.; Scatena, F.N.; Civco, D.L. Rainfall, runoff and elevation relationships in the Luquillo mountains of Puerto Rico. *Caribb. J. Sci.* **1996**, *32*, 413–424.

46. Weaver, P.L. Environmental gradients affect forest structure in Puerto Rico's Luquillo mountains. *Interciencia* **2000**, *25*, 254–259.

47. Weaver, P.L.; Gould, W.A. Forest vegetation along environmental gradients in northeastern Puerto Rico. *Ecol. Bull.* **2013**, *54*, 43–65.

48. R Development Core Team. *R: A Language and Environment for Statistical Computing*; R Foundation for Statistical Computing: Vienna, Austria, 2014.

49. Bond-Lamberty, B.; Bolton, H.; Fansler, S.; Heredia-Langner, A.; Liu, C.; McCue, L.A.; Smith, J.; Bailey, V. Soil respiration and bacterial structure and function after 17 years of a reciprocal soil transplant experiment. *PLoS ONE* **2016**, *11*, e0150599. [CrossRef] [PubMed]

50. Kirschbaum, M.U.F. The temperature dependence of organic-matter decomposition—Still a topic of debate. *Soil Biol. Biochem.* **2006**, *38*, 2510–2518. [CrossRef]

51. Sayer, E.J.; Heard, M.S.; Grant, H.K.; Marthews, T.R.; Tanner, E.V.J. Soil carbon release enhanced by increased tropical forest litterfall. *Nat. Clim. Chang.* **2011**, *1*, 304–307. [CrossRef]

52. Müller, T.; Höper, H. Soil organic matter turnover as a function of the soil clay content: Consequences for model applications. *Soil Biol. Biochem.* **2004**, *36*, 877–888. [CrossRef]

53. Ruan, H.H.; Zou, X.M.; Scatena, E.; Zimmerman, J.K. Asynchronous fluctuation of soil microbial biomass and plant litterfall in a tropical wet forest. *Plant Soil* **2004**, *260*, 147–154. [CrossRef]

54. Wang, S.; Ruan, H.; Wang, B. Effects of soil microarthropods on plant litter decomposition across an elevation gradient in the Wuyi mountains. *Soil Biol. Biochem.* **2009**, *41*, 891–897. [CrossRef]

55. Malhi, Y.; Grace, J. Tropical forests and atmospheric carbon dioxide. *Trends Ecol. Evol.* **2000**, *15*, 332–337. [CrossRef]

56. Clark, D.A. Sources or sinks? The responses of tropical forests to current and future climate and atmospheric composition. *Philos. Trans. R. Soc. B Biol. Sci.* **2004**, *359*, 477–491. [CrossRef] [PubMed]

57. Plante, A.F.; Conant, R.T.; Carlson, J.; Greenwood, R.; Shulman, J.M.; Haddix, M.L.; Paul, E.A. Decomposition temperature sensitivity of isolated soil organic matter fractions. *Soil Biol. Biochem.* **2010**, *42*, 1991–1996. [CrossRef]

58. Conant, R.T.; Drijber, R.A.; Haddix, M.L.; Parton, W.J.; Paul, E.A.; Plante, A.F.; Six, J.; Steinweg, J.M. Sensitivity of organic matter decomposition to warming varies with its quality. *Glob. Chang. Biol.* **2008**, *14*, 868–877. [CrossRef]

59. Von Lützow, M.; Kögel-Knabner, I. Temperature sensitivity of soil organic matter decomposition—What do we know? *Biol. Fertil. Soils* **2009**, *46*, 1–15. [CrossRef]

60. Davidson, E.A.; Janssens, I.A. Temperature sensitivity of soil carbon decomposition and feedbacks to climate change. *Nature* **2006**, *440*, 165–173. [CrossRef] [PubMed]

61. Conant, R.T.; Ryan, M.G.; Ågren, G.I.; Birge, H.E.; Davidson, E.A.; Eliasson, P.E.; Evans, S.E.; Frey, S.D.; Giardina, C.P.; Hopkins, F.; et al. Temperature and soil organic matter decomposition rates—Synthesis of current knowledge and a way forward. *Glob. Chang. Biol.* **2011**, *17*, 3392–3404. [CrossRef]

62. Balser, T.C.; Wixon, D.L. Investigating biological control over soil carbon temperature sensitivity. *Glob. Chang. Biol.* **2009**, *15*, 2935–2949. [CrossRef]

MDPI

St. Alban-Anlage 66

4052 Basel

Switzerland

Tel. +41 61 683 77 34

Fax +41 61 302 89 18

www.mdpi.com

Forests Editorial Office

E-mail: forests@mdpi.com

www.mdpi.com/journal/forests

www.ingramcontent.com/pod-product-compliance
Lightning Source LLC
Chambersburg PA
CBHW041215220326
41597CB00033BA/5910